IET TELECOMMUNICATIONS SERIES 71

Understanding Telecommunications Networks

Other volumes in this series:

Understanding Telecommunications Networks

2nd Edition

Andy Valdar

The Institution of Engineering and Technology

Published by The Institution of Engineering and Technology, London, United Kingdom

The Institution of Engineering and Technology is registered as a Charity in England & Wales (no. 211014) and Scotland (no. SC038698).

© The Institution of Engineering and Technology 2017

First edition published 2006
Second edition published 2017

The Institution of Engineering and Technology
Michael Faraday House
Six Hills Way, Stevenage
Herts, SG1 2AY, United Kingdom

www.theiet.org

British Library Cataloguing in Publication Data
A catalogue record for this product is available from the British Library

ISBN 978-1-78561-164-3 (hardback)
ISBN 978-1-78561-165-0 (PDF)

Typeset in India by MPS Limited
Printed in the UK by CPI Group (UK) Ltd, Croydon

To
Melanie Johnston and
Mathew Valdar

Contents

Foreword

Although everyone is familiar with fixed and mobile telephones and the ability to dial anywhere in the world, not many people understand how calls are carried, or how the various networks link together. Similarly, the workings of the Internet and all the different data and broadband services can be equally mysterious to many, as is the role of the myriad of underground and overhead cables in the streets. More importantly, many people now think that in any case all the telephone networks will be replaced by the Internet! This book aims to address such aspects.

The primary purpose of this book is to describe how telecommunications networks work. Although the technology is explained at a simple functional level, emphasis is put on how the various components are used to build networks – fixed ('wired'), mobile ('wireless'), voice and data. I have tried to pitch the explanations at a level that does not require an engineering knowledge. Indeed, it is hoped that this book will be helpful to the wide range of people working in the telecommunications industry: managers who specialise in marketing, customer service, finance, human resources, public relations, investor relations, training and development, legal and regulation. However, many of the engineers in the industry may also appreciate an understanding of aspects beyond their specialisations that this work aims to provide.

I have also written the book with a view to making the broad-based coverage and emphasis on networks valuable background reading for students on Undergraduate and Graduate University courses in telecommunications. The majority of the material for the book has been based on my lecturing experience over the last 18 years in my role of visiting professor of telecommunications strategy at the Department of Electronic and Electrical Engineering of UCL. But my understanding of the business of planning and operating networks is based on the previous 30 years of wide-ranging experience at BT.

Finally, this is a rapidly changing industry – one of the reasons why it is so fascinating – with new technology constantly hitting the scene. But public networks do not change overnight, and the principles of networks and how technology can be used as put forward in this treatise should endure for many years. So, I hope that this book has a long and useful life.

ARV (2005)

Foreword to the second edition

I have become increasingly aware that, although the principles covered in the original version of this book still hold, the significant changes to the technology, marketplace and social needs for telecommunications networks means an update is now necessary. In this fully revised version I have tried to include essential new material, as well as making appropriate updates to the existing text and Figures.

Another important reason for this new edition is that I am pleased to say that a companion book – Understanding Telecommunications Business – was published in 2015. This book, co-authorised with my colleague Ian Morfett, provides the commercial, marketing, regulatory, planning, operations and business management framework for the various networks described in this book. Therefore, throughout this new edition I have cross-referenced to the companion work. I hope that the pair of books provides a comprehensive underpinning of today's fascinating world of telecommunications.

ARV (2017)

Acknowledgements

Writing this book as a personal project has been a great source of enjoyment for me. But, of course, this was not done in isolation and many people have been kind enough to give me help in the form of providing information and offering critical reviews.

Firstly, I would like to acknowledge the late Professor Gerry White, who gave the initial impetus for this book. Although he was able to give early direction and comments, Gerry's sad and untimely death in 2004 unfortunately truncated his involvement.

However, I have been fortunate indeed in having the willing help of many friends and colleagues from the telecommunications industry and UCL, whom I am pleased to be able to acknowledge here. Firstly, I am grateful to Chris Seymour, a veteran of BT and now a consultant, for his early guidance on the structure of the book. I am especially indebted to my colleague and long-time mentor Professor Keith Ward, another veteran of both BT, and more recently UCL, for his support, comments and permission to use his ideas and diagrams as a basis for some of the content. Several other colleagues deserve a special mention for their helpful inputs and guidance: Tony Holmes and Tim Wright of BT on Chapters 10 and 11, respectively; and Professor Izzat Darwezah of UCL on Chapter 9. Also, the help from Professor John Mitchell and Dr Bob Sutherland, both of UCL, is acknowledged.

Furthermore, I have also been lucky enough to have several of my graduate students – notably, David Schultz, Kevin Conroy, and Peter Weprin – who, despite their studies and full-time work for BT, found time to critique my draft chapters and I have done my best to incorporate their views. I am also grateful to Liam Johnston of Fujitsu for his support, as well as Maria Pajares and Sorrel Packham for preparing the Appendix 2 and the Abbreviations, respectively.

But, this set of acknowledgements must finish with a big thank you from me to my wife, Su, for not only putting up with me working on the book at all hours, but for her helpful and diligent reviewing of all the draft text and, of course, her continuous encouragement.

ARV (2015)

Additional acknowledgements with the new edition

I am most grateful to the following people who have been kind enough to help me with guidance and suggestions on the new material included in this edition: Peter Leonhardt of BT on Chapters 4, 5 and 8; Colin Phillips of BT on Chapter 8; Sheik Sahul Hameed, Prof Izzat Darwazeh and Dr Clive Poole of UCL, and Andy Sutton of BT on Chapter 9.

Finally, I owe a big vote of thanks to my wife, Su, for kindly reviewing the draft text, putting up with me beavering away at all hours on the revision work, and always being supportive.

ARV (March 2017)

Abbreviations

AAA	Authentication, Authorisation and Accounting
AAL	ATM Adaptation Layers
AAR	Automatic Alternative Routeing
ABR	Adaptive Bit Rate
AC	Alternating Current
ACD	Automatic Call Distribution
A/D	Analogue-to-Digital (conversion)
ADM	Add-Drop Multiplexer
ADSL	Asymmetric Digital Subscriber Line
AESA	ATM End-System Address
AFI	Authority and Format Indicator
AM	Amplitude Modulation
AMPS	Advance Mobile Phone System
AP	Access Point
API	Application Programming Interface
APON	ATM over Passive Optical Network
ARP	Address Resolution Protocol
ARPANET	US defence department project: the forerunner of the Internet
AS	Autonomous System
AS	Application Server (re IMS)
ATM	Asynchronous Transfer Mode
AU	Administration Unit
AUC	Authentication Centre
BHCA	Busy Hour Call Attempts
B-ISUP	Broadband ISDN User Part
BORSCHT	Battery, Overload protection, Ringing, Supervision, Codec, Hybrid, Test
BPON	Broadband Passive Optical Network
BRA	Broadband Radio Access
BRAS	Broadband Remote Access Server

BS	Base Station
BSC	Base Station Controller
BSI	British Standards Institute
BSS	Business Support Systems
BTC	Base Transmitter Controller
BTS	Base Transceiver Station
C7	see SSNo.7
CATV	Community Antenna Television
CB	Citizens' Band (Radio)
CBR	Constant Bit Rate
CCS	Common Channel Signalling
CCU	Cross Connection Unit
CDM	Code-Division Multiplex
CDMA	Code-Division Multiple Access
CDPD	Cellular Digital Packet Data
CDN	Content Distribution Network
CDR	Charging Data Record
CEPT	Conférence Européenne des administrations des Postes et des Télécommunications
CER	Cell Error Ratio
CIDR	Classless Inter-Domain Routeing
CG	Charge Group
CLI	Caller Line Identification
CM	Connection Memory
CN	Communication Network
COS	Class of Service
COTS	Commercial Off-The-Shelf (software)
CPE	Customers Premises Equipment
CPU	Central Processing Unit
CSCF	Call/Section Control Function
CSFB	Circuit-Switched Fall Back
CSMA/CA	Carrier-Sense Multiple-Access/Collision Avoidance
CSMA/CD	Carrier-Sense Multiple-Access/Collision Detection
CTD	Cell Transfer Delay
CTS	Core Transmission Stations
D-AMPS	Digital-Advanced Mobile Phone System

Datel	Data over Telephone
DC	Direct Current
DC	Data Centre
DCC	Data Country Code
DDF	Digital Distribution Frame
DDI	Direct Dialling In
DECT	Digital Enhanced Cordless Telecommunications
DHCP	Dynamic Host Configuration Protocol
DJSU	Digital Junction Switching Unit
DLE	Digital Local Exchange
DMSU	Digital Main Switching Unit
DMT	Discreet Multitone
DNS	Domain Name Server
DP	Distribution Point
DPNSS	Digital Private Network Signalling System
DRM	Digital Rights Management
DSL	Digital Subscriber Line
DSLAM	Digital Subscriber Line Access Multiplexor
DSP	Digital Signal Processor
DSP	Domain Specific Part
DUP	Data User Part
DWDM	Dense Wave Division Multiplex
DXC	Digital Cross Connection (unit)
E1	2 Mbit/s PDH digital transmission block
E2	8 Mbit/s PDH digital transmission block
E3	34 Mbit/s PDH digital transmission block
E4	140 Mbit/s PDH digital transmission block
EC	Echo Canceller
EDGE	Enhanced Data Rate for Global Evolution
EFS	Error-Free Second
EHF	Extra High frequency
EGP	Exterior Gateway Protocol
EIR	Equipment Identity Register
eNodeB	Evolved NodeB (LTE Mobile)
EPC	Evolved Packet Core (LTE mobile)
EPON	Ethernet Passive Optical Network

eTOM	Extended Telecommunications Operations Map
ETSI	European Telecommunications Standards Institute
E-UTRAN	Evolved UMTS RAN (LTE mobile)
FAB	Fulfilment, Assurance, and Billing
FAS	Frame Alignment Signal
FCC	Federal Commission for Communications
FDD	Frequency-Division Duplex
FDM	Frequency-Division Multiplex
FDMA	Frequency-Division Multiple Access
FEC	Forward Error Control
FEI	Federation of Electrical Industries
FEXT	Far End Cross Talk
FM	Frequency Modulation
FSAN	Full Service Access Network
FSK	Frequency-Shift Keying
FTP	File Transfer Protocol
FTTH	Fibre to the Home
FTTK	Fibre to the Kerb
FTTO	Fibre to the Office
2G	Second Generation (of mobile networks)
2½G	Step between second and third generation (of mobile networks)
3G	Third Generation (of mobile networks)
4G	Fourth Generation (of mobile networks)
5G	Fifth Generation (of mobile networks)
GGSN	Gateway GPRS Support Node
GMSC	Gateway Mobile Switching Centre
GOS	Grade of Service
GPON	Gigabit/s Passive Optical Network
GPRS	General Packet Radio System
GPS	Global Positioning Satellite
GSM	Global System for Mobile (originally, Groupe Spécial Mobile)
GTP	GPRS Tunnelling Protocol
HD	High Definition
HDF	Hand-over Distribution Frame
HDSL	High-Bit-Rate Digital Subscriber Line
HF	High Frequency
HFC	Hybrid Fibre-Coax

HLR	Home Location Register
HSCD	High-Speed Circuit-Switched Data
HSS	Home Subscriber Server
HTTP	Hypertext Transfer Protocol
Hz	Hertz (cycles per second)
IAM	Initial Address Message
ICANN	Internet Corporation for Assigned Names and Numbers
ICD	International Code Designator
ICT	Information and Communications Technology
IDI	Initial Domain Indicator
IDN	Integrated Digital Network
IDP	Initial Domain Part
IEE	Institution of Electrical Engineers
IEEE	Institution of Electrical and Electronic Engineers
IETF	Internet Engineering Task Force
IGP	Internet Gateway Protocol
IMS	IP Multimedia Subsystem
IN	Intelligence Network
INAP	Intelligent Networks Application Part
IND	International Network Designator
INDB	Intelligence Networks Database
INX	Internet Exchange
IOT	Internet of Things
IP	Internet Protocol
IPTV	IP Television
IPVPN	IP Virtual Private Network
ISC	International Switching Centre
ISDN	Integrated Services Digital Network
ISO	International Standards Organisation
ISP	Internet Service Provider
ISP	Intermediate Service Part
ISUP	ISDN User Part
IT	Information Technology
ITU	International Telecommunication Union
IWF	Interworking Function
JT	Junction Tandem
LAN	Local Area Network

LD	Loop Disconnect (signalling)
LE	Local Exchange
LED	Light Emitting Diode
LEO	Low Earth Orbit (satellite)
LER	Label Edge Router
LF	Low Frequency
LMDS	Local Multipoint Distribution Service
LOS	Line of Sight
LPWA	Low Power Wide Area
LSR	Label Switch Routers
LTE	Line Terminating Equipment
LTE	Long-Term Evolution (4G mobile)
LTU	Line Terminating Unit
MA	Multiple Access
MAC	Medium-Access Control
MAN	Metropolitan Area Network
MAP	Mobile Application Part
MDF	Main Distribution Frame
ME	Mobile Exchange
MEO	Medium Earth Orbit (satellite)
MF	Multi-Frequency (signalling)
MF	Medium Frequency (radio)
MG	Media Gateway
MGC	Media Gateway Controller
MGCF	Media Gateway Controller Function
MGCP	Media Gateway Control Protocol
MIMO	Multiple-Input Multiple-Output
MME	Mobility Management Entity
MMI	Man–Machine Interface
MNO	Mobile Network Operator
MPEG	Moving Picture Experts Group
MPLS	Multi-Protocol Label Switching
MRF	Media Resource Function
MRFC	MRF Controller
MS	Mobile Station
MSAC	Multi-Service Access Code
MSAN	Multi-Service Access Node

MSC	Mobile Switching Centre
MS-DPRING	Multiplex Section – Dedicated Protection Ring
MSO	Multiple Service Operators
MSP	Multi-Service Platform
MSP	Mobile Service Provider
MS-SPRING	Multiplex Section – Shared Protection Ring
MT	Mobile Terminal
MTP	Message Transfer Part
Mux	Multiplexer
MVA	Mega Volt-Amperes
MVNO	Mobile Virtual Network Operator
NANP	North American Numbering Plan
NEXT	Near End Cross Talk
NFC	Near-Field Communications
NFI	Network Function Infrastructure
NFS	Network File System
NFV	Network Function Virtualisation
NGN	Next Generation Network
NGS	Next Generation Switch
NIC	Network Interface Card
NLOS	Non-Line of Sight
NMT	Nordic Mobile Telephone
NSAP	Network Service Access Point
NT	Network Termination
NTE	Network Terminating Equipment
NTN	Network Terminal Number
NTP	Network Termination Point
NTTP	Network Termination and Test Point
NTU	Network Terminating Unit
OA&M	Operations, Administration and Maintenance
OC	Optical Carrier
OFDM	Orthogonal Frequency Division Multiplex
OFDMA	Orthogonal Frequency Division Multiple Access
OLT	Optical Line Termination
OMAP	Operations, Maintenance and Administration Application Part
OMC	Operations and Maintenance Centre
ONP	Open Network Provision

ONT	Optical Network Termination
ONU	Optical Network Unit
OSI	Open Systems Interconnection
OSPF	Open Shortest Path First
OSS	Operations Support Systems
OTT	Over-the-Top
OXC	Optical Cross Connect
PABX	Private Automatic Branch Exchange
PAD	Packet Assembler-Disassembler
PAM	Pulse Amplitude Modulation
PAN	Personal Area Networking
PAPR	Peak to Average Power Ratio
PBX	Private Branch Exchange
PC	Private Circuit
PCM	Pulse Code Modulation
PCN	Personal Communication Network
PCP	Primary Connection Point
PCRF	Policy and Charging Rules Function (re IMS)
PCS	Personal Communication Service
PCU	Packet Communications Unit
PDA	Personal Digital Assistant
PDCP	Packet Data Convergence Protocol
PDH	Plesiochronous Digital Hierarchy
P/GW	Packet data network GateWay
PID	Packet IDentity
PIG	PSTN-to-Internet Gateway
PLMN	Public Land Mobile Network
PM	Phase Modulation
PNO	Public Network Operator
POI	Points of Interconnection
PON	Passive Optical Network
POP	Point of Presence
POP	Post Office Protocol
POTS	Plain Old Telephone Service
PPM	Pulse Position Modulation
PPP	Point-to-Point Protocol
PSK	Phase-Shift Keying

PSS	Packet Switch Stream
PSTN	Public Switched Telephone Network
PVC	Permanent Virtual Circuit
QAM	Quadrature Amplitude Modulation
QOS	Quality of Service
RAN	Radio Access Network
RARP	Reverse Address Resolution Protocol
RC	Routeing Code
RCU	Remote Concentrator Unit
REN	Ringing Equivalent Number
RET	Re-Transmission
RLC	Radio-Link Controller
RLR	Received Loudness Rating
RNC	Radio Network Controller
RRC	Radio Resource Controller
RSVP	ReSource Reservation Protocol
RTP	Real-Time Protocol
SCCP	Service Connection Control Part
SCE	Service Creation Environment
SCP	Secondary Connection Points
SCP	Service Control Point
SCTP	Stream-Control Transmission Protocol
SD	Standard Definition
SDH	Synchronous Digital Hierarchy
SDN	Software-Defined Network
SDSL	Symmetric Digital Subscriber Line
SG	Signalling Gateway
S/GW	Serving Gateway
SGSN	Serving GPRS Support Node
SHF	Super High Frequency
SHR	Self-Healing Ring
SIM	Subscriber Information Module
SIP	Session Initiation Protocol
SLF	Subscriber Locator Function
SLP	Switch Label Path
SLR	Sending Level Rating
SMC	Service Management Centre

SMDS	Switched Megabit Data Service
SME	Small-to-Medium Enterprise
SMS	Short Message Service
SMSC	Short Message Service Centre
SMTP	Simple Mail Transfer Protocol
SNMP	Simple Network Management Protocol
SONET	Synchronous Optical Network
SPC	Stored-Programme Control
SPN	Service Protection Network
S/R	Sender/Receiver
SRE	Sending Reference Equivalent
SS No7	Signalling System Number 7 (also known as 'CCITT SSNo.7' or 'C7')
SSC	Sector Switching Centre
SSP	Service Switching Point
STB	Set-Top Box
STD	Subscriber Trunk Dialling
STM	Synchronous Transport Module
STP	Signal Transfer Point
SU	Signal Unit
SVC	Switched Virtual Circuit
T1	1.5 Mbit/s PDH digital carrier (USA)
T2	6 Mbit/s PDH digital carrier (USA)
T3	45 Mbit/s PDH digital carrier (USA)
T4	274 Mbit/s PDH digital carrier (USA)
TACS	Total Access Communications System
TAS	Telephone Application Server
TAT	Trans-Atlantic Telecommunications (cable)
TC	Trans-Coder
TCAP	Transaction Capabilities Application Part
TCP	Transmission Control Protocol
TDD	Time-Division Duplex
TDM	Time-Division Multiplex
TDMA	Time-Division Multiple Access
TE	Trunk Exchange
TETRA	Terrestrial Trunked Radio

TLD	Top Level Domain
TMF	Telecommunications Management Forum
TPON	Telephony over Passive Optical Network
TRS	Transmission Repeater Station
TS	Time slot
TS	Transport Stream (IPTV)
T-S-T	Time–Space–Time (switch block)
TU	Transmission Unit
TUG	TU Group
TUP	Telephone User Part
UE	User Equipment
UDP	User Datagram Protocol
UDRP	Uniform Domain-Name Dispute Resolution Policy
UHDTV	Ultra-High Definition TV
UHF	Ultra-High Frequency
UMTS	Universal Mobile Telecommunications System
URL	Uniform Resource Locator
UTRAN	UMTS Terrestrial Radio Access Network
VBR	Variable Bit Rate
VC	Virtual Circuit
VC	Virtual Container (SDH)
VDSL	Very High-Bit-Rate Digital Subscriber Line
VF	Voice Frequency
VHF	Very High Frequency
VLF	Very Low Frequency
VLR	Visitor Location Register
VLSI	Very Large-Scale Integration (Integrated Circuits)
VNF	Virtual Network Functions
VOATM	Voice over ATM
VOD	Video on Demand
VOFR	Voice over Frame Relay
VOIP	Voice over IP
VOLTE	Voice over LTE
VP	Virtual Path
VPN	Virtual Private Network
VT	Virtual Terminal

WADM	Wavelength Add-Drop Multiplexor
WAN	Wide Area Network
WAT	Wide Area Tandem
W-CDMA	Wide-Band Code-Division Multiple Access
WDM	Wave-Division Multiplex
WiMAX	Worldwide Interoperability for Microwave Access
WLAN	Wireless Local Area Network
WRC	Wireless Regulation Congress
WWW	World Wide Web
XDM	XML Document Management
XDR	eXternal Data Representation
xDSL	generic Digital Subscriber Line
XML	eXtendible Mark-up Language

Metric prefixes used in this book

Name	Symbol	Factor	Power of 10
Less than One:			
milli	**m**	0.001 (one thousandth)	(10^{-3})
micro	**μ**	0.000001 (one millionth)	(10^{-6})
nano	**n**	0.000000001 (one billionth)	(10^{-9})
pico	**p**	0.000000000001 (one trillionth)	(10^{-12})
Greater than One:			
Kilo	**k**	1,000	(10^{3})
Mega	**M**	1,000,000	(10^{6})
Giga	**G**	1,000,000,000	(10^{9})
Tera	**T**	1,000,000,000,000	(10^{12})

Binary multipliers used in this book

Measure	Symbol	Factor	Power of 2
Kilo	**k**	1,024	(2^{10})
Mega	**M**	1,048,576	(2^{20})
Giga	**G**	1,073,741,824	(2^{30})
Tetra	**T**	1,099,511,627,776	(2^{40})

The above description shows the normally accepted multipliers when applied to bits or bytes of data. For example, *2 kilobytes (kB)* means 2,048 bytes rather than 2,000 bytes.

Chapter 1

An introduction to telephony

1.1 Introduction

Telecommunications are today widely understood to mean the electrical means of communicating over a distance. The first form of telecommunications was that of the telegraph, which was invented quite independently in 1837 by two scientists, Wheatstone and Morse. Telegraphy was on a point-to-point unidirectional basis and relied on trained operators to interpret between the spoken or written word and the special signals sent over the telegraph wire. However, the use of telegraphy did greatly enhance the operations of railways and, of course, the dissemination of news and personal messages between towns. This usefulness of telecommunications on the one hand and the limitation of needing trained operators on the other led to the aspiration for a simple means of bidirectional voice telecommunications that anyone could use. Alexander Graham Bell met this need when he invented the telephone in 1876. Remarkably soon afterwards, the world's first telephone exchange was opened in 1878 in New Haven, Connecticut, United States. Since then, telephony has become the ubiquitous means of communicating for mankind.

Of course, telecommunications are now global and include not only telephony over fixed and mobile networks, but also data – the World Wide Web, e-mail, broadcast TV, video, social networking, etc. In setting out how all of this works it is helpful to understand how the basic telephone networks are structured since they form the basis of all today's telecommunications networks. We will then progressively address the full range of networks and how they link together. This first chapter introduces the principles of telephony, covering the operation of a telephone and the way that telephones are connected via a fixed (wireline) network. We will be considering telephony over mobile and data networks in subsequent chapters.

1.2 Basic telephony

Figure 1.1(a) illustrates a basic simple one-way telephone circuit between two people. The set-up comprises of a microphone associated with the speaker, which is connected via an electrical circuit with a receiver at the remote end associated with the listener. A battery provides power for the operation of the microphone and receiver. (The concept of an electrical circuit is given in Box 1.1.) During talking, variations in air pressure are generated by the vocal tract of the speaker. These

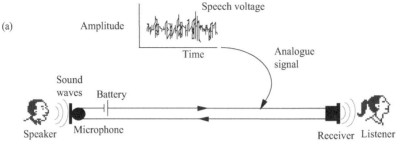

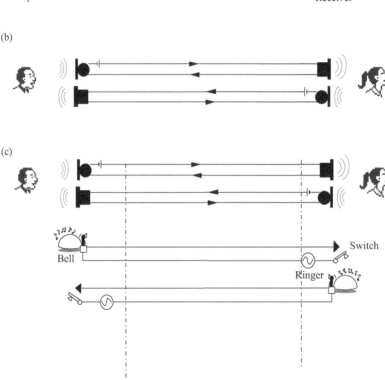

Figure 1.1   (a) Simple one-way speech over two wires. (b) Both-way speech over four wires. (c) Both-way speech and alerting over eight wires. [Ward]

variations in air pressure, known as sound waves, travel from the speaker to the microphone, which converts them into an electrical signal varying in sympathy with the pattern of the sound waves (see Box 1.2 for more explanation). Indeed, if you are to look at the electrical signal on the circuit leaving the microphone, as illustrated in Figure 1.1(a), the level of the electrical signal varies with time, with an average value set by the voltage of the battery and with modulations above and below this level, representing the variation in sound pressure hitting the microphone. This electrical signal is an analogue signal because it is an analogue of the sound wave variations in air pressure. (Later, in Chapter 3, we consider how an analogue signal is converted into a digital signal within a telephone network.)

Box 1.1   Electrical circuits

Consider an electrical circuit comprising of a power source, for example, a battery, and a length of wire linking both terminals of the power source to some device, say a lamp. Whilst the circuit is complete the lamp will glow and so a switch is normally inserted in to the arrangement to control the light on and off. In this simple example, the voltage applied by the battery can be viewed as forcing an electric current to flow around the circuit from its positive terminal to its negative terminal. This flow is referred to a 'direct current (DC)'. The lamp contains a coil of special wire that provides an obstacle to the flow of current, known as 'resistance'. The greater the resistance of the lamp the less current the battery can force to flow through the whole circuit. This gives rise to the simple relationship, known as 'Ohm's Law' in which the resistance (measured in ohms or Ω) is given by the voltage (measured in volts or V) divided by the current (measured in amps or A).

An alternative form of electrical voltage is one which cyclically varies from zero up to a maximum positive value, drops to zero and then goes to an equal but opposite maximum negative value and then back to zero. The shape corresponds to the sinusoidal waveform shown in Figure 1.2. This so-called 'alternating voltage' creates a corresponding 'alternating current (AC)' [1]. The electrical main supply is typically at 240 V alternating current (240 V AC), with the cycles occurring at 50 times/s (50 Hz) in the United Kingdom and Europe and at 120 V AC cycling at 60 Hz in the United States.

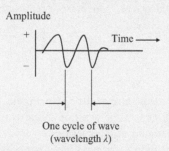

Figure 1.2   Sinusoidal waveform

The continuous cycling of the alternating electrical current causes additional changes to the flow of electricity when passing through a circuit. The first phenomenon (capacitance) causes the waveform to be delayed; the second phenomenon (inductance) causes the waveform to be advanced. The results of these effects, known collectively as impedance, are that the AC current is out of step with the applied AC voltage. These effects are used throughout telecommunications and electronic equipment, for example: inductance forms the basic mechanism exploited in hybrid transformers and loading coils, as mentioned in this chapter.

Box 1.2   How a microphone (transmitter) and receiver (earpiece) work [2]

The microphone used in a telephone handset is really an electro-acoustic transducer, a device for converting acoustic energy to electrical energy. There are several types of transmitter used for telephones, for example, carbon granule, electro-dynamic and electret. Each performs the conversion of acoustic or vibration energy to electrical energy in different ways. For example, in the electro-dynamic type of microphone a flexible diaphragm is made to vibrate when in the path of a stream of sound waves. The diaphragm's movements are transferred to a coil of wire in the presence of a magnetic field, thus inducing a current in the coil through a phenomenon known as 'electromagnetic induction'. This varying electric signal has a voltage pattern that is a replica of the sound wave pattern impinging on the microphone, i.e. it is an analogue signal.

The telephone receiver (earpiece) is also an electro-acoustic transducer, but works in the opposite direction to the microphone. For example, with an electro-dynamic device at the receiving end of the circuit the analogue electrical signal is passed through a coil attached to an electromagnet, which is attached to a permanent magnet. The varying electrical signal in the coil induces a magnetic field (electromagnetic induction) to vary similarly, which reacts against the bias field made by the permanent magnet, thus causing a diaphragm in the vicinity to vibrate in sympathy. In this way, the diaphragm generates a set of sound waves, which are a reasonable reproduction of the original sound of the speaker.

At the receiving end of the circuit the analogue electrical signal energises the receiver (i.e. an earpiece), generating a set of sound waves, which are an approximate reproduction of the sound of the speaker.

Obviously for conversation to be possible it is necessary to have transmission in both directions, and therefore a second circuit operating in the opposite direction is required, as shown in Figure 1.1(b). Thus, a basic telephone circuit comprises of four wires: one pair for each speech direction. This is known as a '4-wire circuit'. In practice, of course, a telephone system would need to include a mechanism for the caller to indicate to the recipient that they wished to speak. Therefore, we need to add to the assembly in Figure 1.1(b) a bell associated in a circuit with a power source and a switch. The electrical current flowing in a circuit used to ring a bell in a telephone is known as 'ringing current'. Again, one such arrangement is required in each direction. This argument brings us to the conclusion, illustrated in Figure 1.1(c), that a set of eight wires, four pairs, is needed to provide bidirectional telephony service between two people.

In a practical telephone network, the most important requirement is to mini-mise the cost associated with connecting each customer. Clearly, the provision of eight wires for each connection between users in a telephone network would be extremely costly. However, some ingenious engineering has enabled significant economy to be achieved through the reduction of the numbers of wires from eight to two, i.e. one pair. This is achieved using the 4-to-2-wire conversion (and vice versa), and the time-sharing of functions, as described below.

(i)   The 4-to-2-wire conversion. The two directions of speech circuits, show in Figure 1.1(c), can be reduced to a single circuit carrying speech currents in both directions, using a device known as a hybrid transformer, as shown in Figure 1.3 (see Box 1.3 for a brief explanation of how a hybrid transformer works).

(ii) Time-sharing of functions. The need for two pairs of wires to be dedicated to ringing circuits can be eliminated by exploiting the fact that ringing does not occur during the speaking phase of a telephone call. Therefore, the single pair provided for speech can instead be used at the start of a call to carry ringing current in either direction, as necessary. Once the call is answered, of course, the single pair is used only to carry the two directions of speech current.

Figure 1.4 shows that, for our simple two-person scenario, the telephone instrument at each end needs to comprise a handset with microphone and receiver; a hybrid transformer; a bell; and a means to send ringing current to the far end. The two telephones need to be connected by a single pair of wires and a battery.

We can now extend this basic two-person scenario to the more general case of several people with phones wishing to be able to talk to each other. For example, the logical extension to a four-telephone scenario is shown in Figure 1.5. In this case, six links (i.e. 2-wire circuits) are required in total, each telephone terminating three links. Not only does each telephone need to terminate three links rather than just one, but it also needs a 1-to-3 selection mechanism to choose which of the links should be connected to converse with the required telephone. (Not only does this

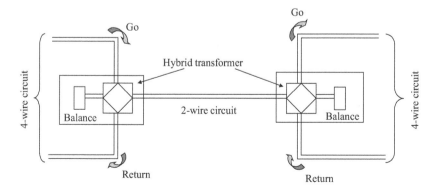

Figure 1.3   The 4-to-2 wire conversion

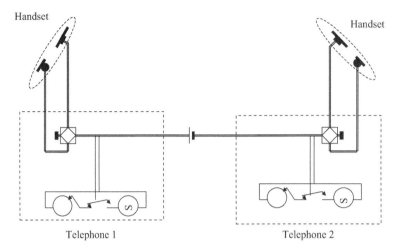

Figure 1.4   A simple two-phone system

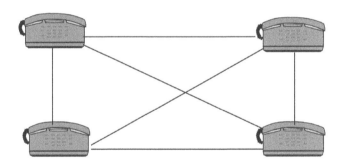

For four phones, number of links required = 6
Generally: For *n* phones, number of links required = $n(n-1)/2$

Figure 1.5   Direct interconnection of several phones

involve a selection switch within each phone, but the arrangement also needs each phone to be designated with a name or number, as discussed later in Chapter 10.)

Whilst the arrangement shown in Figure 1.5 is quite practical for networks of just a few phones – indeed, many small office and household telephone systems are based on such designs – it does not scale up well. In general, the number of links to fully interconnect *n* telephones is given by $n(n-1)/2$. As the number of phones becomes large, the number of directly connected links approaches $n^2/2$. Clearly, providing the necessary 5,000 direct links in a system serving just 100 telephones would not be an economical or practical design! (In addition, the complexity of the selection mechanism in each telephone would increase for it to be capable of switching 1-out-of-99 lines.) The solution to the scaling problem is to introduce a

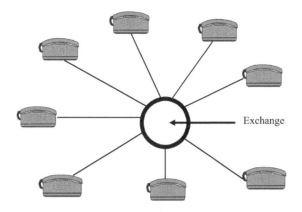

For *n* phones, number of links to each exchange = *n*

Figure 1.6   Interconnection: Use of an exchange

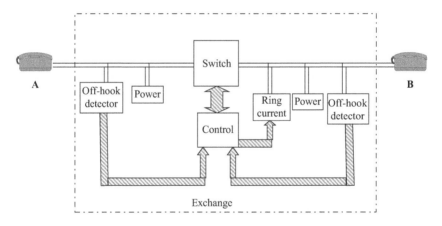

Figure 1.7   The functions of a telephone exchange

central hub (commonly called an exchange or central office) onto which each phone is linked directly, and which can provide connectivity between any two phone lines, on demand (Figure 1.6). With a single exchange serving *n* phones only *n* links are required; a good solution, which in practice scales up to about 50,000 telephone lines with modern telephone exchanges.

We can now deduce the role of a telephone exchange. Figure 1.7 gives a block schematic diagram of the basic functions required to connect two exchange lines. In the example shown it is assumed that telephone A is calling telephone B. The first requirement is that both telephones A and B need to have an appropriate power source. Although a battery could be provided inside each telephone, indeed in the early days of telephony this was in fact done, it is far more practical to locate the battery centrally at the exchange, where the telephone company (usually known as a 'Telco') can maintain it. When telephones are connected to their exchange by a pair of metallic wires, usually made of copper, the power for the phones can conveniently be passed over that pair. This arrangement is convenient for the telephone user because they have no need to either manage the charging of batteries in their premises, nor are they dependent on the reliability of the local electricity supply.

However, there are situations where power cannot be passed to the telephone from the exchange. For example, this is not possible when optical fibre is used to connect telephones because glass does not conduct electricity! The other notable example is that of a mobile network, where a radio path is used to link telephones to the exchange, as described in Chapters 2 and 9.

The first step the exchange must undertake in managing a call is to detect that the calling telephone (i.e. telephone A in Figure 1.7) wishes to make a call. The simplest method for conveying such an indication from a telephone, and the one that is still most commonly-used today, is for advantage to be taken of the fact that the pair of wires between the exchange and the telephone can be closed at the telephone end, thus creating a loop. This looping of the pair by the telephone causes a current to flow, which operates a relay at the exchange. (A relay is a device that when activated by an electrical current flowing through its coil causes one or more switches to close. The latter in turn then pass electrical current on to other circuits or devices, hence the term 'relay'.) In the case of the original manual exchanges, the closure of the relay switch caused a lamp to glow, hence alerting a human operator to the calling state of the line. For a modern electronic automatic exchange the energising of the relay by the loop current causes changes in the state of an electronic system, which is subsequently detected by the exchange control system.

The loop is closed in the telephone by a switch activated by the lifting of the handset off the telephone casing; this causes a small set of hooks to spring up. The lifting of the handset creates a condition known as 'off-hook'. In the case of a cordless phone this off-hook switch is in the base unit (attached to the copper pair termination in the premises) and is controlled remotely from the radio handset when the subscriber presses the 'dial' or 'send' button, often indicated on the button by an image of a handset being lifted.

On the outgoing side of the exchange attached to the line for telephone B there is a power source, a generator to send ringing current, and a relay to detect the 'off-hook' condition when the called telephone answers. (For simplicity, Figure 1.7 assumes that telephone A is calling telephone B so the ringing current generator is shown attached only to the line B, but of course any telephone can initiate calls, so in practice all lines have power supply, off-hook detector, i.e. relay, and a ringing current generator.)

When telephone exchanges were first introduced the method of connecting two telephone lines together was through a human operator using a short length of a pair of wires across a patching panel. Each telephone line is terminated on the panel in a socket with an associated small indicator lamp. The human operator was made aware that telephone A wished to make call by the glowing of the relevant lamp, activated by telephone A going 'off-hook'. On seeing the glowing lamp the operator started the procedure for controlling a call by first talking to caller A and asking them to which number they wished to be connected. The operator then checked the lamp associated with the called line, if this was not glowing then that line was free, and the call could be established. The next step was for a ringing current to be applied to telephone B. The operator did this by plugging a line connected to a special ringing generator into the socket for telephone B. It was important for the operator to monitor B's lamp to ensure that the ringing was stopped as soon as B answered, otherwise there would be a very annoyed person at the end of the line! The operator would then make the appropriate connection using a jumper wire across the patching panel. Finally, the operator needed to monitor the two lamps involved in the call so that the connection could be taken down (by removing the connecting cord between the two sockets) as soon as one of the lamps went out. The call had then been terminated.

In making a call connection, the operator had to follow certain procedures, including writing on a ticket the number of the caller and called lines, the time of day and the duration of the call, so that a charge could be raised later. It is important to remember that an exchange needs to serve many lines and that at any time there will be several calls that need to be set up, monitored, or cleared down. The human operators had to share their attention across many calls; each operator typically was expected to be able to deal with up to six calls simultaneously.

Generally, today telephone exchanges are fully automatic. However, there are still occasions when a human intervention is required, for example, providing various forms of assistance and emergency calls, and special automanual exchanges with operators provide such services. For convenience, the telephone exchanges considered in the remainder of this book will be only the fully automatic types. The description above is based on a fully manual exchange system because it enables the simple principles of call connection to be explained in a low-technology way – yet, all the steps and the principals involved are followed in automatic exchange working [3,4].

In an automatic exchange, as shown in Figure 1.7, the role of the operator is taken by the exchange control (representing the operator's intelligence), which in modern exchanges is provided by computers with the procedures captured in call-control software, and the switch or 'switch block', usually in the form of a semiconductor matrix, which performs the functions of the connecting cords and patch panels. Chapter 6 describes modern exchanges in more detail.

1.3   A telephone network

A telephone exchange serves many telephone lines, enabling any line to be connected to any other (when they are both free). In a small village, all the lines could easily be connected to a single telephone exchange, since the distances are short. However, if telephone service needs to be provided to a larger area, the question arises as to how many exchanges are needed. This is illustrated in Figure 1.8, where a region of the country has a population of telephones to be served. They could all be served by one central large exchange or by several smaller exchanges. Obviously,

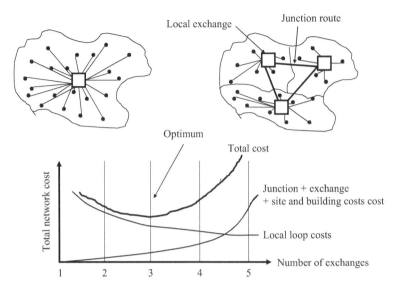

Figure 1.8   Network optimisation [Ward]

the lengths and hence costs of the telephone lines reduce as the number of exchanges increases, but this saving is offset by the increase in costs of exchange equipment and buildings. In addition, a link (known as a 'junction route') needs to be established between each exchange to ensure full connectivity between all telephones in the region. The trade-off between the cost of the telephone lines and the costs of the switching and buildings and junction-route costs plotted for various numbers of exchanges, *n*, to serve the population of telephones in the region shows a typical bath-curve shape [5]. In this example of Figure 1.8, the optimum cost is achieved with three exchanges.

Of course, it is not only the number of exchanges that contributes to the optimum costs, but also the location of the exchanges within their catchment areas. The optimum total cost for the region being achieved when the exchanges are at the centre of gravity (i.e. the location where the sum of all the line lengths is the minimum) of the population of telephones served. In practice, network operators locate telephone exchanges as close to this centre of gravity as possible, within the constraints of the availability of suitable sites within a town.

There are also practical limitations on the lengths of telephone lines which constrain the size of the catchment area of lines dependent on one exchange. These limits are set by the electrical characteristics of the lines, predominantly the resistance of the loop (see Box 1.1 for an explanation of resistance). Typically, this resistance needs to be less than 2,000 ohms (written as '2,000 Ω') to ensure that sufficient current flows for the 'off-hook' condition to be detected by the exchange, and also to ensure that the loudness of the call is acceptable. There are several ways in which the telephone lines can be kept within the electrical limits, including the use of thicker gauge wire (more expensive, but having less resistance) on the longer telephone lines. Also, in the United States, where the terrain requires larger catchment areas, inductors (i.e. devices comprising tightly coiled wire), known as 'loading coils', are added to long lines to reduce the signal loss. The majority of telephone lines in the United Kingdom are below 5 km, whereas in the United States there are many lines more than 10 km and exchange catchment areas can be as large as 130 square miles [6]. These aspects are considered in more detail in Chapter 5.

Thus, the primary design requirement for a network operator is to achieve a cost optimised set of exchanges, each of which is located at the centre of gravity for the catchment area, whilst keeping within the electrical limits on telephone line lengths.

We can now consider extending this logic to a telephone network serving several regions, or even the entire country. Given that each exchange needs to be able to connect to every other exchange (via junction routes), a further network optimisation issue arises. For example, for an area comprising 26 exchanges the number of fully interconnected junction routes would be $26(26 - 1)/2 = 325$, as shown in Figure 1.9(a). (For a country like the United Kingdom with some 6,000 local exchanges the number of directly connected junction routes would be an impossible 17,997,000!) As before, this situation is alleviated using a central exchange within each region, known as a trunk exchange. These trunk exchanges also need to have links (i.e. 'trunk routes') between them to ensure full connectivity across all regions. In the example of Figure 1.9(b) this arrangement requires just 15 trunk links and 26 junction links, but with the added cost of six trunk exchanges. Further network optimisation is possible by the addition of a single trunk tandem exchange to provide connectivity between each of the trunk exchanges, reducing the number of trunk links to six, but with the added cost of a trunk tandem exchange (see Figure 1.9(c)).

The arrangement of exchanges and links can be drawn as a traffic routeing hierarchy, as shown in Figure 1.9(d). (The term 'traffic', which is used to describe the flow of telephone calls, is described in more detail in Chapter 6.) The bottom of the hierarchy is the exchange serving its catchment area of telephone lines. These exchanges are known as 'local exchanges' in the United Kingdom (and 'Class 5 central offices' in the United States [6]). Similarly, telephone lines are known as 'local lines' or 'subscriber lines', and the collection of local lines as the 'local network'. At the second level of the hierarchy are the trunk exchanges, and at the top of the hierarchy is the trunk tandem exchange. This parenting of trunk exchanges on higher-level exchanges can continue further. However, in practice telephone networks generally have no more than three levels of trunk exchanges: primary, secondary and tertiary. Note that only the local exchanges have subscriber lines attached, the various levels of trunk exchanges switch only between trunk and junction routes to and from other exchanges. The significance of this difference between local and all other types of exchanges will be considered further in Chapter 6.

As an example of a typical public switched telephone network (PSTN), we will consider the simplified representation in Figure 1.10. This shows the PSTN for British Telecommunication (BT), which covers all the United Kingdom (except for

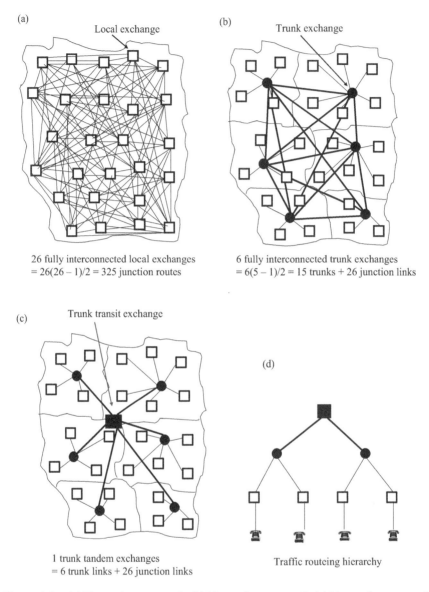

Figure 1.9   (a) Network structure 1. (b) Network structure 2. (c) Network structure 3. (d) Network structure 4 [Ward]

the town of Hull). The structure is shown as a hierarchy, with the exchanges serving subscriber lines forming the lowest level. For reasons that will be explored in Chapter 6, there are three ways of connecting subscriber's lines to the network:

Type (i), where subscribers are connected directly to the central local exchange (known as a 'processor node').

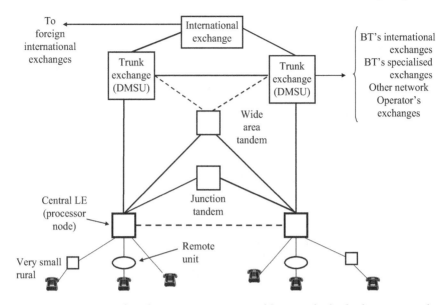

Figure 1.10   British Telecommunication's public switched telephone network

Type (ii), where subscribers are connected to a remote unit (known as a 'remote concentrator'), which in turn is parented onto and controlled by the central local exchange.

Type (iii), where subscribers are connected to an autonomous very small rural exchange, which in turn is connected to the central local exchange.

In BT's PSTN there are about 800 of the central local exchanges (processor nodes), about 5,200 remote units (remote concentrators), and some 480 very small rural exchanges. In total, these exchanges and units serve about 29M subscriber's lines [7]. The (central) local exchanges are parented onto their trunk exchange (known as a digital main switching unit (DMSU)). This hierarchy has only one level of trunk switching, with all trunk exchanges having direct routes to all others. This arrangement ensures that any subscriber can be connected to any other on this network and that there will be a maximum routeing of two trunk exchanges between the originating and terminating local exchanges. Calls between local exchanges, which are in the same locality and where there is sufficient demand, are carried over direct links, known as 'junction routes', as shown in Figure 1.10. The number of junction routes is minimised in large urban areas, such as London, Birmingham and Manchester using junction tandem exchanges (known as digital junction switching unit (DJSU)). Further network optimisation is achieved by off-loading the trunk exchanges of calls flowing just within the region by switching these at regional tandems (known as wide area tandem (WAT)). BT's PSTN has some 15 junction tandems, about 20 regional tandems and around 80 trunk exchanges.

In addition to providing full interconnectivity between all subscribers on its network, BT's PSTN also provides access to international exchanges within the United Kingdom for calls to other countries. As Figure 1.10 shows the international exchange is above the trunk exchange in the hierarchy. BT is one of several network operators in the United Kingdom that have international exchanges (known as international switching centre (ISC)). The routeing of international calls is considered in more detail in Chapter 2.

The PSTN of BT also acts as a collector of calls from its subscribers to the networks of other operators (fixed and mobile) in the United Kingdom, and similarly as a distributor of calls from other operator's networks to subscribers on the BT network (Figure 1.10). The points of interconnection between BT's PSTN and the other networks are at the trunk exchanges. Similarly, the trunk exchanges also act as collectors and distributors of traffic between PSTN subscribers and the variety of specialised BT exchanges covering functions such as operator services and directory enquiries, Centrex and virtual private network (VPN) service for business customers, and network intelligence centres for the variety of calls that need more complex control than the PSTN exchanges can provide. Chapter 2 considers how these specialist exchanges are linked to the PSTN, whilst the intelligence, data and mobile networks are fully described in Chapters 7, 8 and 9, respectively.

1.4   How does a network set up a call connection?

A fundamental requirement of routeing through any telecommunications network, whether it is to carry a voice call or a packet of data, etc., is that each termination (or endpoint) on the network has an address, which can be used to indicate the desired destination. In the case of telephony, each subscriber's line has a permanent unique number, which is used by the exchanges in the network as its address.

We are now able to consider a call set up across the PSTN. Figure 1.11 illustrates the routeing of a national call within the United Kingdom from a subscriber in Penzance in the South West tip to a subscriber in Brighton on the South coast. As an example, we assume that the subscriber in Penzance, *01736 89YYYY* (i.e. Area Code *01736*, exchange code *89* and subscriber number *YYYY*), dials their friend in Brighton, *01273 34XXXX* (i.e. Area Code *01736*, exchange code *34* and subscriber number *XXXX*). There are several local exchanges in the Brighton area, two of which (*34* and *39*) are shown in Figure 1.11.

On receiving the dialled digits from the caller, the control system of the local exchange in Penzance (the 'originating exchange') examines the first five digits, as shown by the underlining of the numbers in Figure 1.11. The initial '0' indicates to the control system that a national number has been dialled. Since the control system does not recognise *1273* as either its own code or that of any other exchanges to which it has direct routes, the call is switched through to its parent trunk exchange in Plymouth. This is achieved by the sending of signals between the control systems in Penzance and Plymouth exchanges (shown by dotted arrows), conveying the required destination number. The number information sent in the forward direction between the exchanges is shown between curly brackets in Figure 1.11.

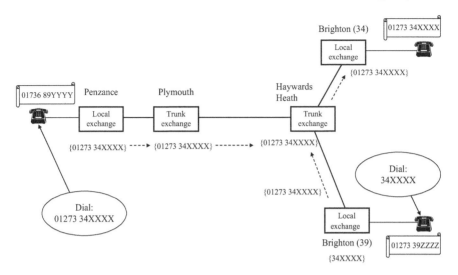

Figure 1.11   An example of a call routeing

The control system in Plymouth exchange examines digits *01273* and concludes that this is not one of its dependent Area Codes, but is owned by Haywards Heath trunk exchange. In the United Kingdom network all BT trunk exchanges are fully interconnected and so the control system of Plymouth exchange selects a direct route to Haywards Heath and signals the full destination number.

Haywards Heath exchange control system recognises *01273* as one of its dependent Area Codes and so it examines the next two digits, *34*, to determine the destination local exchange. The Hayward's Heath trunk exchange then routes the call to the Brighton-34 exchange. At the destination exchange, the control system on recognising *0123 34* then examines the final *XXXX* to determine called subscriber number.

In the case of a local call from the other Brighton exchange, 39, the set-up sequence is simpler, as shown in Figure 1.11. The calling subscriber dials only the local number since the destination number has the same Area Code. The control system of Brighton-39 on examining the first two digits, *34*, recognises that the call is destined for one of the other exchanges dependent on the Hayward's Heath trunk exchange, and so routes the call accordingly. Notice that, in this case of a local call, the control system inserts the appropriate Area Code and signals the full destination national number to its trunk exchange. Hayward's Heath trunk exchange then uses the standard decoding procedure to determine that the call is destined for Brighton-34 exchange; the call is then completed as before.

The dialled number is not only used by the exchanges to route the call through to its destination, but importantly the number is also used to determine the appropriate charge rate for the call. In both examples above, it is the parent trunk exchange that determines the charge rate for the call and indicates that this to the originating local exchange, which then times the call and records the incurred charge in its control system memory. This is later downloaded to produce an entry on the calling subscriber's telephone bill.

1.5   Waveforms

It is important at this point to ensure an understanding of the basic concept of waveforms and how they will be used throughout the remainder of the book.

In general, all communication systems are characterised by the range of frequencies that can be carried, usually termed the 'bandwidth' of the system. Figure 1.2 shows the classical sinusoidal waveform showing the regular oscillating pattern that occurs often in nature, for example, a swinging pendulum. The time between successive peaks or troughs in the amplitude of the waveform (which might be length, voltage, power, etc.) is known as the 'period'. Each period comprises of one full cycle of the waveform. The distance travelled in this time is called the wavelength, given the symbol λ. The number of periods per unit time is known as the frequency of the waveform, measured in cycles per second (cps), or to use the international standard, Hertz (Hz) [1].

The sinusoidal waveform of Figure 1.2 in the case of a sound wave would represent a single tone. For example, a single sinusoidal sound wave of 256 Hz produces the well-known musical note of middle C. However, speech and music are a mixture of many different tones, each with different frequencies and amplitudes, resulting in complex waveform shapes, as shown in Figure 1.1(a).

Within this book two fundamentally different forms of waves are considered. The first is that of *sound waves*, which are moving physical vibrations in a substance. These can be heard and felt by humans. Thus, sound waves are carried through the air in the form of vibrating air molecules, for example, from a speaker to a listener, or a guitar string being plucked. Sound waves can also be carried as vibrations through a wooden door, through water, or as vibrations along a railing which someone at the far end is banging, or as vibrations along the railway lines as the distant train approaches. Sound is generated by vibrating the air, as is apparent when a finger touches a loud speaker of a HiFi system at full volume! Vibrations carried along the surface of the sea (i.e. sea waves) are another example of sound waves.

The other forms of waves are *electromagnetic waves*. Light is the most obvious example of electromagnetic radiation. Light travels as very high-speed waves of electromagnetic forces. Other examples of this type of waveform are radio waves (carrying radio and TV broadcasts, etc.), microwaves and X-rays. They are all examples of the phenomenon of electromagnetic radiation, but at different frequencies. Thus, within telecommunications networks, it is electromagnetic waveforms that are carried as electricity over metallic wires, or as radio waves through the air from radio masts to a mobile handset, or as light waves through optical fibre cables. The full range of the electromagnetic radiation spectrum is illustrated in Figure 1.12.

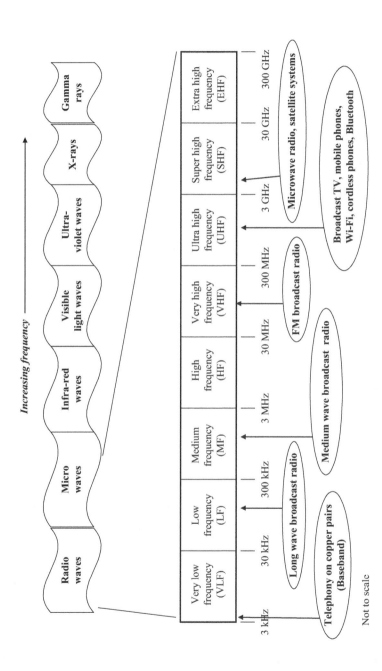

Figure 1.12   The electromagnetic radiation spectrum

Box 1.3   How a hybrid transformer works [2]

The Hybrid transformer is a system comprising of four input terminals, two for the Go signal and two for the Return signal of the 4-wire circuit, and two output terminals for the 2-wire signal. The role of the Hybrid system is to ensure that the Go signal is conveyed to the 2-wire circuit only and the signal from the 2-wire circuit is conveyed to the Return terminal only. This separation is achieved using a pair of transformers, each with two sets of windings, as shown in Figure 1.13. The windings are arranged such that the induced currents from the Go signal, which would otherwise appear at the Return terminals, are equal and opposite and so cancel, with the power dissipated in a resistance device (the 'balance') matching the electrical characteristics of the 2-wire circuit. The signal from the Return circuit, which would otherwise go into the Go circuit, is similarly cancelled. In this way, the Hybrid transformer ensures that the two directions of electrical signal on the 2-wire circuit are transferred to the appropriate Go and Return parts only of the 4-wire circuit.

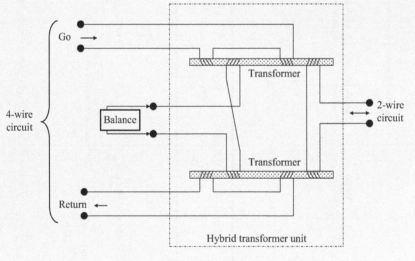

Figure 1.13   Hybrid transformer

The key point is that the sound waves emitting from a speaking person are converted into electromagnetic waves for the conveyance through the telecommunications network. (Thus, in Figures 1.1 and 1.3, the sound waves are carried over the pair of wires as an electromagnetic signal over the wires, and converted back to sound waves at the far end.) The sound or speech waveform (occupying the frequency range of 0–4 kHz as described in Chapter 3) can be carried by an electromagnetic wave at any appropriate frequency level. Thus, if carried straight over a copper pair the sound waveform is converted to an electrical analogue signal occupying the same frequencies (0–4 kHz); or they may be carried at higher frequencies within multiplex signal (as described in Chapter 3), or at radio frequencies if carried over long wave, medium wave or FM radio, or at even higher frequencies in the electromagnetic spectrum if carried as part of a TV broadcast, and so on up to the highest frequencies when the speech is carried as light waves over an optical fibre cable.

1.6   Summary

In this opening chapter, we introduced the basics of telephony, in which economies are gained using of a single pair of wires to convey a two-way telephone call over the link to the serving exchange. We also examined the role of local, trunk and international exchanges, noting that they can be manual or automatic. This led to an understanding the basic structure of a PSTN. A simple example of a call across the country was examined to illustrate the sequence of events involved in progressing a telephone call. This highlighted the roles of numbering and addressing, call control, and the need for signalling between the control systems of exchanges involved in the call. We also introduced the concept of the two forms of waveforms used in telecommunication networks. Subsequent chapters develop these basic concepts.

References

[1]   Langley, G. and Ronayne, J.P. *Telecommunications Primer*, 4th edn. Boston, MA: Prentice Hall; 1993, Chapter 6.

[2]   Bigelow, S.J., Carr, J.J., and Winder, S. *Understanding Telephone Electronics*, 4th edn. Boston, MA: Newnes; 2001, Chapter 2.

[3]   Redmill, F.J. and Valdar, A.R. 'SPC digital telephone exchanges'. *IET Telecommunications Series No. 21*. Stevenage: Peter Peregrinus Ltd. on behalf of the Institution of Electrical Engineers, 1995, Chapter 1.

[4]   Anttalainen, T. *Introduction to Telecommunications Network Engineering*. Norwood, MA: Artech House; 1999, Chapter 2.

[5]   Flood, J.E. *Telecommunications Switching, Traffic and Networks*. Harlow: Prentice Hall; 2001, Chapter 1.

[6]   Bigelow, S.J., Carr, J.J., and Winder, S. *Understanding Telephone Electronics*, 4th edn. Boston, MA: Newnes; 2001, Chapter 1.

[7]   France, P.W. and Spirit, D.M. 'An introduction to the access network'. In France P.W. (ed.), *Local Access Network Technologies. IET Telecommunications Series No. 47*. Stevenage: The Institution of Electrical Engineers, 2004, Chapter 1.

Chapter 2
The many networks and how they link

2.1 Introduction

In Chapter 1, we considered how a public switched telephone network (PSTN) establishes telephone call connections between subscribers on its network. However, this is only part of the story. Firstly, there may be several PSTNs in any one country, each owned by different operators, although some may have a limited geographical coverage or serve only certain (typically corporate business) customers. Many countries also have cable TV operators who provide telephone service, in addition to broadcast TV distribution. Then, of course, nearly all countries now have one or more mobile telephone networks, which are separate from the so-called 'fixed' PSTNs. All these networks need to interconnect so that their subscribers can call subscribers on any of the other networks. This chapter, therefore, briefly describes the various telecommunication networks and how they all interconnect.

Although the fixed PSTN and mobile networks were originally primarily designed for voice communication, they do also enable subscribers to connect their computers to the Internet (a network primarily designed for data service). Thus, to complete the interconnect story, this chapter briefly introduces the concept of the Internet and how access to it is provided by the telephone networks.

The second aspect of the story is the need for a set of specialist networks which enables a typical national telephone operator to provide a wide range of services beyond simple telephone calls. This chapter briefly reviews these specialised networks and uses a simple model to help explain how they link and support each other.

So, clearly there are many networks in each country all linking appropriately to provide a variety of types of call connections between subscribers across the World. (Not to mention the interconnection of the various nonvoice or data networks! This, for simplicity, is not covered until Chapter 8.) We will now try to understand this complex picture.

2.2 Other forms of telephone networks

2.2.1 Mobile networks

The concept of a cellular mobile telephone network is shown in Figure 2.1. The principle of operation is like that of the PSTN, as described in Chapter 1, although

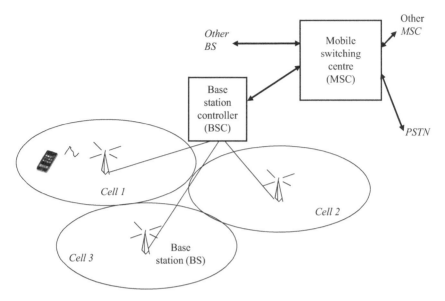

Figure 2.1   Mobile cellular network concept

there are some important differences. Obviously, the first difference is that a two-way radio link is used to connect the mobile handset to its exchange, rather than a copper pair of wires, so that the user has freedom of movement. This ability of the handset to move around means that the serving exchange must have a system, to identify a calling handset and another system to keep track of where a particular handset is at any time, so that it can send or receive calls.

The basis of the 'access network' is a set of cells. These cells form an area ranging from about 1 km to 10 km radius and have a centrally-located radio transmitter/receiver (also known as a 'transceiver'), collocated with a base station (BS). A two-way radio link is potentially available to all handsets in the cell area. Groups of radio-channel pairs (Go and Return), each provided by a separate set of frequencies, are preassigned to the cell, and a pair of radio channels is allocated on demand by the base station controller (BSC) to a mobile handset wishing to make or receive a call.

A BSC serves a catchment area of several cells with their associated BSs. The BSs are connected to their BSC by a fixed transmission link carried over either a point-to-point microwave radio system or an optical fibre cable.

During the call, the hand set is free to move within the cell. However, if the handset travels towards the boundary of its cell during a call the weakening signal from the BS and the strengthening radio signal from the adjacent cell's BS is detected by the BSC and a 'handover' between the two is managed without interruption to the call. This requires changing the send and received channels to those of the new cell. (If a spare set of radio channels is not currently available within the new cell, the call drop out.) At the end of the call, or if the terminal moves into another cell's area, these radio channels are available for use by other calls.

The mobile switching centre (MSC), which is similar to a large local PSTN exchange (but without the subscriber terminations), performs the switching of the mobile calls between all handsets operating within its catchment area of BSs or to other MSCs on its network. The MSC is also associated with control systems providing terminal-location management and authentication [1]. The terminology in the above description relates to GSM (global system for mobile communication), subsequent generations of mobile systems use different terminology but the principles of cellular operation is broadly similar. A more detailed account of mobile networks is given in Chapter 9.

2.2.2   Cable TV networks

Cable TV networks (often referred to as just 'Cable Networks') are primarily based around a cable-distribution system, often deployed in only certain areas of a town, which delivers many TV broadcast channels primarily to the residential market. In some countries, for example, the United Kingdom, these networks also provide telephone service to the premises taking their broadcast TV service. Figure 2.2 illustrates the principle of such a network. Two separate feeds serve each customer's premises: A coaxial cable (like TV aerial cable) terminating on a set-top box (STB) linked to the television set provides TV broadcast service, and a copper pair of wires provides telephone service to the household. The coaxial cable is part of a tree and branch arrangement radiating from a street cabinet, serving customers located along one or a few roads in the vicinity. A separate pair of copper wires back to the street cabinet is required for each household served. At the street

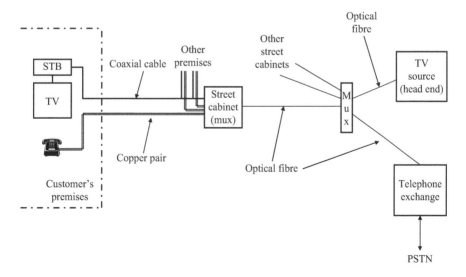

Figure 2.2   Cable TV network

cabinet, the TV distribution and the telephone lines are combined, i.e. multiplexed (see Chapter 3), onto an optical fibre cable back to the network centre. Here the TV channels are provided from the TV source, known as the head end, which may receive the TV channels from a satellite or other terrestrial links to TV broadcast companies and possibly a TV channel switching and control centre. At this network centre the telephone channels are connected to the cable company's telephone exchange [2].

The telephone portion of the cable TV network is similar to the standard PSTN. However, the catchment areas of the exchanges are usually much larger than those of the PSTN. In addition, the cable companies tend to combine the functions of local and trunk switching in each exchange. One or more of the exchanges act as the gateway to the PSTN and possibly mobile and other networks.

2.3   Interconnection of networks

2.3.1   *International calls*

As shown in Figure 1.10 of Chapter 1, international exchanges (known as 'international switching centres' (ISCs)) are above the trunk-exchange level in the PSTN routeing hierarchy. Several network operators in the United Kingdom have ISCs linked to their PSTN. A subscriber wishing to call a number in another country first dials the international prefix (internationally recommended to be *00*) to indicate to its network that this is an international call. The next one, two or three digits, which form the country code indicating the destination country (see Chapter 10), are then examined by the gateway ISC serving that PSTN. Traffic between ISCs in different countries is directly routed where there is a high volume of calls between the two countries, otherwise it is routed via one or more transit ISCs located in intermediate countries. A simple summary of these international routeing options is shown in Figure 2.3. This illustrates an example in which Country 1 provides international transit switching for calls between Country 3 and Country 4, and calls between Country 1 and Country 2 are directly routed.

In addition to determining the appropriate route to the destination country, direct or to a transit ISC, the originating gateway ISC determines the appropriate rate to be charged for the call. This charging information is passed back to the originating local exchange to be associated with the calling subscriber's billing records.

The cost of building and maintaining the transmission links between ISCs is shared between the two operators concerned. In the case of a trans-border land cable, ownership extends from the ISC to the border. Where there is a sea boundary, e.g. between the United States and the United Kingdom, a hypothetical boundary exists midway along the subsea cable, thus both operators own a so-called 'half circuit'.

Each operator pays the other for the cost of accepting traffic and completing the calls in the destination country. In practice, this is done by a periodic reckoning for each ISC (Country A)-to-ISC (Country B) route of the net flow of calls

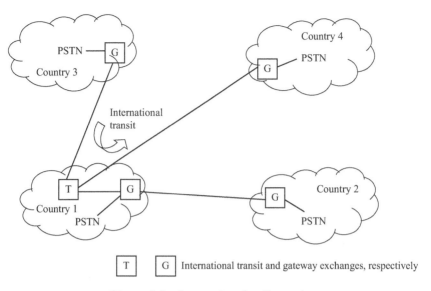

Figure 2.3   International call routeing

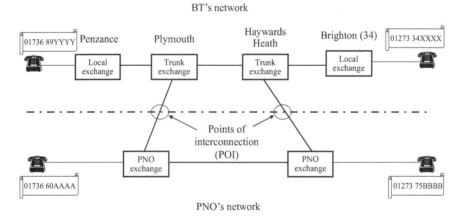

Figure 2.4   Interconnection with a PNO's network

(the difference between A-to-B and B-to-A), which is then charge at a determined tariff. Thus, the operator who sends the higher amount of traffic pays the other operator for the net termination charge.

2.3.2   Interconnection of a PSTN and a PNO's network

Figure 2.4 shows the situation in the United Kingdom where BT's (British Tele-communications) PSTN interconnects with several hundred other network operators, for example, other PSTN, cable TV network or mobile network in the same country.

These are generically referred to as public network operators (PNOs). We take as our example the situation described in Figure 1.11 of Chapter 1, but with the addition of two points of interconnection (POI) with the network of a PNO. Note that the subscriber on the PNO exchange in Penzance has the same area code, *1736*, as the BT subscriber, but has a different exchange code, i.e. *60*; similarly, the PNO exchange code at Brighton is *75* compared to *34* for the BT exchange. The two networks are connected at various POI, usually at the trunk exchange level in BT's network. All calls from BT subscribers destined for a PNO's network are routed to the destination exchange via the designated POI. Similarly, calls originating in the PNO network are routed to the destination BT exchange via the appropriate POI. As Figure 2.4 shows, there may be a choice of POIs to use for a call: the routeing either staying on the originating network until the last available POI – known as 'far-end handover', or interconnecting at the nearest POI – known as 'near-end handover'.

As an example, let us consider the routeing of a call from the BT subscriber in Penzance (*01736 89YYYY*) to the subscriber on the PNO network in Brighton (*01272 75BBBB*). The control system of the BT local exchange in Penzance, on examination of the nonlocal Area code *01273* passes the call onto its parent trunk exchange at Plymouth. The control system of Plymouth trunk exchange identifies *01273* as an Area code parented on Haywards Heath Trunk exchange and switches the call accordingly. The Hayward Heath trunk exchange, on recognising *01273*, examines the exchange code *75* and routes the call to the designated POI for the serving PNO, which in this case is a route from the BT trunk exchange to the Brighton PNO exchange. The call is passed to the PNO exchange, the signalling from the control systems of Haywards Heath giving the full dialled number (*01273 75BBBB*), so that the call can be completed within the PNO network – in this case directly to the called subscriber's line.

A similar arrangement applies to calls from a PNO to BT's fixed network.

The flow of calls is measured between the BT trunk exchanges and each of the interconnected PNO exchanges so that interconnect conveyance charges can be determined. Each network operator must pay the receiving operator for the conveyance and completion of calls passed over the POI. Normally, this follows the principle of a receiving operator charging on a call-minutes basis, according to the amount of their network traversed in delivering the call (either to its subscriber or onto another network for completion). The rate of charge for traversing a network depends on the type of network involved, the mobile networks being rated higher than a fixed network. Like most aspects of interconnect, these call-completion-charge rates are usually subject to scrutiny or determination by the national regulator [3,4].

2.3.3   Mobile-to-mobile call via the PSTN

Figure 2.5 illustrates the general situation for calls between two different mobile networks in the same country. It is interesting to note that there are two basic scenarios for needing such a configuration. The first is a call between a terminal

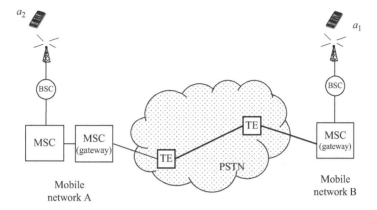

Figure 2.5   Mobile-to-mobile call via PSTN

belonging to (i.e. registered with a service provider using) mobile network A and a terminal belonging to mobile network B. The other is a call between two terminals belonging to mobile network A, but the second terminal has moved temporarily onto (i.e. visiting) mobile network B – a process known as 'roaming'.

Although calls between mobile networks in the same country may be delivered over direct links between MSCs on their respective networks, generally it is more economical to route calls to other mobile operators, as well as the wide range of fixed PNOs, via the national incumbent's PSTN, with its widespread coverage and full interconnectivity with all the operators in the country. Each mobile network has at least one MSC that acts as a gateway to the PSTN, providing the required technical interface between the two networks, as described in more detail in Chapter 9.

Before we consider the case of roaming, although technically possible it may not be allowed within a country for regulatory reasons (as is the case in the United Kingdom). However, roaming is allowed between operators in most countries. Let us consider the case where a terminal, a_1, has roamed from network A to network B (see Figure 2.5), and a call is to be made to it from another terminal, a_2, on network A. The location-control system of the MSC on Network B has already advised the location-control system of network A that terminal a_1 is currently on its network, so network A knows that it needs to pass the call from terminal a_2 over to network B [1,5]. On receiving a call completion request from network A the control system of the MSC on Network B allocates on a temporary basis (for the duration of the call) one of a small pool of PSTN numbers it holds, and advises the Control system of the originating MSC on network A of this number. The originating MSC then routes the call via its nominated gateway MSC to the PSTN, using the allocated PSTN number as the destination address. The MSC of network B can then associate the incoming call from the PSTN with the required BS currently serving terminal a_1. The call connection between a_2 and a_1 can then be completed. Interconnect charging applies across the boundary of network A to the PSTN, and across the boundary between the PSTN and network B [4]. In the case where the roaming is between countries the PSTN cloud shown in Figure 2.5 would be two PSTNs linked by an international call connection.

2.4   The Internet

The Internet, that widely-known entity, is really a super constellation of many networks around the world – a network of networks. Indeed, the name 'Internet' is derived from the term 'internetworking'. Each of these networks is based on a special form of switching designed specifically for handling data, using a standard way of packaging and addressing the data: the so-called Internet protocol (IP). The enormous utility of IP is due to the way that it enables data to be easily transferred between programs and applications run on different types of computers. (The subject of IP and data networks is covered in more detail in Chapter 8.) However, it was the introduction of the World Wide Web (WWW), enabling general wide-spread access to sections (i.e. 'pages') of data stored in people's computers that really caused the Internet to spread throughout the world and to become the phe-nomenon that it is today. A huge range of applications by business and residential users now exploits the Internet. The most common of these applications are social networking, e-mails and other peer-to-peer applications such as gaming, informa-tion gathering from web pages and Internet shopping.

Figure 2.6 presents a simplified schematic diagram of the Internet. Each of the component networks in the Internet constellation are owned and operated by

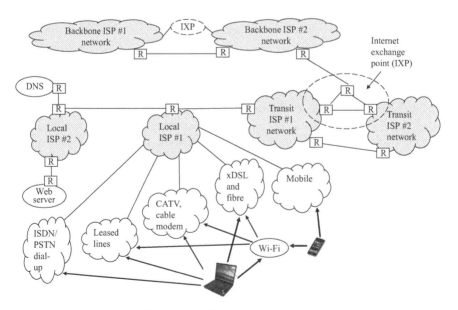

Figure 2.6   The Internet concept

Internet service providers (ISPs), shown shaded. Users and providers of information (in the form of web pages) subscribe to an ISP for their Internet service. ISPs networks usually comprise of several IP switching units – known as 'routers' (shown as *R* in Figure 2.6) located at various points in the country and linked by high-capacity transmission circuits leased from a network operator ('leased lines', as described in Section 2.5.5 later in this chapter). As the diagram shows, there is a hierarchy of ISPs, with Local ISPs providing a gateway to the Internet for its set of subscribers. Local ISPs are parented on one or more Transit ISPs, who provide links across the country. Full interconnectivity across the country is provided by cabling links between the Transit ISP's routers collocated in Internet exchange points (IXPs). The Transit ISPs also have connections to Backbone ISPs with their international coverage, who in turn have direct links to other Backbone ISPs, so spreading the Internet worldwide. The largest IXP in the United Kingdom is in the City of London: 'Telehouse'. This provides participants with full interconnectivity within the United Kingdom and to links going to the United States, Asia and mainland Europe.

Figure 2.6 also shows how a user gains access to a website, with a web server shown linked to Local ISP #2. It also shows a domain name server (DNS) connected to that ISP; this important device acts as a directory listing Internet addresses for web names, as described in Chapter 8.

Users access their ISPs over links provided by one of a range of fixed networks (PSTN, integrated services digital network (ISDN), leased line, xDSL (digital subscriber line; DSL) broadband and Cable TV), and mobile networks, as described in the Section 2.5 of this chapter. Figure 2.6 shows that for wires-free convenience users may link their terminals – laptops, tablets, etc. over Wi-Fi, to the fixed broadband links to the ISPs. Also, users with smartphones may access the Internet either over a mobile network at home or on the move or via the Wi-Fi in premises (e.g. residence, office, café, airport or hotspot) and hence over the fixed network to their ISP.

2.5   Access to the Internet

2.5.1   Dial-up via the PSTN and ISDN

The original way of accessing the Internet is through a telephone call, thus taking advantage of the full interconnectivity of the PSTN, its ubiquity and its ease of use. The procedure is in two stages: the first is a telephony call to the ISP; the second stage is a data transfer session between the computer and the Internet via the ISP. The arrangement is illustrated in Figure 2.7.

For the first stage, a standard call is set up from the computer, which is connected to a telephone line in the normal way. A modem (a shortening of 'modulator-demodulator') card in the computer makes it behave like a telephone instrument, providing the equivalent of lifting the receiver (off-hook), detecting dial tone and sending out the dialled digit tones. The modem also enables data to be passed over the standard local telephone line. The ISP is connected to the PSTN over a standard

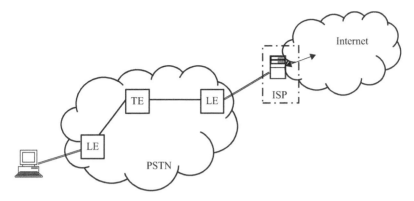

Figure 2.7   Dial-up access to the Internet

local line or by an ISDN local line, but in either case standard telephone numbers are allocated.

Once through to the ISP the second stage begins, the call connection through the PSTN now just acts as a two-way path between the computer and the ISP. An interchange of data occurs so that the ISP can verify the identity of the user and establish a data session to the Internet. IP is used between the computer and the ISP's router, as described in Chapter 8, passing over the transparent PSTN call-connection path. The ISP then routes the data between the computer and the Internet for the duration of the session. When the user indicates that the session is over, the modem card in the computer creates the equivalent of the 'on-hook' condition so that the telephone exchange clears the call to the ISP in the normal way.

Generally, the exchange handles the call to the ISP as standard telephony, in terms of charging and routeing. However, the users want to be able to run their often-lengthy sessions to the Internet, involving periodic flows of data as web pages are accessed or e-mails sent or received, without being concerned about the dura-tion and paying telephony charges on a timed basis. One solution is for the ISP to take a special service from the PSTN operator using *0800* numbers, which enables the users to have free or reduced rate call charges, the ISP being able to recover the charges through subscriptions from the registered users. In addition, the PSTN operator may offer special flat rate (i.e. un-timed) charges to subscribers for their calls to the Internet. Similarly, the incumbent PSTN operator may be required by the national regulator to make such special flat rate tariffs available as a wholesale interconnect service to other operators (PNOs) or ISPs, for example, FRIACO (flat-rate Internet access call origination) [6].

In addition to the needing special tariffs, the carrying of Internet-bound calls causes congestion for PSTN telephone operators due to the Internet sessions tend-ing to last longer than telephone calls, typically some 45 minutes compared to 3 minutes. Thus, telephone exchanges whose capacity is dimensioned assuming 3-minute average calls can experience severe congestion if the proportion of

Internet-bound calls becomes significant. The most effective way of avoiding this problem is the total segregation of telephony and Internet (data) traffic right at the source, i.e. the user's computer. Such segregation is provided by the xDSL (so-called 'broadband') and cable modem access systems as described later; which provide users with a high-speed Internet access by routeing their data traffic to the ISP independently of the PSTN, the latter handling just the telephony traffic.

Accessing the Internet via ISDN is essentially the same as for the PSTN, except that the interface of the ISDN line to the computer is digital (as described in Chapter 3) rather than analogue so the modem card does not convert the computer output to analogue and consequently higher speeds up to 64 kbit/s or even 128 kbit/s are possible. Although dial-up access to the Internet is now largely obsolescent, with broadband fixed access provided mainly by xDSL or cable modem, the ISDN service is still used by some business users as an alternative standby system. The ISDN service is discussed in more detail in Chapter 6.

2.5.2   Over xDSL broadband links

There are now a family of transmission systems being deployed to provide fixed broadband access using the existing copper pair telephone wires, a technique known as DSL. Figure 2.8 illustrates the main two members of the 'xDSL' family: asymmetric DSL (ADSL) and very high-speed DSL (VDSL).

Both ADSL and VDSL enable a high-speed data service to be carried in addition to telephony over the copper pair of a standard telephone line, as described

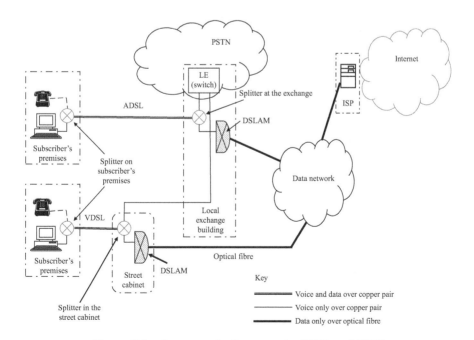

Figure 2.8   Access to the Internet via ADSL and VDSL

in Chapter 4. At the subscriber premises, the internal telephone line and the computer are separately connected to a 'splitter' within the ADSL termination equipment. This device directs the computer output to the high-speed data channel on the line, and similarly in the reverse direction; while enabling the telephone signal to be carried in the normal way on the local line. The ADSL splitter and terminating equipment located at the exchange similarly separates the telephony and high-speed data signal [7]. The telephony is carried over a copper pair to the input of the local exchange switch. The data stream carrying the Internet-bound traffic is connected to the DSLAM (digital subscriber line access module) equipment, which combines the data from many such lines onto a high-speed transmission link to a data network. This network carries the Internet-bound data in an efficient way, interleaving it with data streams from other sources, to the connecting points of the various ISPs in the country. A corresponding process applies in the reverse direction from the Internet to the computer.

Charging for access to an ISP (and hence the Internet) may be purely on a subscription basis, or in the case of high-capacity users there may also be usage or throughput charges.

The VDSL is an important enhancement of this DSL concept in which higher data speeds are achieved by keeping the copper pair short, terminating it on a splitter and DSLAM in a nearby street cabinet [8]. The high-speed data channels is carried over an optical fibre to the data network, leaving the telephony-only signal to be carried over the remaining copper line back to the telephone exchange, as shown in Figure 2.8.

Chapter 4 describes the xDSL technologies in more detail, together with a further development of the concept known as G.fast.

2.5.3   Over a cable modem

Cable TV networks provide a high-speed data link to an ISP over one of the spare TV channels carried through the network. The technique requires a so-called 'cable modem', which converts the output of the computer into a signal compatible with the TV distribution system. As Figure 2.9 shows, the computer is connected to splitter-like device which segregates the received TV signal from the broadband data stream at the subscriber's premises, where both are carried as a composite TV signal over the coaxial cable from the street cabinet. Since many subscribers are sharing the data modem channel on the coaxial cable tree, some form of combining of the many signals is required; the street cabinet therefore performs a similar aggregating function as the DSLAM, described above, following the DOCSIS (data over cable service interface specification) industry standard for cable TV networks [9]. As described for xDSL access, the Cable TV operator routes the aggregated data traffic over a data network to the connecting points of the appropriate ISPs. Charging for access to an ISP may be on a subscription basis or usage basis.

The telephony is carried over a copper pair alongside the coaxial cable back to the street cabinet, which the voice channel is combined with the data channel and

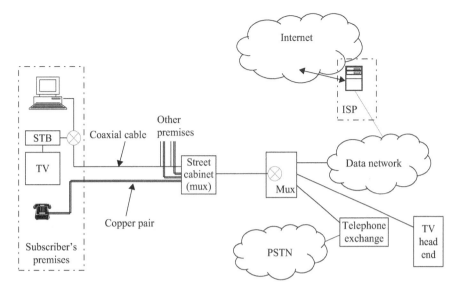

Figure 2.9   Access to the Internet via cable modem

carried over the optical fibre and separated out at the far end and linked to one of the cable TV operator's telephone exchanges, as shown in Figure 2.9.

2.5.4   Over optical fibre

There are several broadband access systems that use optical fibre technology all the way from subscriber's premises to the serving exchange building, so ensuring higher data speeds and usually a lower maintenance liability than systems using copper cables. The set of optical fibre systems is described in Chapter 4.

2.5.5   Over leased line access

Leased lines are primarily designed to provide a dedicated end-to-end commu-nications channel (narrow band or broadband) between two subscribers' premises to carry either voice or data traffic privately. However, leased lines may also be used to provide Internet access over high-capacity dedicated links from large business premises (e.g. office block, university campus) to their ISP. The concept of leased lines is further explained later in the Section 2.6.4.

2.5.6   Over mobile network

Of course, a popular way to access the Internet following the advent of smartphones and tablets is via a mobile network connection, as shown in Figure 2.6. The access from the phone or tablet may be via the mobile network or via a local Wi-Fi connection to the fixed broadband service using one of the configurations described in the preceding sections. The various mobile network architectures and config-urations for voice services and Internet access are described in Chapter 9.

2.6   The specialist networks associated with a PSTN

Network operators, particularly the national incumbents, need to provide a range of services beyond the standard telephony of the PSTN. These extra services, which may be grouped as those available to all customers and those aimed at business customers, are provided by specialised networks owned and run by the operator and closely linked to their PSTN. Most of the non-incumbent network operators in a country also have one or more of these specialised networks, usually aimed at the business customers. These networks comprise of special equipment located in exchange buildings. However, the transmission between the customer's premises and the exchange building uses either the copper Access Network used by the PSTN or the capacity in optical fibre cables if these already exist, or else a cable is provided specially. Links between the exchange-located equipment are provided over common capacity provided by the Core Transmission Network (optical fibre cable and microwave radio), linking all exchange buildings, as described in Chapter 5.

This section introduces the most common specialised networks; Sections 2.6.1 and 2.6.2 describe networks that provide services available to all, while Sections 2.6.3–2.6.6 describe networks that provide service aimed at business customers only.

2.6.1   *Operator-services network*

This network comprises of several specialist telephone exchanges around the country, like those described in Chapter 1, but which support a suite of operator consoles (typically 10–30), instead of subscriber lines. The operators have more control of the calls than is provided to normal subscribers; thus, such exchanges are normally referred as 'automanual'. The operator-services network provides call assistance (dial *100*) and access to the emergency services (dial *999* in the United Kingdom, the European standard *112*, United States standard *911*, etc.), directory enquiry, and blind and disabled special assistance services. Calls to this network are routed either directly from the PSTN originating exchange or from the parent trunk exchange. There are also special charging arrangements for such services; either free, as in the case of assistance and emergency calls, or at a special rate, as for directory enquiries.

2.6.2   *The intelligent network*

The so-called intelligent network (IN) comprises of several centres around the country containing control systems and databases that provide a variety of advanced switched services. Typically, these services are based on the translation of the number dialled by the subscriber to another number to complete the call, together with some special charging arrangements (e.g. the recipient rather than the caller pays) and tariffs, as is the case of '0800' services. The translations of the dialled numbers might be fixed or, for example, vary by the time of day, day of the week, the location of the caller, or the availability of assistants in the case of a company providing a service from various call centres. The IN concept is described in Chapter 7.

Calls are routed to an IN centre when the PSTN local or, more usually, trunk exchange detects that the number dialled requires translation. This might be on the basis that the number is from a special range (e.g. beginning *08* in the United Kingdom) or it is in the standard range but listed as special in the exchange-control system (e.g. for number portability purposes, see Chapter 10).

2.6.3   Business-services network

There are a range of telephony services designed for businesses which give direct desk-to-desk dialling using private numbering schemes and special charging arrangements between offices and factory sites, etc. – examples being Centrex and VPN (see Boxes 2.1 and 2.2). These services can be provided by the local PSTN exchange. However, it is usually more economical to use separate dedicated special exchanges for these services, which form a business-services network. Since most calls on this network are within the businesses, only a few calls need to 'break out' to (or 'break in' from) the PSTN. Interconnection is provided between one or more of the business-services exchanges and a trunk exchange in the PSTN.

Access to these services is separate from standard PSTN, using private or combined public/private numbers (known as direct dialling in (DDI), as discussed in Chapter 10) and possibly using special telephone instruments. Subscriber lines from the business customers' sites (e.g. an office block, factory or warehouse) are provided over the copper pairs of the local telephone network or over specially provided optical fibre cables or microwave radio links.

2.6.4   Leased-lines services network

The leased-lines-service network provides point-to-point un-switched links, known as 'leased lines or private circuits', between business premises (e.g. offices, factories or warehouses) [10]. Although the leased-line service provided to the business customers is non-switched, the network may use automatic digital cross-connection units (DXC), a form of switching equipment for leased-lines circuits, as a way of automating the connection of circuits between transmission links, conceptually a form of electronic jumpering. The DXCs are technically exchange switch blocks but without the exchange call-control system. Connections once set up are held for the duration of the lease of the circuit, typically several years. Such circuits are used for linking private telephone systems (i.e. voice) or alternatively, linking data terminals and computers (i.e. data) in different business premises.

Figure 2.10 illustrates the outline arrangement for the leased-line services network. Two examples are given. Figure 2.10(a) shows how a leased-line circuit is provided simply by the manual jumpering at the local exchange between the copper pairs from premises A and B. Longer leased-line circuits are obtained by jumpering of transmission links in concatenation from the customer's premises at A to

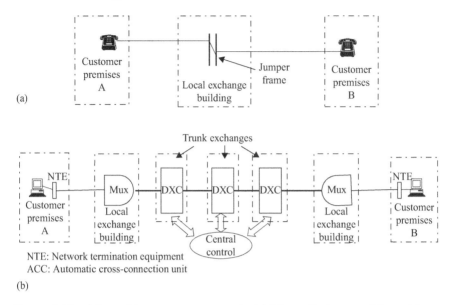

Figure 2.10   A leased-line services network. (a) Manually-jumpered leased-line circuit. (b) Leased-line circuit using digital cross-connection units

their premises at B. In Figure 2.10(b) gives a block schematic diagram of leased lines provided using DXCs to establish the required path across the network. The connection across the DXCs is controlled and monitored from a central network management centre [11].

2.6.5   Data services networks

There are a variety of specialised networks providing a range of data services using the packet switching technique, rather than the circuit switching technique normally used for telephony, as described in Chapter 1. The most common of these data services are: ATM (asynchronous transfer mode), Ethernet, Frame Relay (FR), IP and MPLS (multiprotocol label switching). Typically, each of these data services is provided from a separate network, i.e. five data networks would be needed for the above list of services.

Business customers use the data services provided by the network operators to connect terminals and computers from one set of premises to another. In addition, businesses now use data networks within their premises, such as LANs (local area networks); the network data service then acts as a link between distant LANs, forming wide-area networks (WANs). Data services and networks are described further in Chapter 8.

2.6.6   Telex network

Telex is a basic text messaging service which was used extensively throughout the World by businesses. It is now very much a legacy service, being replaced by fax and e-mails. However, it is still used within many developing countries and internationally to their trading partners. The Telex Network is quite separate from the PSTN, with its own numbering scheme for the telex terminals and its own exchanges. However, these are like telephone exchanges and the Telex Network is structured with a routeing hierarchy similar to the PSTN. Also, the Telex subscriber lines are connected to their exchanges (which are or were in the same buildings as the PSTN exchanges) over the local copper Access Network [12]. BT closed its inland Telex network in 2008, although it still retains access to international Telex service.

2.7   A model of the set of a Telco's networks

It is useful to picture the above-described range of specialised networks associated with a PSTN operator as a set of building blocks, as shown in Figure 2.11. This simple model portrays the various types of customers at the top and a stack of networks, shown as building blocks, arranged as a set of horizontals with a set of verticals sandwiched in between. The top horizontal blocks represent the fixed and the mobile (cellular) Access Networks, which provides the means of linking customers (users or subscribers) to the various service networks. The fixed Access Network is made up of the copper lines (local loop) as described in Chapter 1, providing links to the PSTN and several of the specialised networks. The

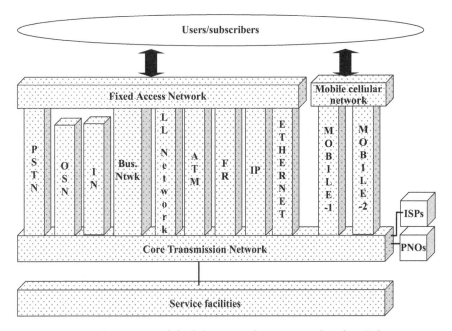

Figure 2.11   A model of the networks associated with a Telco

specialised networks, arranged as vertical blocks, comprise a range of switching and intelligence equipment. The fixed Access Network also contains optical fibre cables providing broadband access to business premises, and point-to-point microwave radio providing an alternative means of linking subscribers to many of the specialised networks. Some of the specialised networks cannot be directly linked to subscribers through the Access Network, but are linked instead via one of the other vertical networks, usually the PSTN. For example, calls to operator services are first switched at the PSTN exchange and then routed over the Core Transmission Network to the Operator Services Network (OSN) (shown as OSN in Figure 2.11).

The set of mobile networks (e.g. 2G, 3G, LTE (long-term evaluation)) shown for convenience as two verticals gain access to the subscribers via the cellular networks, as shown.

All the above service networks (vertical blocks) are supported by the (horizontal block) Core Transmission Network, which comprises of a set of transmission links, usually carried over optical fibre cables and microwave radio paths, together with automatic cross-connection and add/drop terminal equipment forming the nodes. (The technology for the Core Transmission Network is described in Chapter 4.) As its position in the model suggests, the Core Transmission Network acts as a common resource for all the service networks. Since the service networks, i.e. the verticals in the model, do not contain any transmission links all their nodes need to use the Core Transmission Network for connectivity.

In practice, the exchange buildings would house not only a local switching unit (local exchange), but also potentially a trunk switching unit (Trunk Exchange), and several of the non-PSTN specialised units [13]. To illustrate this point we will refer again to the example used in Chapter 1 where the transmission capacity between the exchange buildings in Haywards Heath and Penzance would also provide the required links between the Private Circuit Network, Business Service Network or OSN specialised units, respectively, located in these buildings.

Access to the ISPs for the Internet, and interconnection of voice and data services to the networks of other operators are shown at the right-hand side of the Core Transmission Network block. Finally, the bottom horizontal block represents the various service facilities for the networks, such as Cloud services from Data Centres, or IPTV (IP television) and video streamed services.

Other models of the various telecommunication networks are introduced in Chapter 11, where the concept of network architecture is examined in more detail. Also, the commercial aspects of these various networks and their services are described in detail in the companion book (*Understanding Telecommunications Business*). Not only does a Telco's network provide services to the subscribers, but their networks also support many other services/applications provided by third parties – a concept known as 'over-the-top (OTT)' – examples being Netflix and WhatsApp [14,15].

Box 2.1   PBXs and VPN

Businesses which have more than a few telephones use a private branch exchange system, known as a PBX, to provide call connections between each telephone (which become 'extensions') and links into the PSTN [10]. The PBX is really a small version of the PSTN exchanges, typically ranging in sizes from 10 up to 5,000 extensions. A private numbering scheme is required to enable extension-to-extension dialling, also special codes (e.g. 'dial 9') are required to enable calls to be made to the PSTN. Incoming calls from the PSTN must be answered by a receptionist or operator at a manual console so that the appropriate (privately numbered) extension can be contacted. Alternatively, the extension numbers can form part of the public numbering scheme (see Chapter 10) so that calls from the PSTN can be directly switched by the PBX to the required extension (known as direct dial in (DDI)), so avoiding the need for manual intervention where the caller knows the number of the wanted extension. Only the calls to the PSTN are charged. The corporate customer owns and pays for its PBX.

In the case where a company extends over two or more sites (e.g. office or factory buildings) the PBXs on each site can be linked by leased lines, thus enabling calling between all the extensions. This is known as a 'private corporate network' (or just 'private network'). In this case, the private numbering scheme extends across all the PBXs and usually each PBX is linked to the PSTN. Charging applies only to calls leaving the private network for the PSTN, although, of course, a rental charge is made by the network operator for the rental of the leased lines.

A virtual private network (VPN) provides an alternative to the use of leased lines between each PBX, as shown in Figure 2.12(a). The VPN exchange switches calls between the PBXs connected to it (e.g. between PBXs 'a' and 'b') as well as to trunk links to the other PBXs in the VPN corporate network. However, the VPN is provided over public exchanges, either special business exchanges or as part of the PSTN. Each VPN exchange switches the private network calls of several private corporate networks, although each operates in isolation, using its own numbering scheme. Thus, each corporate network appears to have the benefits of a private set of links between their PBXs, even though connectivity is provided over public exchanges, hence the use of the word 'virtual' in VPN. The VPN customer is charged a subscription for the VPN service based on a certain level of inter-PBX calls (traffic); there are usually charges made when the level of calls between any two PBXs exceeds the agreed threshold.

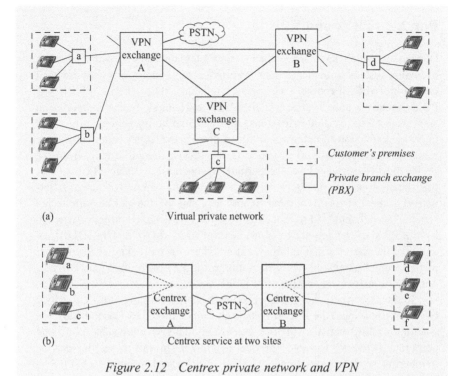

(a)   Virtual private network

(b)   Centrex service at two sites

Figure 2.12   Centrex private network and VPN

Box 2.2   Centrex service

Centrex is the generic name of a service in which extension-to-extension calls within a customer's site are switched by the public exchange, thus eliminating the need for a PBX [10]. This requires that each extension from the building be carried over the access network to the Centrex exchange, which serves many Centrex customers, each with their own private extension numbering scheme. Again, calls between the different Centrex groups are kept isolated within the exchange, and charging only relates to calls that go out to the PSTN. The service can extend across several Centrex exchanges (known as 'networked Centrex'), as required to serve the company's private corporate network, as shown in Figure 2.12(b). (Note that networked Centrex is not the same as a VPN since the role of the PBXs is taken by the Centrex exchanges.)

2.8   Summary

This chapter introduces the concept of the many other telecommunications networks that exist today and how they all interlink. Firstly, we look briefly at the two other types of network, in addition to the PSTN, that carry telephone calls, namely: mobile networks and cable TV networks. It is noted that all three networks use essentially the same type of switching system, but that the mobile networks use radio links and mobility management in place of the fixed wire access of the PSTN and cable TV networks. The necessary interconnection of all these types of networks within a country and to networks in other countries is described.

We also introduce the concept of the Internet and how subscribers gain access to it via the PSTN, cable modems over Cable TV networks, xDSL broadband, mobile networks or over leased-lines using optical fibre.

Finally, the various specialised networks associated with a PSTN Telco are introduced. Namely:

- Operator services network;
- Business services network;
- IN;
- Leased-line-services network;
- Frame relay network;
- ATM network;
- IP network;
- MPLS network;
- Telex network.

Also, many PSTN Telco's also have mobile networks. The chapter concludes with a simple model, which is introduced to help position the specialised networks with the fixed Access and Mobile Access and the common Core Transmission networks.

References

[1] Schiller, J.H. *Mobile Communications*, 2nd edn. Boston, MA: Addison-Wesley, 2003, Chapter 4.

[2] France, P.W. and Spirit, D.M. 'An introduction to the access network'. In France P.W. (ed.), *Local Access Network Technologies. IEE Telecommunications Series No. 47.* Stevenage: The Institution of Engineering and Technology; 2004, Chapter 1.

[3] Buckley, J. 'Telecommunications regulation'. *IEE Telecommunications Series No. 50.* Stevenage: The Institution of Engineering and Technology; 2003, Chapter 5.

[4]   Valdar, A. and Morfett, I. 'Understanding telecommunications business'. *IET Telecommunications Series No. 60*. Stevenage: The Institution of Engineering and Technology; 2016, Chapter 2.

[5]   Anttalainen, T. *Introduction to Telecommunications Network Engineering*. Norwood, MA: Artech House; 1999, Chapter 5.

[6]   Valdar, A. and Morfett, I. 'Understanding telecommunications business'. *IET Telecommunications Series No. 60*. Stevenage: The Institution of Engineering and Technology; 2016, Chapter 8.

[7]   Foster, K.T., Cook, J.W., Clarke, D.E.A., Booth, M.G., and Kirkby, R.H. 'Realising the potential of access networks using DSL'. In France P.W. (ed.), *Local Access Network Technologies. IEE Telecommunications Series No. 47*. Stevenage: The Institution of Engineering and Technology; 2004, Chapter 3.

[8]   Clarke, D. 'VDSL: The story so far'. In France P.W. (ed.), *Local Access Network Technologies. IET Telecommunications Series No. 47*. Stevenage: The Institution of Engineering and Technology; 2004, Chapter 6.

[9]   Tassel, J. 'TV, voice and broadband IP over cable TV networks'. In France P.W. (ed.), *Local Access Network Technologies. IEE Telecommunications Series No. 47*. Stevenage: The Institution of Engineering and Technology; 2004, Chapter 15.

[10]   Bell, R.K. 'Private telecommunication networks'. In Flood, J.E. (ed.), *Telecommunications Networks*, 2nd edn. *IEE Telecommunications Series No. 36*. Stevenage: The Institution of Engineering and Technology; 1997, Chapter 11.

[11]   Marshall, J.F., Adamson, J., and Cole, R.V. 'Introducing automatic cross-connection into the KiloStream network'. *British Telecommunications Engineering*. 1985;**4**(3):124–128.

[12]   Drewe, C. 'BT's telex network: Past, present and future?' *British Telecommunications Engineering*. 1993;**12**(1):17–21.

[13]   Valdar, A. and Morfett, I. 'Understanding telecommunications business'. *IET Telecommunications Series No. 60*. Stevenage: The Institution of Engineering and Technology; 2016, Chapter 6.

[14]   Valdar, A. and Morfett, I. 'Understanding telecommunications business'. *IET Telecommunications Series No. 60*. Stevenage: The Institution of Engineering and Technology; 2016, Chapter 1.

[15]   McCarthy-Ward, P. 'The growth of OTT services and applications'. *Journal of the Institute of Telecommunications Professionals*. 2016;**10**(4):9–14.

Chapter 3

Network components

3.1   Introduction

The first two chapters have considered the way that a telephone call is set up and how the various networks providing telephone service are interconnected, as well as how broadband access to the Internet is provided. Also, Chapter 2 introduced the roles of the other types of network: specialised voice networks and data networks. This chapter considers the basic components that go together to make up these networks. In fact, these components can be treated as part of a set of building bricks, each of a different shape and size; all the telecommunication networks described in this book are made up using a suitable mixture of these bricks. Subsequent chapters describe the networks and systems in more detail, assuming the reader has an understanding of the basic components described in this chapter.

3.2   Network topologies

Any network is a system of nodes and links. There are many examples of networks which are encountered in everyday life. We talk of the road network, where the junctions form the nodes and the stretches of road in between are the links of the network. Similarly, the rail network comprises of the rail links joining the station nodes. A further example is that of the airline's network, where airports provide the nodal functions and the airline routes provide the links. Networks can also be on far smaller scale, for example, the electrical circuitry in a television set with its physical wires or printed circuit rails linking the nodal electronic components. The various telecommunication networks are similarly made up of a variety of nodal functions and transmission links.

The pattern of links between the nodes in a network, that is, its topology, determines the possible routeings between any two nodes – either direct if a link between the two nodes exists or connected through one or more intermediate nodes, known in general as a 'tandem routeing'. Figure 3.1 shows the possible set of network topologies. A star configuration provides a direct link between all nodes and a hub node, all routeings between nodes need to transit via the hub. A typical example of a star topology is the local network with dependent exchanges linked to the parent exchange. With the ring configuration, routeings

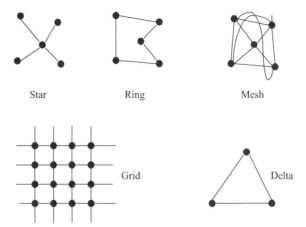

Figure 3.1   Network topologies

between all nonadjacent nodes must transit around the ring. The local and regional transmission networks tend to use this topology because it enables a dispersed set of exchanges (nodes) to be linked by cables which follow the road network and usually requires the least overall length of cable. In contrast, the mesh configuration uses a direct link between each node, thus no tandem routeings are required. The top level of a routeing hierarchy (see Chapter 1) is configured as a mesh. The grid or matrix arrangement is useful within switching and similar equipment (see later in this chapter) because it enables any horizontal link to be connected to any vertical link. The delta is shown for completeness, although it is really a special case of a mesh. In practice, telecommunication networks usually comprise of a combination of topologies to minimise the cost, while producing an appropriate level of resilience. For example, a star network may be used to achieve minimum link costs, but some meshing would be added between certain nodes to give the network a degree of resilience to link failures.

The main service features are provided in the nodes of telecommunication networks. It is for this reason that network operators (other than virtual network operators) need to own and control all their nodes, while none of the transmission links needs to be owned and can instead be leased from other network operators. The following sections introduce the most common component functions that are used to provide the service features in the network nodes.

3.3   Nodal: concentrator switching

The concept of telephone switching is introduced in Chapter 1, where the switch is shown in Figure 1.7 providing connections between an input circuit to a required output circuit for the duration of the call. Actually, the single box marked 'switch'

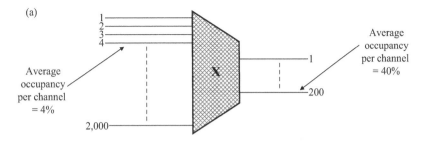

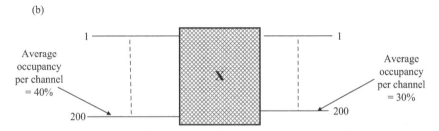

Figure 3.2   (a) Concentration switching. (b) Route switching

comprises of two types of switch, that is, concentrator and route (also known as 'group' or 'interconnect'), as described in more detail in Chapter 6. The purpose of concentration in a network is to achieve cost savings in equipment by taking calls (or traffic) from lightly-loaded subscriber lines – over which only a few calls a day may be made – and switching these on to fewer, but consequently more highly loaded lines. Figure 3.2(a) illustrates the case of the calls from 2,000 lines on the input side of the switch, each line having an average occupancy of 4%, being concentrated onto 200 lines with 40% occupancy. This concentration ratio of ten to one is typical for a telephone concentrator switch.

Although the concentrator switch described above has improved the occupancy of the lines by a factor of ten, it has introduced the possibility of subscribers not being able to get through since calls from only 200 of the 2,000 subscribers can be carried at any one time. For the time that any calls are unable to get through, the switch is deemed to be in 'congestion' and these calls are considered as 'lost'. Clearly, the probability of lost calls may be determined statistically. Normally, this probability is low, typically only about 1% or 2% during the busiest time of day. The network operator makes an economic trade-off between keeping the acceptable probability of lost calls as low as possible against gaining as much cost saving in equipment and network capacity by having as high a concentration ratio as possible.

3.4   Nodal: route switching

Route switching provides the interconnection of calls from the set of input circuits to an equal number of circuits on the output side, as shown in Figure 3.2(b). The role of the switch is to provide connectivity at an exchange between either internal lines with high occupancy (40% in the example of Figure 3.2(b)) or external lines, also with high occupancy, going to other exchanges. Trunk, junction tandem, and international exchanges comprise of only route switches, since they do not have any subscriber lines with their low occupancy, and therefore no concentration switching is needed.

Route switches, which are sometimes known as 'group switches', introduce no possibility of losing calls (or traffic) if working within the dimensioned load, that is, normal conditions.

3.5   Nodal: packet switching and routeing

Data is fundamentally different to voice in its nature and origin. The primary difference is that data is information represented in the form of letters and numbers which are generated, stored, and displayed by devices, for example, a computer; whereas voice is generated as a sound by humans. The consequence of this as far as telecommunication networks are concerned is that voice originates as an analogue signal (i.e. a varying waveform that copies the sound pressure pattern of the speaker) and data originates as a digital signal (i.e. a string of numbers). We look at the important distinction between analogue and digital signals and their attributes in more detail later in this chapter.

The need to convey data may be on the basis of a slow trickle, as in the case of a telemetry signal (e.g. the remote measurement of the height of water in a reservoir), or a very high-speed voluminous flow, as in the case of the transfer of a file between two computers. The need for data communication between two points may be continuous during a session, or intermittent and in bursts, or a single one-off enquiry. Furthermore, much of the data is not time critical, and it can therefore tolerate some delay before it is delivered to the far end, for example, e-mails.

Data can be conveyed over a telephone network, if appropriate modems are used to convert the digital signal to an analogue signal for transfer over the telephone lines. Indeed, the first form of public switched data services did use the public switched telephone network (PSTN), in which a telephone call was established between a terminal and a computer. This service was known in the United Kingdom as 'Datel', meaning 'data over telephone'. Similarly, as described in Chapter 2, data sessions with the Internet may also be provided over the PSTN between the user and the Internet service provider (ISP). However, the continuous nature of the call connection provided by the PSTN, in which a circuit is held for the duration of the call (defined as 'circuit switched') is not an efficient way of conveying data which might be burst and intermittent. Also, the call set-up time of

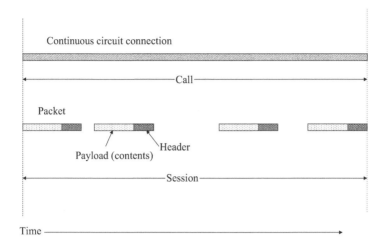

Figure 3.3   Circuit and packet switching

circuit switching may be proportionately too long in the case of short rapid-enquiry type data transfer. These issues, together with the fact that the charging and numbering arrangements of the PSTN are optimised for telephony voice service, mean that the alternative use of packet networks for data is usually preferred.

In a packet network, the data to be conveyed is split into conveniently-sized segments, each associated with an address, to form packets. These packets, which being digital are in the form of a string of 1's and 0's (i.e. 'binary'), may be of fixed or variable length and sent on a regular or on an as-and-when basis, depending on the type of packet system. Figure 3.3 illustrates the difference between a circuit and a packet connection. Note that the circuit is held for the duration of the call, irrespective of whether either person is speaking, while in contrast the packets need be sent only when data is present during a session between two terminals/computers. The advantage of a packet system compared to the continuous call connection system of a PSTN is that packets from different users can be interleaved on each circuit between two packet switches, thus giving improved utilisation and hence lower cost than the equivalent single occupancy of a circuit-switched (i.e. PSTN) connection.

Figure 3.4 illustrates the principle of a packet switch. The input line to the switch, which comes from another Switch A, carries a stream of interleaved packets. The addresses in the header of each packet are read in turn and the packet is passed to the appropriate output line. In the example of Figure 3.4 all packets with addresses 'x' and 'z' are sent out to the line to Switch C, while all packets with address 'y' are passed to the line to Switch B. The way that the packets are sent through the network differs according to whether a predetermined path is set up for all packets belonging to a session between two endpoints, known as a virtual path, or whether each packet follows an independent routeing. The former is usually

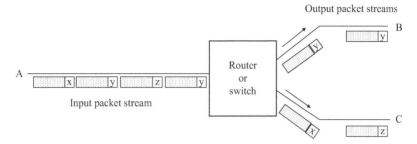

Figure 3.4   Packet switch or router

referred to as 'packet switching', the latter as 'packet routeing'. Chapter 8 describes data networks and the roles of packet (or cell) switches and routers more fully.

3.6   Nodal: control (computer processing and storage)

There is a category of nodal functions associated with the control of calls or the provision of additional service features which can be used by callers. These functions are provided by systems using computer processing and storage. Such systems, which are referred to as 'network intelligence', are usually located at several centres in a national network, generally housed in exchange buildings. Calls requiring the use of such facilities are handled in the normal way in the PSTN until one of the control systems of the exchanges detects, for example, from the number dialled, that more advanced call control and service features are required. That exchange will then request further information and instructions from the network intelligence so that the call can be completed. Chapter 7 describes the role of these systems in more detail.

3.7   Nodal: multiplexing

One of the most common nodal functions of a telecommunications network is that of multiplexing, that is, the carrying of several tributary streams over one high-capacity bearer. Figure 3.5 illustrates the example of ten tributaries, each with one channel's worth of capacity being multiplexed onto a ten-channel bearer. In general, a multiplexed system comprises an m:1 multiplexing node, a multiplexed bearer (link) and at the far end a 1:m de-multiplexing node. The multiplexed bearer may be a transmission link between two cities, extending to several hundreds of km or a bus (i.e. a common highway) within an exchange switching system, extending just a few metres. Either way, the multiplexing system provides important reduction in the total network cost due to the economies of scale of higher-capacity bearers. The important characteristic of a multiplexed system is that the capacity of the multiplexed bearer equals that of the input and output tributaries, so there is no concentration and hence no loss of traffic.

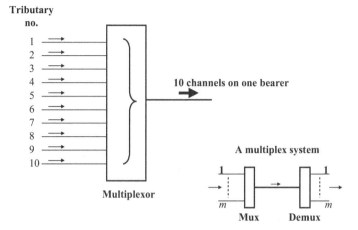

Figure 3.5   Multiplexing

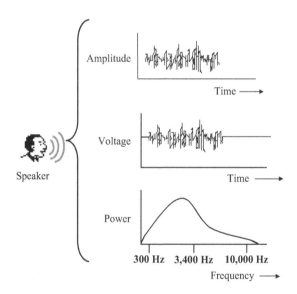

Figure 3.6   Waveform analysis of speech

There are many ways in which multiplexing may be achieved. The most commonly used methods in telecommunications are described below, using the example of telephony. The specification of any multiplexing system depends on the frequency characteristics of the signals to be carried (e.g. voice, video, music-quality sound, broadcast colour television, data). The concept of frequency and waveforms is introduced in Chapter 1. In the case of telephony, the sound power of the human voice throughout the frequency range is characterised in Figure 3.6. The sound wave power spectrum is low at frequencies below 400 Hz, and rises to a peak around 3–4 kHz, and progressively drops off with increasing frequency. Of course, the exact shape varies between individuals, with louder speakers having more power in their voice and hence the curve is higher up the Y-axis. Also, female speakers have their frequency-power curve shifted to the right due to the higher pitch of their voice, while male speakers have their curve shifted to the left of the average.

It has been agreed worldwide that a band of frequencies between 400 Hz and 3.4 kHz should be allocated within a PSTN to each telephone connection. This band represents a compromise between capturing sufficient of the voice power spectrum to ensure acceptable reproduction of the speaker's voice on the one hand, and the cost of providing the capacity on the other. In practice, an absolute band of 4 kHz (0 Hz–4 kHz) is actually allocated to each channel, with the 3.4 kHz point representing the level where the output power is halved and the attenuation rapidly approaches cut-off at higher frequencies. While passing this relatively narrow band of the voice frequencies manages to capture the bulk of the power spectrum, the absence of the frequencies above 4 kHz causes not only some loss to the quality of the voice, but also more importantly some of the consonants become indistinct. Thus, speakers over the telephone are often required to say 'S for sugar', etc., to overcome the lack of precision caused by the loss of the higher frequencies.

In describing the commonly used types of multiplexing in telecommunications, it is useful to consider how the dimensions of time and frequency are allocated to the channels within the multiplexed system.

3.7.1   Frequency-division multiplexing

With frequency-division multiplexing (FDM) the frequency dimension is divided into 4 kHz bands, each channel occupying one band. Figure 3.7 illustrates the FDM concept, showing that each of the five channels occupies just its allocated band of frequencies for all time. The bandwidth of this FDM system is 5×4 kHz = 20 kHz. Inter-exchange transmission systems introduced from the 1950s onwards (and still in use in many parts of the World) used FDM systems based on a 12-channel format, known as a 'group'. The FDM technique is also used for

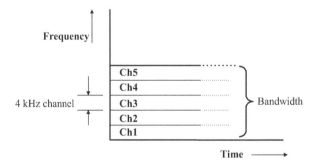

Figure 3.7   Frequency-division multiplexing

radio and TV broadcast, with each radio or TV channel allocated to a different carrier frequency; the tuning mechanism on the radio or TV set enables the user to select the carrier frequency and the associated band of frequencies for that channel.

3.7.2   Time-division multiplexing

With time-division multiplexing (TDM) only small slices of time are allocated to each channel and during these times the full bandwidth of the TDM highway is made available to that channel. The concept is illustrated in Figure 3.8. Each of the five channels is sampled sequentially and passed through the multiplexed system, and thereafter the five channels are repeatedly sampled in sequence. The time period for all five samples is called the 'timeframe', and the time allocated to each sample is known as a 'time slot'. Slicing up a voice call into these periodic samples, to be carried over their respective time slots, will not introduce any impairment to the quality of the voice heard at the receiving end if the samples are taken frequently enough. The rule for the required sampling rate, attributed to Nyquist and also known as the 'Sampling Law', states that the minimum rate of sampling is twice the highest frequency of the signal to be carried on each channel [1].

For telephony, where each channel is frequency-band limited to an absolute top frequency of 4 kHz, as described above, the necessary Nyquist rate of sampling is 8 kHz (i.e. 8,000 times/s) or once every 125 μs (millionths of a second, microseconds

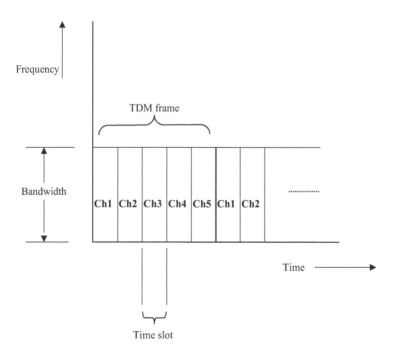

Figure 3.8   Time-division multiplexing

are written as 'μs'), that is, 1/8,000th second. Thus, for the five-channel TDM system shown in Figure 3.8, the second sample of Channel 1 needs to be made 125 μs after the first sample; and subsequent samples need to be sent at 125 μs intervals thereafter. Similarly, the samples for Channels 2, 3, 4, and 5 also need to be sent at this rate. The timeframe, therefore, for all telephony TDM systems, irrespective of the number of channels carried, is always 125 μs long. In this example of a five-channel TDM system, each time slot occupies a fifth of the time frame, that is, 25 μs.

The principle of sampling and TDM is illustrated in more detail in the three picture sequence of Figure 3.9, in which, for visual clarity and simplicity, a three-channel TDM system is used as an example. The voice signal on Ch1 is taken through a low-pass filter to eliminate all frequencies above 4 kHz before entering the sampling gate, as shown in Figure 3.9(a). A string of pulses, P1, each separated by 125 μs opens the sampling gate at the sampling intervals and the stream of samples of the band-limited voice on Ch1 pass onto the TDM highway. Notice how each of the samples on the highway equals the height of the Ch1 waveform at the respective instant of sampling. At the output of the TDM highway a second sampling gate is opened at the appropriate instants by another stream of P1 pulses to allow the sequence of samples of Ch1 to pass. These samples are then taken through a low-pass recovery filter which, in effect, smooths the sequence of samples into a continuous waveform, a replica of the input to the sampling gate at the beginning of the TDM system.

Now we can consider adding the second channel to the system, as shown in Figure 3.9(b). The Ch2 input waveform after being band-limited to 4 kHz is sampled under the control of a string of pulses P2, which are 125 μs apart but positioned a third of a timeframe (i.e. 41.7 μs) later than the P1 set of pulses. The samples from Ch2 are then passed to the TDM highway, interleaved in the gaps between the samples from Ch1. Application of the second stream of P2 pulses then opens the output sampling gate at the appropriate instants so that the samples of Ch2 may be taken through the output low-pass filter in order to recover the Ch1 waveform.

Finally, Figure 3.9(c) shows how the full three-channel TDM system works. The stream of P3 pulses, also 125 μs apart, but staggered by 41.7 μs from the P2 stream of pulses, operates the sampling gate to allow the appropriate samples of the band-limited Ch3 waveform onto the TDM highway interleaved between the samples of Ch2 and Ch1. At the end of the TDM highway the second stream of P3 pulses opens the output sampling gate allowing the Ch3 samples to pass to the recovery filter.

Clearly, for this TDM system to operate without interference between channels the input and output pulse trains must be precisely timed. Thus, for example, the output set of P2 pulses must occur at the same instants as the input P2 pulses otherwise some portion of the samples of either the preceding Ch1 or the following Ch2 will pass through the CH2 output sampling gate and corrupt the final result of the Ch2 output. TDM systems in general, therefore, need a mechanism to

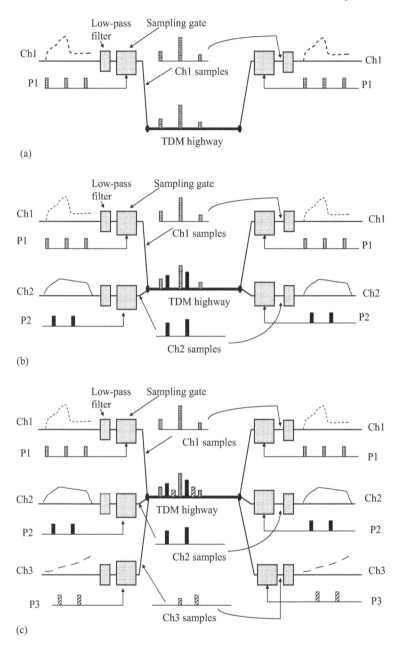

Figure 3.9   Principle of a time-division multiplexing system

synchronise and align the corresponding streams of input and output sampling pulses. TDM transmission systems are described in more detail in Chapter 4.

TDM is employed by the majority of transmission systems used in today's telecommunication networks (with the TDM multiplexers at the network nodes) as well as within the exchange switching (nodal) equipment. A derivative of TDM, known as time-division multiple access (TDMA) is used by telecommunication satellite systems and mobile phone networks to allocate calls to available radio channels, as described in Chapter 9.

3.7.3 Code-division multiplexing

With code-division multiplexing (CDM) the full frequency and time capacity of the system is made available to all the channels, as illustrated in Figure 3.10. This technique is more difficult than either FDM or TDM to explain. The best way to understand the principle of operation is to consider the analogy of a cocktail party. At such parties there are many people talking together in small groups around the room. Anyone at the cocktail party can join a group and, despite the general high level of background chatter, they would be able to concentrate on the conversation of their group by tuning in to the particular voices involved. During the conversation the person would also be able to monitor what is being said elsewhere in the room by momentarily tuning in to the various groups within earshot – just in case something more interesting is being said! Similarly, CDM allows all the channels to talk at once on the CDM highway, but a different binary code is first applied to each input channel, the required output channel is selected at the far end by applying the appropriate code, all other channels appear as unintelligible background noise. Just as with the cocktail party, there is a limit to the number of channels a CDM system can convey because the increasing background noise will reach a point where it swamps any individual channel (i.e. the 'signal-to-noise ratio' becomes unacceptable).

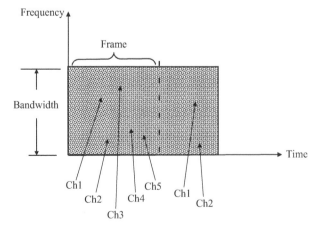

Figure 3.10 Code-division multiplexing

Like the other multiplexing systems, the CDM system needs to synchronise the binary codes used at the input and outputs of each channel. Also, each input channel is band limited to 4 kHz before being converted to digital (see below) prior to entering the CDM system. Multiple access systems based on CDM, that is, CDMA, are used extensively in second-generation digital mobile phone systems [2], particularly in the United States, and a wideband variant is used by the third-generation (3G) phone systems around the World, as described in more detail in Chapter 9.

3.8   Nodal: grooming

The grooming function is an important component in minimising network transmission cost where several services are carried over lines from many different sources, for example, as in the subscriber access network. Figure 3.11 illustrates the principle of grooming. Multiplexed bearers from nodes **i**, **ii**, **iii**, and **iv** all terminate at a node at which the grooming system is located, each bearer carries circuits (Service A or Service B) which are destined for either **v** or **vi** nodes, respectively. Typically, nodes **i** to **iv** would be big customer sites and Service A could be private circuits dealt with by node **v**, and Service B might be a data service dealt with by node **vi**. The grooming system permanently assembles all the channels carrying Service A (21 channels in total) and passes them to the output line to **v**, and assembles all the channels carrying Service B (29 channels in total) and passes them to the line to **vi**. Note that the total input capacity is equal to the total output capacity, so there is no traffic loss. Groomers operate on a semi-permanent basis, where the reconfiguration of channels is usually controlled by a local equipment-management system within the exchange building.

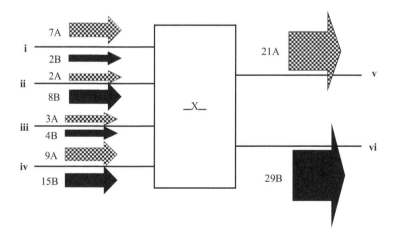

Figure 3.11   Grooming

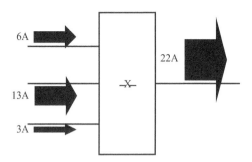

Figure 3.12   Consolidation

3.9   Nodal: consolidating

The consolidation function, as the name suggests, is the packing of the occupied circuits on several tributaries onto one better-filled output line transmission system. As Figure 3.12 illustrates, the consolidation of the occupied capacity of three transmission systems onto a single output system carrying all 22 circuits would provide network economies. The Core Transmission Network of a typical telecommunication operator undertakes consolidation where ever possible at the transmission nodes, as described in Chapter 5. This act of consolidation is also common in many of the other types of networks, for example, the consolidation of containers on cargo ships, which are restacked onto other ships at strategic ports around the world.

3.10   Link component

All the components described so far in this chapter relate to functions undertaken at the nodes of a telecommunication network. We will now consider a generalised description of the link transmission component. The link comprises three parts, as shown in Figure 3.13. The transmitter (Tx) converts the electrical (usually) output from the sending node into the format required for the type of transmission link. Examples of such formats are special digital signals designed for propagation over a copper line or coaxial cable, the digital pulsing of light for transmission over optical fibre, or the generation of a radio signal for a radio (mobile or fixed) or satellite transmission link. The transmission medium part comprises of the copper cable, transverse-screen cable, coaxial cable, optical fibre, or in the case of a radio system, the air, over which the telecommunication takes place. This part may also include repeater and regenerator equipment (located below ground, above ground, or even in outer space) along the route of the transmission medium which boosts and corrects the signal, compensating for the loss and distortions incurred. At the far end, the receiver (Rx) performs the reverse of the Tx, converting from the format required for the transmission medium to the format of electrical signal required by the terminating node.

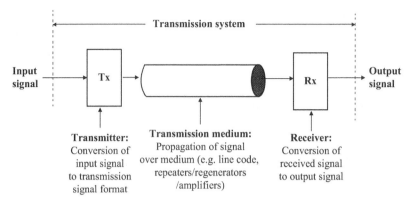

Figure 3.13　Link (transmission) functions

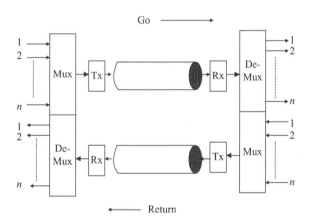

Figure 3.14　A transmission system incorporating the nodal function of multiplexing

Since the cost effectiveness of a transmission link system is higher the larger its capacity, in terms of the unit cost per circuit (provided the capacity is used), practical transmission systems are usually associated with multiplexing equipment. Figure 3.14 illustrates the generalised arrangement for a transmission system incorporating the nodal functions of multiplexing. It should be noted that because electronic devices are unidirectional, the combined transmission/multiplexing arrangement is a four-wire system, with separate Go and Return paths.

The configuration of Figure 3.14 can be optimised by using a further stage of multiplexing so that the Go and Return transmission signals are sent over the same transmission medium. So, for example a single optical fibre (which is a bidirectional medium) could carry both the Go and Return transmission signals by using FDM.

However, it should be noted that if this optical fibre system had a repeater using unidirectional electronics, the Go and Return signals would need to be separated and recombined at the entry and exit of the repeater.

3.11   Analogue-to-digital conversion

3.11.1   The advantages of digital networks

Chapter 1 introduced the concept of the analogue signal generated by the microphone in a telephone handset receiving air pressure variations from a speaker and the subsequent conversion back to air pressure variations by the earpiece at the receiving telephone handset. Figure 3.6 considered earlier in this chapter also shows the characteristics of an analogue signal. Originally, the PSTNs around the World were fully analogue; they comprised of analogue transmission systems in the Access and Core parts of the network, using a variety of FDM-based systems to gain economies, together with analogue switching in the local and trunk exchanges. Digital transmission systems (pulse-code modulation (PCM), as described later) using TDM, were first introduced around the mid-1960s into the shorter-distance inter-exchange routes (so-called 'junction' and 'toll' routes), as a way of providing multiple circuits over copper pairs. Later, digital transmission was introduced onto the long distance coaxial-cable routes.

Digital transmission not only provides a cost-effective multiplexing and transmission technique, but it also greatly improves the quality and clarity of the circuits, providing immediate benefit to the telephone users. This quality improvement is due to the ability of digital systems to regenerate the binary 1's and 0's where necessary in the network to compensate for the deterioration of the conveyed signal, unlike analogue systems where use of amplification to boost the signal equally affects the background noise, causing the impairments to increase with distance. We are familiar in everyday life with the merits of the quality of digital systems compared to the analogue equivalents, for example, (digital) CDs and (analogue) audio cassette tapes; and (digital) DVDs and (analogue) VHS video cassette tapes.

However, it was not until the late-1970s and early-1980s that digital switching was used to replace the analogue switching systems initially in the trunk then later the local exchanges. The primary driver for introducing digital switching was to be able to make fully semiconductor electronic exchanges, with the consequential reduction in manual operational costs, lower capital costs, less accommodation required, improvements in clarity of the connections, and the potential for new services and features. Progressively, the analogue switching and transmission systems have been replaced by digital systems so that since the mid-1990s nearly all PSTNs in the developed world are entirely digital [3].

A PSTN with digital transmission and switching is known as an integrated digital network (IDN). There are several major advantages of an integrated digital PSTN, compared to an equivalent analogue PSTN, as summarised below [4].

- Capital cost savings due to the all-electronic nature of the equipment enabling the cost and performance benefits of semiconductor computer technology to be exploited. In addition, the integration of the switching and transmission technology (all digital) eliminates the need for costly interworking equipment.
- Operational cost savings resulting from the significant reduction in floor space taken up by digital electronic equipment compared to that required for the bulky electro-mechanical and semi-electronic analogue switches. This leads to a reduction in the accommodation costs for the network operator.
- Operational cost savings, primarily in manpower, due to the virtual elimination of manual adjustments to the equipment and the reduction in the overall maintenance load.
- A high level of transmission quality, in terms of clarity for the user, lack of background noise and a constant loudness irrespective of the distances involved or the number of exchanges in the connection.
- The ability to easily mix different types of services on the same line by virtue of the use of digital conveyance for both voice and data, that is, 'integrated services'.

Figure 3.15 shows the concept of an IDN. Despite the merits of digital working, the ubiquity, cost effectiveness, and adaptability of the copper local loop means that most PSTN subscriber lines are still analogue. Thus, the IDN begins at the entry to the digital local exchange where the analogue-to-digital (A/D) conversion is provided and extends to the exit of the local exchanges, where the analogue line is connected via a D/A conversion.

The provision of integrated services digital network (ISDN) service adds a digital line system over the local loop together with a digital termination at the subscriber's premises, thus in effect extending the IDN to cover the full end-to-end connection between subscribers' terminals. The term ISDN means 'integrated services digital network', a name derived from 'IDN' and the attribute (e) above. The advantages of the IDN listed above then extend directly from the customer's

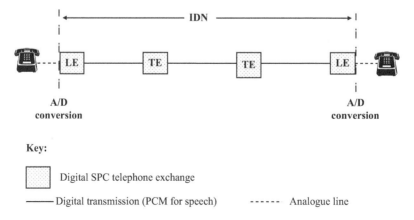

Figure 3.15   Integrated digital network

terminal equipment, for example, a computer connected digitally to the ISDN termination at the desktop.

3.11.2   The analogue-to-digital conversion process

As previously discussed, an analogue electrical waveform or signal is, by definition, an analogue of an input waveform, that is, the air pressure variation caused by a speaker. An analogue waveform also has another defining attribute, namely: that it may take any value within the permitted range. In contrast, a digital signal is a series of numerical representations of an input waveform and, importantly, the digital signal can take only a limited number of values, that is, it has a fixed repertoire – in the case of a binary digital signal, this number is two. (It is the use of binary values that enable digital transmission to achieve such good quality performance irrespective of distance, since the signal can be periodically cleaned up and regenerated by comparing the impaired signal with a just one threshold, unlike the analogue signal in which impairments accumulate with distance and cannot be corrected.)

Figure 3.16 shows a simple illustration of the process of A/D conversion. Firstly, the analogue input signal must be passed through a low-pass filter to limit the maximum frequency of the waveform, as described above for the process of TDM. The waveform is then sampled at or above the Nyquist rate to produce a series of analogue samples or pulses, whose heights represent the waveform at the sampling instants. (This stream of pulses is known as 'pulse-amplitude modulation' (PAM).) These analogue samples are then converted to digital by the comparison of

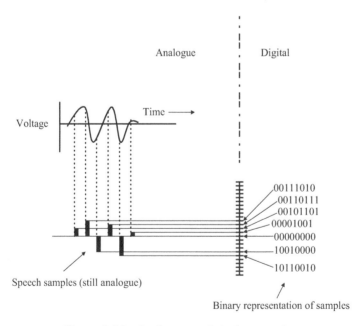

Figure 3.16   Analogue-to-digital conversion

each sample with a digitalising ruler. The digital representation is the numerical value of the graduation on the ruler at or just above the PAM pulse height, as shown in Figure 3.16. This coding process generates a binary number for each PAM pulse, the digital output of the digital-to-analogue (D/A) converter.

Conversion back to analogue from digital operates in reverse to the D/A described above. However, in converting each of the binary numbers to the PAM height given by the graduation on the ruler (i.e. decoding) there is an inherent error introduced when the regenerated PAM sample is compared to the sent original PAM sample. This is because of the finite number of graduations on the ruler, the binary number being the next closest reading to the input PAM sample in the A/D process, yet the binary number will generate a PAM sample exactly at the graduation level on the ruler during the D/A process. The amount of difference between the sent and regenerated PAM sample clearly varies between zero and the distance between ruler graduations. This is known as the quantisation error, and it is perceived by a listener as 'quantisation noise'.

In practical systems, the quantisation noise is kept acceptably low by using many steps, or quantisation levels, on the ruler. In addition, the quantisation noise is kept proportional to the level of input signal by bunching the levels closer at the lower end of the ruler and wider towards the top, following a logarithmic progression. The A/D process described above is performed by a digital coder and the D/A process is provided by a decoder; often the equipment needs to deal with the Go and Return direction, and so a combined digital coder/decoder or 'codec' is used.

The complete A/D-and-D/A process for a telephone network, which is known as PCM, is shown in Figure 3.17. This figure plots the various representations of the signal, from analogue waveform, to PAM samples, to digital numbers or PCM 'words' and back. Notice, that once converted to digital (PCM) the signal may pass through any number of digital transmission systems and switching exchanges across the network.

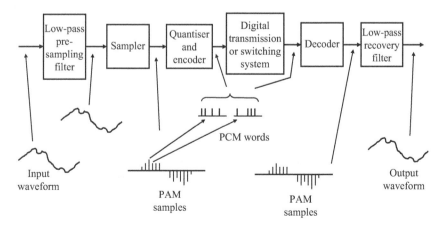

Figure 3.17   The pulse-code modulation process

The A/D process is used for many applications in everyday life in addition to telephone networks. As mentioned, earlier music is digitally coded onto CDs, and films are scanned and coded onto DVDs. Similarly, digital TV and radio (DAB) broadcasts are also increasingly being used in preference to analogue transmissions. Later in this book we will consider how the trend towards the use of digital in fixed and mobile telecommunications, entertainment, information and computing creates possibilities for convergence and what this means for the future generations of networks.

3.12   Summary

In this chapter, we looked at the basic components used in the links and nodes of telecommunication networks. Beginning with the nodes, we considered the two types of circuit switching:

1. Concentrator switching, dealing with low-occupancy subscriber lines
2. Route switching, interconnecting routes between concentrator units and with other exchanges.

Then, we considered the alternative form of switching, generally known as packet switching, which is the preferred way of conveying data.
The remaining nodal functions considered include:

*   Control nodes
*   Multiplexing (FDM, TDM, and CDM)
*   Grooming
*   Consolidation

A generalised view of transmission links was then described, noting that within the switched and Core Transmission networks these are four-wire systems, with Go and Return channels, whereas in the copper Access Network two-wire circuits are used for economy. Then we considered the advantages of an IDN and how it can be extended over the Access Network to the subscribers using ISDN. Finally, the important A/D conversion process was described.

References

[1]   Redmill, F.J. and Valdar, A.R. 'SPC digital telephone exchanges', *IEE Telecommunications Series No. 21*. Stevenage: The Institution of Engineering and Technology; 1995, Chapter 4.

[2]   Anttalainen, T. *Introduction to Telecommunications Network Engineering*. Norwood, MA: Artech House; 1998, Chapter 5.

[3]   Valdar, A.R. 'Circuit switching evolution to 2012', *Journal of the Institute of Telecommunication Professionals*. 2012;**6**(4):27–31.

[4]   Redmill, F.J. and Valdar, A.R. 'SPC digital telephone exchanges', *IEE Telecommunications. Series No. 21*. Stevenage: The Institution of Engineering and Technology; 1995, Chapter 21.

Chapter 4

Transmission systems

4.1 Introduction

This chapter describes the wide variety of transmission systems used in tele-communication networks today. In doing this, we will be drawing upon the concepts of multiplexing and analogue-to-digital (A/D) conversion introduced in the previous chapter. Earlier chapters have also introduced the important distinction between the Access Network, which serves the subscribers, and the Core Transmission Network, which provides links between network nodes only. In general, the transmission systems described in this chapter may be deployed in both the Access and the Core Transmission Networks – although they are usually more appropriate to one or the others, and this will be indicated. Chapter 2 introduced the model of the networks associated with the public switched telephone network (PSTN) (see Figure 2.11), in which the Access and Core Transmission Networks act as a common utility for providing circuits to the various specialised networks as well as the PSTN. However, we will leave the description of how the various transmission systems are deployed in the Access and Core Transmission Networks until Chapter 5.

4.2 Transmission bearers

In this section, we look at the general principles of transmission and the range of practical systems used in a telecommunication networks.

4.2.1 Transmission principles

The purpose of a transmission system is to provide a link between two distant points or nodes. The link may be unidirectional, as in the case of radio broadcast where transmission is only from the transmitter to the radio receivers, or bidir-ectional, as in the case of a telephone connection with its two-way conversation. In telecommunication networks, the links between nodes are either 2-wire or 4-wire circuits, as described in Chapter 1. With 4-wire circuits, the separate 'Go' and 'Return' paths are considered as transmission channels, as explained in the generalised view of a multiplexed transmission system in Figure 3.14 of Chapter 3.

In general, a transmission system takes as an input the signal to be conveyed (which may be a single channel or a multiplexed composite of channels) and converts that signal into a format suitable for the transmission medium, propagates the transmission signal to the far end, where after conversion from the transmission signal a reproduction of the input signal is produced (see Figure 3.13 of Chapter 3). The actual conveyance of the channel over the transmission system is achieved by 'modulating' an electrical bearer signal which travels over the transmission medium. It is the modulation or modification of the bearer signal that constitutes the transmitted information, not the bearer itself – indeed, the bearer signal is usually called a 'carrier' for this reason.

There are many forms of modulation [1]. The normal form of carrier is a sinusoidal waveform, as introduced in Chapter 1 (see Figure 1.2). Some of the common forms of modulating sinusoidal carrier waveforms are:

- Amplitude modulation (AM), in which the input signal directly varies the height or amplitude of the carrier, shown in Figure 4.1.
- Frequency modulation (FM), in which the input signal directly varies the frequency of the carrier, as shown in Figure 4.1.
- Phase modulation (PM), in which the input signal directly varies the phase (i.e. positioning of the start of the wave) of the carrier.

All three methods described above are used widely in radio and TV broadcasting, where a radio carrier is modulated, and most people would recognise the commonly used terms of 'FM' and 'AM'. They are also used in telecommunication transmission systems, as described later. However, these modulation systems are analogue because the carrier is continuously moderated by the input signal and the carrier can take any value within the working range of the system.

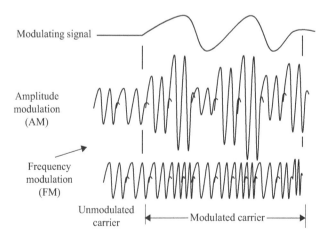

Figure 4.1   Carrier-wave modulation

A digital carrier, in the form of a stream of pulses can also be modulated, taking a fixed number of values, by an input signal. Examples of such modulation schemes include:

- Pulse amplitude modulation (PAM), in which the height of each carrier pulse is moderated by the corresponding bits in the digital input signal.
- Pulse position modulation (PPM), in which the position of each carrier pulse is moderated by the corresponding bits in the digital input signal.
- Pulse-code modulation (PCM), described in Chapter 3, whereby a binary stream is modified to convey the successive digital representations of a set of time-division multiplexing (TDM) (PAM) speech samples.

These digitally modulated streams may themselves also be used to modulate a radio or optical carrier on a transmission system. The input signal, in the form of pulses, is then used to modulate the carrier in a series of step changes between a fixed set of values. This is known as 'keying'. Examples of such systems include: phase-shift keying (PSK) – in which the output signal steps between two or more phases in response to the binary modulating signal; and frequency-shift keying (FSK) – in which the output shifts between two or more frequencies in response to the binary input. The advantage of these keying modulation systems is that the output signal can step between more than just two values, for example, 4, 8, or even 16 levels of phase can be used in what is known as 4-, 8-, or 16-phase PSK, respectively. In the example of the 4-phase PSK system, the input-modulating digital stream is taken 2 bits at a time (i.e. *00, 01, 10,* or *11*) to set the appropriate phase level in the output signal. Similarly, three bits at a time (*000, 001, 010, 011, 100, 101, 110,* or *111*) are taken for 8-phase PSK, four bits at a time for 16-phase PSK, and so on.

All transmission media introduce some impairment to the signal being conveyed due to the physical mechanisms involved. Examples of the most common forms of impairment are:

- Loss of power, due to the absorption, scattering and reflections of the signal within the transmission medium, known as 'attenuation'.
- Delay to the signal, due to the propagation through the medium. This will vary according to the characteristics of the medium (e.g. the purity of the glass in an optical fibre) and external factors such as temperature.
- Phase distortion, due to the different amount of delay introduced by the transmission medium to each of the component frequencies in a multi-frequency transmitted signal. The resulting spread in the arrival time of the waveform is known as 'dispersion'.
- Amplitude distortion, resulting from the different loss of power introduced by the transmission medium for each of the component frequencies in the transmitted signal.
- Noise, the pick-up of unwanted signal generated intrinsically by the medium (e.g. thermal noise), as well as induced noise from outside sources (e.g. crosstalk from adjacent transmission systems).

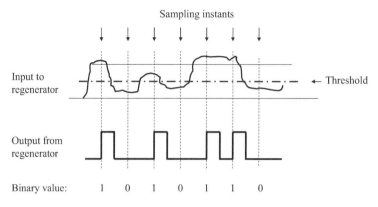

Figure 4.2   Digital regeneration

The design of transmission systems aims to minimise and compensate for the various impairments introduced by the medium. It should be noted that the extent of the above-mentioned impairments may change with the temperature of the transmission medium. There are methods for compensating for phase dispersion and amplitude distortion, known as equalisers, for example. However, these are effective only over a defined operating range. Thus, there will be limits on the length of a transmission system within a network if it is to operate within the accepted range of loss, delay, noise, and phase distortion. There will also be a limit to the bandwidth (range of frequencies carried) that may be used. Normally, the usable length of a transmission system is increased by introducing devices located appropriately along its route to boost the signal strength, compensating for the power loss.

In the case of an analogue transmission system, the signal accumulates the impairments along the length of the medium, so the introduction of an amplifier to compensate for power loss results in not only the wanted signal being boosted, but also that of the added noise. The ratio of the wanted signal to noise (signal-to-noise ratio) gets progressively worse at each point of amplification along the route of the transmission system – so setting a limit on the acceptable number of amplifiers and hence length of transmission link. By contrast, a digital transmission system can reconstitute a noise-free signal at each point of regeneration along the route. This is illustrated in Figure 4.2, where the impaired signal from the line is sampled by the digital regenerator and any signal above the threshold at the sampling instants will result in a clean digit '1' being generated and sent to line. (See also Chapter 3 and Figure 3.15 for a description of how this feature is exploited in an integrated digital network (IDN).)

Figure 4.3 illustrates the components of a generalised digital line transmission system. This is a development of the link functions as shown in Figure 3.13 of Chapter 3. The input digital signal in the 'Go' direction is shown in the left-hand side, this is either a multiplexed assembly of many channels (see Section 4.3.4 for synchronous digital hierarchy (SDH) or Section 4.3.3 for plesiochronous digital

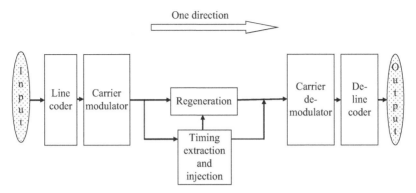

Figure 4.3   Digital line transmission system

hierarchy (PDH)) or one very large capacity single channel. The signal is then converted through the application of a line code to a format that suits the transmission medium. This will produce a signal that contains timing information that can be extracted by the regenerators along the line, to ensure that they stay in synchronism with the sending end. Line coding may also improve the efficiency of the transmission system by converting the two-level binary signal to a multilevel signal for actual transmission over the medium. An example of this was given with the use of four-phase PSK above, where the binary signal was converted to a four-level signal for transmission. This enables the transmission system line rate (known as the 'baud rate') to be half that of the binary input signal. In this way, a transmission system can carry higher capacities by using line codes of increasing numbers of levels. In practice, there is a trade-off between the increased capacity or efficiency of the transmission system and the cost and complexity of the line-coding detection mechanism, which generally increases with the number of levels used. There are many forms of line codes employed in transmission systems, each offering different levels of efficiency against complexity and equipment cost and, of course, suitability for the various transmission media [2,3].

Next, the line-coded signal is used to modulate the carrier for the transmission system. In the case of an optical fibre system, the signal modulates a laser sending a corresponding stream of light pulses down the fibre (i.e. the medium) – undertaking an electro-optical conversion. At each of the regenerators (one only is shown in Figure 4.3) the timing information (inserted by the line-coding process) is extracted and used to drive the sampling gates in order to regenerate the signal, as described above.

At the far end the appropriate demodulation occurs, for example a photo detector in an optical fibre system receives the transmitted light pulses and generates a stream of digital electrical pulses (i.e. opto-electrical conversion). This electrical signal is then transcoded from the line code back to the format of the binary input.

The principles described above apply to all the many forms of transmission systems carried over the different forms of media – namely: metallic cable, optical fibre cable, and radio over free air. Given this wide range of possible transmission options, the choice of system used in a network depends on the capacity to be carried, the existing transmission infrastructure, the distance involved, expected capacity growth potential, and the cost of the equipment. We can now look at these aspects as they affect a telecommunication network operator.

4.2.2   Transmission media

Figure 4.4 illustrates the physical characteristics of the transmission media used in modern networks.

4.2.2.1   Copper pair cable

As described in Chapter 1, copper pair cable has been adopted throughout the world as the means of providing the local loop between the subscriber and the serving telephone exchange. But, the humble copper pair cable can also be used to carry a variety of data high-speed services, and even video, by the addition of appropriate electronics (e.g. asymmetric digital subscriber line (ADSL)) at its ends, as described later in the chapter [4,5].

Copper pair cable is available as a single pair of wires, with plastic insulation, usually deployed between a subscriber's premises and the overhead or underground distribution point (DP), as described in Chapter 5. However, the copper pairs are also provided in multi-pair cable bundles of a variety of sizes (e.g. from 2-pairs up to 4,800 pairs). The thickness or gauge of the copper conductors also comes in a range of sizes, with corresponding electrical resistance values per km. Figure 4.4(a) shows the general construction of a multi-pair cable. The individual pairs have a thin insulating cover of polyethylene and these are grouped into quads in order to minimise adverse electrical interference problems. The coating of the pairs is

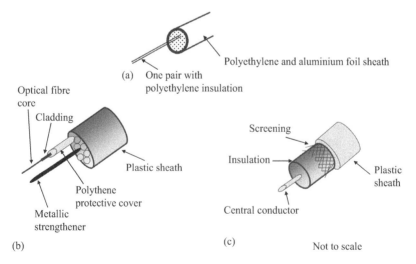

Figure 4.4   Cable transmission media. (a) Copper pair in a multi-pair cable.
(b) Optical fibre in a multifibre cable. (c) Coaxial cable

usually colour-coded to assist their identification by the technician jointing the ends onto other cables and terminals – quite a task when there are tens, hundreds or even thousands of pairs in one sheath! The sheath is constructed from polyethylene for strength and durability, together with aluminium foil to provide a water barrier.

Although not shown in Figure 4.4, special forms of copper cable, arranged with one central conductor surrounded by a copper cylindrically shaped conductor – known as 'transverse screen' cable – were deployed in the United Kingdom and elsewhere during the 1970s and 1980s to provide high-speed digital circuits in the Access Network [5]. Now, optical fibre or digital transmission over standard copper pairs is used in preference.

4.2.2.2 Optical fibre cable

Optical fibres provide a transmission medium of huge potential capacity by constraining pulses of light within a thin core strand of highly pure glass or silica. The light stream is contained within the core by a cladding of glass which has a lower refractive index, causing the light to be totally reflected internally at the core-to-cladding boundary [2]. The light is pulsed at the sending end by solid state electro-optic transducers, such as light-emitting diodes (LED) and lasers. Detection at the distant end is by solid state opto-electric transducers, such as PIN and avalanche photodiodes.

The key requirement for an optical fibre transmission system is the use of a glass (silica) fibre material which offers very low loss to the light passing through. This is achieved through a rigorous manufacturing process in which the glass purity is strictly controlled. In addition, advantage is taken of the transmission characteristics of the glass which offers different attenuation to light of different frequencies, as illustrated in Figure 4.5.

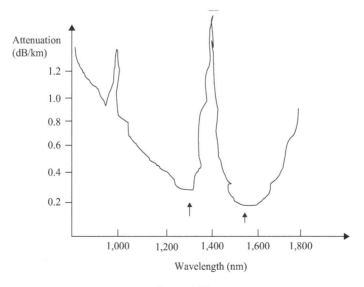

Figure 4.5   Optical fibre operation

Optical fibre transmission has the following characteristics compared to metallic cable transmission:

- Huge potential capacity due to the inherently higher operating frequencies.
- Low power loss (e.g. <0.5 dB/km), resulting in the need for few, if any, repeaters along the route. See Box 4.1 for an explanation of the term dB used as a measure power levels within telecommunication networks.
- The optical nature of the medium means that the transmission is unaffected by electrical inference in the vicinity.
- No escape of energy from the fibre, so that there is no transmission interference between fibres or other lines.
- Physically smaller than metallic cables of comparable capacity.
- Can be difficult to join optical fibres, special splicing techniques are required.

Figure 4.4(b) shows the construction of an optical fibre within a multifibre cable. The cables, which range in sizes (e.g. from 2 to 96 fibres), contain the fibres within a plastic sheath about a central metallic wire, which provides mechanical strength for the otherwise very fragile bundle of optical fibres. Typical diameters of the optical fibres are 50–200 μm core with 125–400 μm cladding for the earlier (multimode) types to 5–10 μm core with 125 μm cladding for the more recent (single mode) types. (A 'μm' is 1 millionth of a meter.) These fine glass threads, which are about the thickness of a human hair, are wrapped in a polythene protective cover for robustness.

Box 4.1   The decibel measure of power

Power levels in telecommunication networks are measured in units known as decibels (one-tenth of a 'Bel'). The important characteristic of dBs is that they represent a logarithmic comparison of the measured power level against some other power level. Thus, if the power output (P1) from a radio was turned up to twice the level (P2), the increase in power is calculated as $10 \log_{10} P2/P1 = 10 \log_{10} 2 = 3$ dB. Since the measure is a logarithmic ratio, it does not matter whether the two powers (P1 and P2) are large or small – a doubling of power is always 3 dB and, by the same logic, a halving of power is -3 dB. The main advantage of the use of this logarithmic measure is that all the increases and decreases of power contributed by the elements of a circuit throughout the network are simply added or deducted from the total to give the overall power level or loudness rating.

However, where an absolute power level is to be measured, for example, the output from some electrical equipment, it is compared to a standard power level, typically 1 mW (milliWatt). Powers are then shown with units of 'dBm'. Thus, a power level of 6 dBm would be equal to 4 mW, that is, double times double (3 dB + 3 dB) the power of 1 mW.

4.2.2.3 Coaxial cable

As the name suggests, coaxial cable comprises of a metallic central core cable encased by a cylindrical conductor. The shielding effect of the outer conductor provides an interference-free transport medium for high-speed electrical signals. Coaxial cable, similar to that used for connecting aerials to a TV set, is used in the Cable TV networks to provide the video connection from the street electronics to an individual household, as described in Chapter 2 (see Figure 2.2). Early forms of long-distance transmission networks for the PSTN employing analogue FDM systems were provided over coaxial cables from the 1960s to the 1980s. Now, telecommunication networks use optical fibre in preference to coaxial cables for the Core Transmission Network.

Even though it is not now deployed in the external telecommunication networks, coaxial cable is still used extensively for interconnecting transmission and switching equipment within exchange and repeater station buildings. This interconnection is usually provided across digital distribution frames (DDFs), as described in the Section 5.3.2 on PDH transmission in Chapter 5.

The structure of a coaxial cable is shown in Figure 4.4(c). For physical flexibility, the shielding or screening conductor is made out of metallic braiding underneath a plastic sheath. The screening is separated from the central core conductor by a soft insulator so that the overall assembly of the cable is robust and flexible.

4.2.2.4 Atmosphere: radio

The medium of the atmosphere is unlike that of the confined environment of metallic cables or optical fibres, and its unbounded nature requires the transmission systems exploiting it to have a high degree of adaptability to the medium's changing characteristics. Radio transmission through the air is achieved in several ways, mainly depending on the frequency of the carrier wave used. The various types of radio paths are illustrated in Figure 4.6 and briefly described below [4].

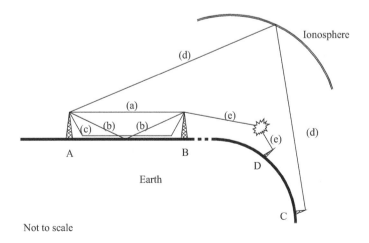

Figure 4.6 Different types of radio paths

Path type (a): direct wave (also known as 'free space wave')
This path uses free space propagation through the air, the radio signal following a straight line of sight between antennas. In order for a sufficient proportion of the transmitted power to be received the two antennas need to be as directional as possible. Since the directional characteristics of antennas increase with frequency, practical systems tend to be at microwave frequencies, i.e. above 1 GHz (1,000,000,000 Hz).

Path type (b): ground reflected wave
In addition to the direct wave there will inevitably be one or more reflected wave paths between the sending and receiving antennas. Normally, these reflected paths cause some degree of interference with the direct signal at the receiver because the reflected signal will experience greater delay, and hence be out of phase with the direct signal. Line-of-sight systems use a variety of methods for reducing the interference from the reflected waves.

Path type (c): ground or surface wave
Ground waves, which follow the surface of the Earth through the process of diffraction, are generated by electrical currents induced in the ground. Radio systems using surface waves are able to extend beyond the line of sight, following the Earth's curvature. The attenuation of surface waves increases with frequency, so practical systems operate in the low-frequency range (30 Hz to 30 kHz) and, thus, are of low capacity.

Path type (d): skywave
This path is created by the reflection of the radio waves off the Ionosphere, the ionised layers belting the Earth. The waves can carry over very long distances through multiple reflections off the Ionosphere and ground. Early long distance, that is, intercontinental, communications were achieved using high-frequency (HF) radio. However, the variable nature of the Ionosphere and the multipath nature of the propagation meant that the transmission was subject to fading and daily fluctuations. Thus, alternative transmission systems (e.g. optical fibre, microwave radio, and Earth satellite) are mainly used in preference to HF radio.

Path type (e): tropospheric scattering
This path, which is created by the scattering of the radio wave by perturbations in the atmosphere, enables communication to a receiver just over the horizon. Practical systems operate at microwave frequencies using very high-transmitting power and large high-gain receiving antennas to compensate for the very high attenuation of tropospheric scattering paths. Such systems are used by the military to provide communication in battle conditions. British Telecommunication (BT) uses tropospheric systems to connect all the oil rigs in the North Sea to mainland United Kingdom, thus extending the PSTN offshore beyond the horizon.

Figure 4.7 summarises the various radio propagation techniques described above and shows the radio spectrum frequency designations.

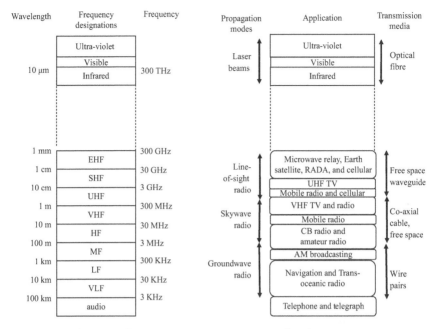

Figure 4.7 The ways that transmission systems use the electromagnetic spectrum

Telecommunication networks today use a mix of radio transmission systems. Invariably, these operate at microwave frequencies in order to exploit the higher bandwidth capacity and hence economy of scale that they provide. With the exception of the tropospheric (or 'troposcatter') systems, these systems work on a line-of-sight point-to-point basis to provide high-capacity links for the Core and Access Transmission Networks, or alternatively in a multipoint configuration in the fixed Access Network linking several customer locations to a central exchange.

Figure 4.8(a) illustrates the configuration for a microwave point-to point relay route, providing transmission capacity between two antennas via an intermediate relay stage (repeater or amplifier for digital or analogue transmission, respectively). The Go and Return channels are conveyed on separate carrier frequencies. Such transmission systems are able to provide routes across a country using many repeater antennas, typically spaced at between 20 km and 40 km apart.

4.2.2.5   Free space: Earth satellites

By locating the microwave radio relay in a communications satellite in space orbit around the World, the distance between a pair of antennas can be extended to intercontinental distances, as shown in Figure 4.8(b). Clearly, the free space loss of the radio signal is correspondingly large because of the long distances between Earth antennas and the satellite relay – typically this so-called 'path loss' up and down totals around 200 dB, depending of the frequency of the carrier wave and the path length, the latter depending on the latitude of the Earth stations. Satisfactory performance requires compensation for this loss by using high-gain repeaters and

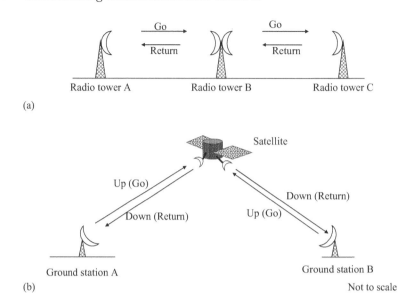

(a)

(b)   Not to scale

Figure 4.8   Free-space transmission media. (a) Microwave radio relay.
(b) Microwave Earth satellite relay

antennas on board the satellite, as well as high-gain antennas on the ground. Since the gain of an antenna increases with size, the gain of the antennas on board the satellite is limited by the weight and diameter of the antennas that can be physically mounted on a space vehicle, so the gain of the antennas on the ground need to be very high – hence the use of the huge satellite dish aerials, which are a familiar sight at satellite Earth stations. Alternatively, smaller capacity transmission links can be achieved from a central large dish antenna at one end of the system and the use of small portable antennas at the other, for example, small-aperture satellite systems as used by news TV broadcasters, and handheld satellite mobile phones used in remote areas of the world.

4.3   Multiplexed payloads

The costs of a telecommunications network are minimised by exploiting the significant economies of scale offered by the above-described transmission systems. In general, the cost per circuit carried reduces as the capacity of the transmission system increases; thus, it pays to multiplex as many channels as possible on to each transmission link. In this section, we consider the standard ways in which channels carrying telephone calls or data communications are assembled into multiplexed composite signals – the payloads – for conveyance over digital transmission systems, that is, copper cable, coaxial cable, optical fibre, microwave radio or satellite, as appropriate. (It should be noted that the multiplexed payloads first used in telecommunication networks were analogue and based upon frequency-division

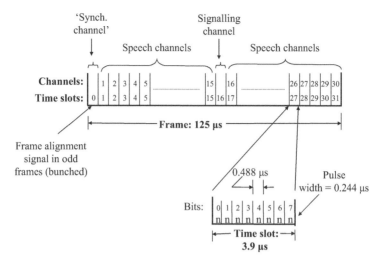

Figure 4.9 The 30-channel pulse-code modulation multiplex frame structure

multiplexing (FDM) – a technique described in Chapter 3.) However, these systems are now obsolete and so they are omitted from this book for brevity. A description of FDM transmission payload systems may be found in [6].

4.3.1 The pulse-code modulation multiplexed payload: the basic building block of digital networks

The primary multiplexed payload created by assembling a group of PCM channels forms the basic building block of all digital transmission and switching networks used throughout the world. Thus, it is an important entity to understand. There have been several versions of the PCM primary multiplex payload, but there are now just two versions standardised by the ITU (previously known as the 'CCITT'): the 30-channel standard, used in Europe, Asia, and elsewhere regionally and on all international links; and the 24-channel DS1 standard, used regionally, mainly in North and South America. Both versions are described below. Chapter 3 introduces the concept of TDM, analogue-to-digital conversion and PCM.

4.3.1.1 30-Channel pulse-code modulation multiplex

In this TDM digital standard, the primary multiplexed structure is based on a timeframe divided into 32 timeslots, as shown in Figure 4.9. The timeframe length is 125 μs, resulting from the sampling rate of 8,000 times a second – 8 kHz (the necessary rate to sample a speech waveform with its frequency content limited to 4 kHz), as explained in Chapter 3. Thus, each timeslot in this frame occupies 1/32 of 125 μs, that is, 3.9 μs. The 32 timeslots of the frame are permanently assigned to 30 speech channels, hence the name of the system, and two non-speech-carrying support channels, as shown in Figure 4.9. The first of these support channels is carried in timeslot 0 (written as 'TS0') and it is often erroneously known as the 'synchronisation' channel, which is used to indicate the start of the frame, as described later. The second support channel is carried in timeslot 16 (TS16) and is used to carry the call-control signalling between the exchanges at either end of the PCM route. (Chapter 7 describes the two sorts of signalling systems that can be carried in TS16.) Timeslots 1–15 are used to carry speech channels 1–15, respectively; timeslots 17–31 are used to carry speech channels 16–30, respectively.

The A/D conversion process (see Chapter 3) within the PCM system encodes each speech sample into an 8-bit binary word. Therefore, each of the 32 timeslots in the multiplex carries 8 bits, making a total of 256 bits per frame. Since there are 8,000 frames transmitted every second, the line rate of the 30-channel PCM multiplex is 8,000 bits/s × 256 bits, i.e. 2,048,000 bits/s. This rate is usually designated as '2,048 kbit/s' or '2.048 Mbit/s' – the latter is usually abbreviated to the more convenient and popular form of '2 Mbit/s'.

Another important figure to remember with the 30-channel multiplex is the line rate provided for each speech channel. This is given simply by the number of bits per sample, 8, times the number of samples per second, 8,000 – that is a line rate of 64,000 bits/s, or 64 kbit/s.

Figure 4.9 also shows how the 8 bits of an individual channel are represented by digital pulses every 0.488 μs (i.e. 3.9 μs/8), each pulse being 0.244 μs or 244 ns (nano equals 1/1,000,000,000) wide.

We are now able to describe the function of the synchronisation channel in TS0. Its role is best explained by considering the terminating end of the digital transmission system which is receiving a stream of digital pulses arriving at the rate of 2 Mbit/s. The timing of the pulses is extracted from the incoming stream and hence sets the receiver sampling rate. However, this stream of pulses is meaningless unless the start of the frame can be identified; thereafter, by counting the bits received the set of eight bits relating to each channel can be located. This frame-start identification is indicated by a special bit pattern, called the 'frame alignment pattern', which is inserted at the sending end into the TS0 of odd frames, as shown in Figure 4.9. At the receiving end the first 8 bits of two frames worth of bits are examined in a digital register. If the frame alignment pattern is not detected, the register shifts one bit and looks at the first 8 bits again. This continues until the pattern is found, the start of the frame has then been detected and the receiver is now in 'frame alignment' with the sender. The spare 8-bit capacity in the even frames of TS0 is available for special purposes – for example, BT used bit 5 to carry network synchronisation control signals for its UK digital network [7].

Digital telecommunication networks have been developed and built on the basis of the 2 Mbit/s building blocks – often referred as '2 Mbit/s digital blocks'. Not only are the blocks used over digital transmission networks, but they are also the entry level into digital exchanges (local, trunk, and international and mobile), as described in Chapter 6. In addition, data can be directly inserted into a 2 Mbit/s multiplex since it is already in digital format. Thus, the 2 Mbit/s digital block may be considered as a payload vehicle capable of carrying 30 speech (or voice) channels or 30 data channels, each of 64 kbit/s, over a digital transmission network.

The 2 Mbit/s block is also used by network operators as the primary rate for a digital stream which is delivered to customer's premises for a variety of business services, including digital leased lines, connections to digital ISDN PABXs, as well as ATM, Frame relay and SMDS data services, as described in Chapter 2. In all these cases the 2 Mbit/s block is carried through the network and presented to the customer as an unstructured stream of 2 Mbit/s, with no implied channel structure.

These 2 Mbit/s digital blocks, whether carry data or voice, may be carried directly over a transmission link or multiplexed together with other blocks to form higher-capacity payloads for PDH or SDH transmission systems, as described below.

4.3.1.2 24-Channel pulse-code modulation multiplex (USA DS1)

Figure 4.10 shows the frame structure for the 24-channel DS1 system. Although the speech is digitally encoded into 8 bits using 8 kHz sampling rate, as in the 30-channel system described above, there is an important difference between the two systems in the way that it is done. This is because the spacing of the graduations on the codec ruler (see Chapter 3) is based on the 'A-law' for the 30-ch. systems and the 'Mu-law' for the 24-ch. system. These two laws, being different ways of spreading the quantum steps using spacing that roughly approximate to a logarithmic scale, ensure a constant signal to quantisation-error ratio over the operating range. Thus, the same waveform would be encoded into a different set of 8 bits by the two systems and a transcoding is required when 30-ch. PCM is connected to 24-ch. PCM links. (E.g. this transcoding takes place at the international gateway exchanges into the United States for calls between the United Kingdom and the United States.) All the 24 timeslots within the 125 μs timeframe are normally used for speech channels (but see below), there being no equivalence of the TS16 or TS0 of the 30-ch. system. However, a single bit is contained at the front of

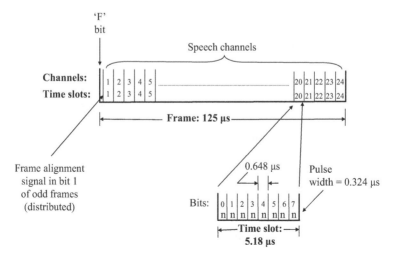

Figure 4.10   The 24-channel pulse-code modulation multiplex structure

the frame before TS1 which is used to carry the frame alignment pattern. The pattern is sent one-bit at a time over odd frames – that is dispersed over eight odd frames, rather than bunched into one frame, as in the 30-ch. system.

The frame comprises of 24 eight-bit time slots plus one bit, totalling 193 bits. The line rate is thus 193 bits per frame, with 8,000 frames/s, giving 1,544 kbit/s – which is usually written as '1.5 Mbit/s'. This rate is also known as the 'DS1' rate in the United States and each of the component 64 kbit/s channels are known as 'DS0'.

There are two ways that signalling can be conveyed over the 24-ch. system. The earlier systems used a method known as 'bit stealing' in which the last bit in each timeslot of every sixth frame was used to carry signalling related to that channel. This periodic reduction in the size of the PCM word from 8 to 7 bits introduced a slight, but acceptable, degradation to the quantisation noise. However, it also prevented the timeslots from being used to carry data at the full 64 kbit/s (i.e. 8 bits at 8,000 frames/s) and the reduced rate of 56 kbit/s (i.e. 7 bits at 8,000 frames/s) was the maximum rate that could be supported. The recently introduced alternative of common-channel signalling avoids the need for bit stealing, since the signalling messages are carried in a data stream carried over the bit 1 at the start of even frames – that is outside of the speech channels. However, this 4 kbit/s (i.e. 1 bit every other frame) of signalling capacity was considered too slow and now the DS1 system has been upgraded to carry common-channel signalling at 64 kbit/s in one of the time slots, leaving 23 channels for speech.

The 1.5 Mbit/s digital block is also offered by network operators as an unstructured primary rate of digital service delivery to customers' premises for business services (leased lines, ISDN, ATM, frame relay, and SMDS). This 1.5 Mbit/s digital block is frequently referred to as a 'T1' system in North America. (The term 'E1' is also used, particularly in North America, to describe the equivalent 2 Mbit/s digital block used in Europe and elsewhere.)

These 1.5 Mbit/s digital blocks, containing data or voice, may be carried directly over a transmission link or multiplexed together with other blocks to form higher-capacity payloads for PDH or SDH/SONET transmission systems, as described later in this chapter.

4.3.2 The time-division multiplexing of digital blocks

Larger payloads may be created by time-division multiplexing several 2 Mbit/s or 1.5 Mbit/s digital blocks. In Figure 4.11 the concept is illustrated using a rotating arm to sample each of four 2 Mbit/s digital block tributaries, each with timeslots 0–31. The 125 μs timeframes of the tributaries is assumed to be aligned so that they all start at the same instant. Since the frame size of the TDM highway is the same as that of the tributaries, the time spent by the wiper on each tributary can be no more a quarter of 125 μs in total and the line rate on the multiplexed highway is 4×2 Mbit/s (i.e. 8 Mbit/s). The wiper therefore must sample each tributary at the rate of 8 Mbit/s. The content of the TDM highway is therefore four times that of each single tributary, that is, 120 traffic channels in 128 time slots, with each time slot

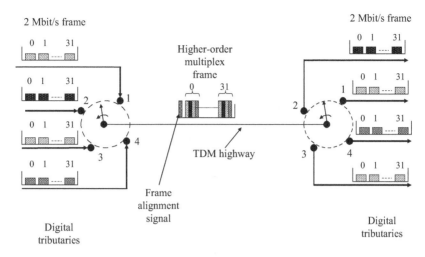

Figure 4.11   Interleaving of digital blocks

occupying 3.9 µs divided by 4, i.e. 0.975 µs. This gives each bit occupying a quarter of 0.488 µs, i.e. 0.122 µs. At the far end of the TDM highway the process is reversed and the appropriate contents are streamed to the corresponding output tributary.

Clearly, for this TDM system to operate accurately the two wipers have to rotate at the same speed and their starting points aligned so that corresponding tributaries at the input and output are sampled simultaneously. The speed of the distant wiper is set by the incoming bit stream, as described above. Whilst the alignment is achieved using a frame alignment pattern (known as a frame alignment signal (FAS)) to indicate the start of the TDM frame, in a similar way to that used to define the start of the PCM frame, described earlier. Figure 4.11 shows the FAS at the start of the TDM frame.

There are two basic ways of sampling the tributaries and interleaving their contents onto the TDM highway: 'bit' interleaving and 'word' or 'byte' interleaving. With the former, one bit at a time is sampled from each tributary in turn; with the latter the full PCM word of 8 bits (also called a 'byte') is sampled in one go from each tributary in turn. Bit interleaving is used in the PDH systems, the original TDM digital payload system, since the design is easier to construct and is readily compatible with digital-to-analogue conversion. Whilst the more recent TDM payload system of SDH, which is designed to carry both data and voice equally, and assumes full digital networking, uses the more complex technique of byte interleaving [8].

The TDM output from the multiplexer can itself be multiplexed again with other tributaries following either the PDH or SDH/SONET formats, and so on to create an appropriately-sized multiplexed payload, as described later in this chapter.

4.3.3   Plesiochronous digital hierarchy system

The plesiochronous digital hierarchy (PDH) was the first internationally standardised form of digital higher-order multiplexing and was deployed over a variety of cable and radio systems, as well as optical fibre cable around the World. Although now largely replaced by SDH systems, there is still some PDH capacity in most incumbent operators' transmission networks. The name 'plesiochronous', meaning nearly synchronous, relates to the situation where the individual 2 Mbit/s tributaries that are to be multiplexed over the PDH system are operating at close-but-slightly-varying rates. This has several consequences. The first consequence is that 'stuffing' or 'justification' bits need to be added to the TDM highway at each stage of multiplexing.

The concept of bit stuffing may usefully be explained by considering a cereal manufacturer who wishes to pack up 12 packets of cornflakes into a large cardboard box for transporting to a grocery shop. The cornflakes packets are nominally sized as 2 cm thick, which means that some boxes will be narrower (e.g. 1.98 cm), whilst others might be wider (e.g. 2.04 cm). Therefore, the size of the cardboard box will need to be greater than 12×2 cm, to allow for this variation in sizes. Once all 12 cornflakes packs are placed into the box the grocer will insert pieces of cardboard as packing to fill up any slack space so that the contents are held tightly. At the receiving shop the cardboard packing will be discarded and the individual 12 packets of cornflakes extracted. Different amounts of cardboard packing will be required in each subsequent box, depending on the mixture of cornflake packet sizes in each batch of 12. The boxes containing 12 cornflakes packets can themselves be packed into larger boxes, again using cardboard spacers to compensate for the variations in box dimensions, and this process can continue with further stages of packing. By analogy bit stuffing in a PDH system is applied to ensure that the TDM frame is filled tightly, irrespective of the actual number of bits sampled from each tributary. There are several ways in which the number and position of the stuffing bits can be indicated to the receiver of the PDH system, so that the appropriate bits can be discarded when the tributaries are extracted.

The European standard for PDH higher-order transmission payloads is shown in Figure 4.12. Four stages of TDM multiplexing are defined, each being a multiple of four. The first stage multiplexes four 2 Mbit/s digital blocks into an 8 Mbit/s stream. The actual line rate of the output of this 2/8 muldex (the name for the

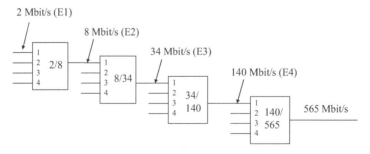

Figure 4.12   Plesiochronous digital hierarchy European standard

multiplexer-to-de-multiplexer assembly, allowing for both directions of transmission) is 8,448 kbit/s. This rate is usually referred to as 8 Mbit/s or E2 line rate. The difference of the E2 rate from four times 2,048 kbit/s results from the addition of an FAS and stuffing bits. The subsequent stages of multiplexing similarly include frame alignment and stuffing bits in the aggregate line rates. Table 4.1 shows all the stages of PDH multiplexing, their nominal and actual line rates, the designation used for line systems at that rate and the number of 64 kbit/s speech channels that could be carried.

The North American PDH standard is shown in Figure 4.13 and Table 4.2. In addition to the DS designations, the line rates when made available for customers' use have the 'T' designations; for example, a 45 Mbit/s leased line is referred to as a 'T3' line system.

Although PDH digital line systems have been successfully deployed since the 1970s around the World, they have proved to be inflexible and cumbersome to manage in large-scale transmission networks due to the so-called multiplexer mountain problem. Figure 4.14 shows that 42 multiplexers are required to provide a fully equipped 140 Mbit/s PDH multiplexed system (for clarity, the line

Table 4.1   European PDH standard

Nominal rate (Mbit/s)	Actual line rate (kbit/s)	Designation	Number of channels (64 bit/s)
2	2,048	E1	30
8	8,448	E2	120
34	34,368	E3	480
140	139,264	E4	1,920
565*	564,148	E5*	7,680

*Not standardised.

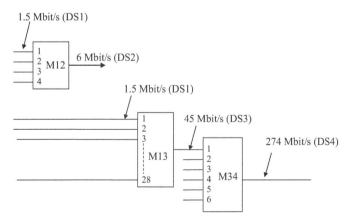

Figure 4.13   Plesiochronous digital hierarchy North American standard

Table 4.2   North American PDH standard

Nominal rate (Mbit/s)	Actual line rate (kbit/s)	Designation	Number of channels (64 or 56 kbit/s)
1.5	1,544	DS1, T1	24
6	6,132	DS2, T2	96
45	44,736	DS3, T3	672
274	274,176	DS4, T4	4,032

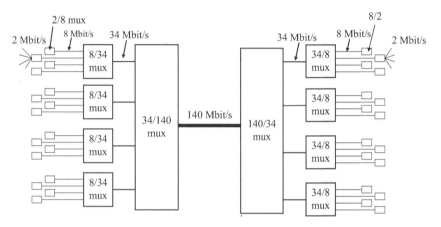

Figure 4.14   A 140 Mbit/s PDH line multiplexed system

transmission equipment is not shown). In order to extract a 2 Mbit/s digital block (e.g. a 30-channel module) from the 140 Mbit/s system, each stage of de-multiplexing is necessary in order to extract the stuffing bits from the 140 Mbit/s, 34 Mbit/s, and 8 Mbit/s frames, respectively. (This is analogous to extracting and discarding the cardboard packing sheets from each nested grocery box until the individual box of cornflakes is obtained.)

Not only does the multiplexer mountain of a PDH network incur many multiplexers, with the consequent cost and potential fault liability, but each item of equipment needs to be connected appropriately. Such connections are usually made manually at the time of setting up a transmission route, using coaxial jumper cables across a distribution frame in the Core Transmission Network station in an exchange or a standalone building, as described in Chapter 5.

4.3.4   SONET and synchronous digital hierarchy system

The PDH standard described above was defined at the start of the introduction of digital transmission networks during the 1970s. However, by the end of the 1980s most of the analogue transmission and switching equipment had been replaced by PDH digital and there was an increasing proportion of optical fibre in use.

A new standard of digital transmission was defined, originally by American National Standards Institute (ANSI) in the United States (see Appendix 1), which was designed to overcome the multiplexer mountain and lack of flexibility of PDH and take advantage of the predominance of digital optical fibre working within the network and the advances made in very large-scale integrated circuits (VLSI) technology. Critically, the system is based upon optical rather than electrical interfaces between transmission and switching equipment in the network. The system is called SONET, derived from 'synchronous optical network'. Later the ITU-T defined a universal standard based upon SONET and extended to cover the European conditions called 'synchronous digital hierarchy' (SDH). The main features of SONET and SDH are:

- Direct access to tributaries through the add/drop multiplexing function (unlike the PDH multiplexer mountain);
- Standard optical interfaces to optical fibres carrying the aggregate signal;
- Compatible to both European and North American standards;
- Reduced operational costs compared to PDH because of the elimination of manual jumpering between the multiplexors;
- Network management features;
- End-to-end performance monitoring capability.

The overall effect of the above features is that SONET and SDH networks use less equipment than the PDH equivalent, with the consequent improvement in operational costs and quality due to the resulting lower fault rate. However, the biggest advantage is that the network management features allow the SONET-SDH transmission to be deployed as part of a coherent managed transmission network, as described in Chapter 5.

As the names SONET and SDH suggest, the system is based upon multiplexing synchronous tributaries, that is they are all operating exactly at one of a fixed set of standard bit rates [9]. This eliminates the need for the use of variable amounts of bit stuffing and the tributaries are time-division multiplexed by directly byte-inter-leaving onto an aggregate highway. The location of a required synchronous tributary within the frame can easily be found by counting the appropriate number of bytes. The system can also multiplex PDH tributaries without the need for bit stuffing by an ingenious use of pointers, which indicate where in the frame the start of the required PDH tributary is to be found.

The tributaries are packed into a basic container, which is known as the synchronous transport module (STM) in SDH or the optical carrier (OC) in SONET. Figure 4.15 illustrates the STM frame structure [10]. Although based on the 125 μs frame, the module actually stretches over nine such frames, creating a matrix of 270 columns and nine rows. Each element in the matrix is 64 kbit/s, that is the capacity of a single voice channel in the basic 2 Mbit/s block, and this is the smallest tributary that can be multiplexed into and extracted from the basic module. The overall rate of the basic STM is 155 Mbit/s. However, higher-speed STMs are created by increasing the number of rows that are contained within the 125 μs frame, each being a multiple 'N' of the basic STM – designated 'STM-N'.

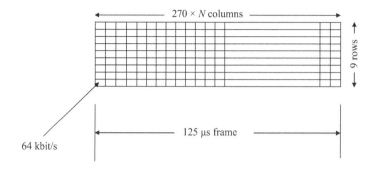

STM–*N* = Synchronous Transport Module, *N* defined for
1, 4, 16 (8 and 12 for USA also)
Gross bit rate = *N* × 155,520 kbit/s

Figure 4.15   Basic STM-N frame structure

Table 4.3   Standard optical interfaces for SONET and SDH

Data rate (in Mbit/s)	SONET optical carrier level (OC-N)	SDH synchronous transport module (STM-N)
51.84 (52)	OC-1	–
155.52 (155)	OC-3	STM-1
622.08 (622)	OC-12	STM-4
1,244.16 (1.2 Gbit/s)	OC-24	–
2,488.32 (2.5 Gbit/s)	OC-48	STM-16
9,953.28 (10 Gbit/s)	OC-198	STM-64

The standard rates for the modules OC-N and STM-N for SONET and SDH, respectively, are shown in Table 4.3.

When packing the tributaries into the modules (i.e. OCs or STMs) other channels in the form of management information as well as the pointers, described above, are also included. This packing process takes place in two stages of multiplexing, as shown in Figure 4.16. In the first stage, tributaries are placed into a lower-order container which, together with an 'overhead' containing transmission-link network management information, is inserted into a virtual container (VC). Several of these VCs are then collected together to form a transmission unit (TU) using pointers to indicate their location. Several TUs form a group (TUG), which is the output of the first stage of multiplexing. In the second stage of multiplexing the TUGs are packed into higher-order containers, which together with an overhead carrying the network management information for the higher-order path – potentially comprising of several links – is inserted into the higher-order VC. Finally, several higher-order VCs may be grouped into an administrative unit (AU) which forms the

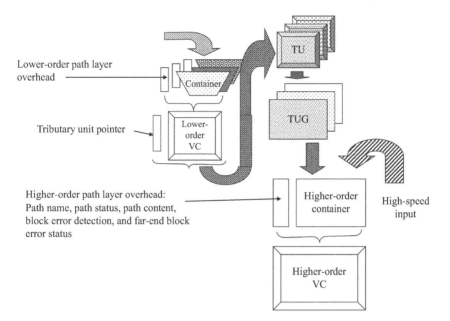

Figure 4.16   SDH packing mechanism

entire payload of the STM-1, again using pointers to indicate the location of the constituent higher-order VCs.

There are a variety of ways that the tributaries may be multiplexed up to the STM-N level, depending on the size of the tributaries and the configuration of the transmission links [10], and these are illustrated in Figure 4.17.

Whilst this multiplexing structure for SDH and SONET might appear complicated, the basic concept used is relatively straight forward. It can be simply explained by considering the synchronous transmission modules (STM) as containers on large transport lorries, which are trundling in a continuous stream down the (SDH/SONET) highways between warehouses (transmission network nodes). Figure 4.18 shows this simplified view of the concept with a lorry containing a single container (STM), in which two different freight companies each have a share – thus forming two AUs – separated by a net curtain inside the container. Within each AU are several large crates (transport unit groups) each containing several smaller crates (transport units). It is these transport-unit crates that contain the various sized parcels for delivery (i.e. virtual containers). The location of the parcels within the crates is indicated by information on written sheets (i.e. 'dockets' or 'inventory sheets'), corresponding to the pointers in the SDH multiplexing system. STMs-4, -16, and -48 are analogous to proportionately larger containers on the lorries or possibly lorries towing a series of container trailers.

As described above, one of the big advantages of equipment based on SONET or SDH is the ability to inject and extract a tributary from a multiplexed aggregate

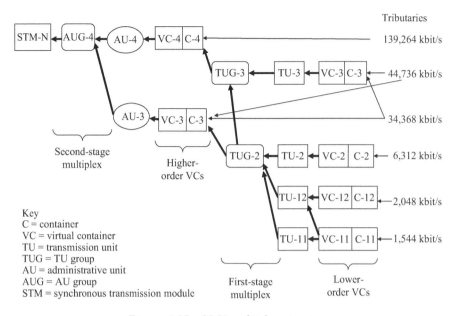

Figure 4.17   SDH multiplex structure

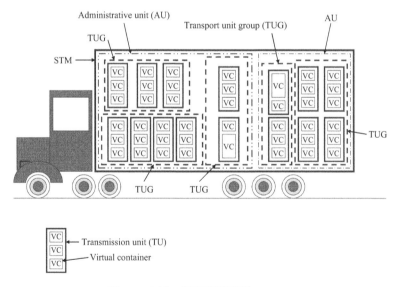

Figure 4.18   SDH/SONET concept

bearer using just one component. The functions of such a component, known as an 'add-drop multiplexer' (ADM) is shown in Figure 4.19, which lists the potential optical interfaces on the input and output ports to the optical fibre bearers, and the electrical interfaces to the set of tributaries.

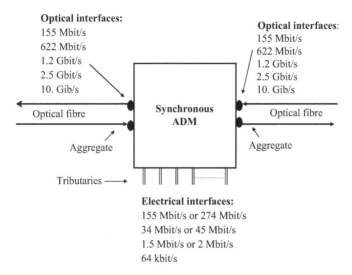

Optical interfaces:
155 Mbit/s
622 Mbit/s
1.2 Gbit/s
2.5 Gbit/s
10. Gib/s

Optical fibre

Aggregate

Tributaries ⟶

Optical interfaces:
155 Mbit/s
622 Mbit/s
1.2 Gbit/s
2.5 Gbit/s
10. Gib/s

Synchronous
ADM

Optical fibre

Aggregate

Electrical interfaces:
155 Mbit/s or 274 Mbit/s
34 Mbit/s or 45 Mbit/s
1.5 Mbit/s or 2 Mbit/s
64 kbit/s

Figure 4.19   A synchronous add-drop multiplexor

The network aspects of a SONET or SDH transmission network are described in Chapter 5.

4.4   The range of transmission systems

In this section, we consider the range of transmission systems currently used in telecommunication networks. Each of these systems is carried over an appropriate medium with the required capacity provided by a direct or multiplexed payload, as described in Sections 4.2 and 4.3. The range of transmission systems is wide. Network operators need to choose which systems to deploy based on the costs, the extent of existing infrastructure, that is whether cables or ducts are already in place, the nature and capacity of transmission channels required, as well as future growth expectations, etc., as described in Chapter 5. In addition, of course, the technologies available for transmission systems are continually being improved, in terms of performance and costs, which can further complicate the choices for the network operator.

4.4.1   Metallic-line systems

Originally all transmission systems used in telecommunication networks were based on metallic – predominantly copper (but, also aluminium) – lines, carried as single pairs or bundled into cables, slung overhead between poles. These pairs carried just one telephone circuit as payload, either from the subscriber's premises (i.e. in the Access Network) or between exchanges providing 'trunk' and 'junction' routes. Today, single circuit payload copper pairs still constitute most of the subscriber lines in the PSTNs of the World. In addition, copper line systems – usually overhead – are still used in some networks, particularly in rural areas, between local exchanges and between a local exchange and its trunk exchange.

However, there has been a progressive use of, first analogue then digital, multiplexed payloads to provide extra capacity over the single copper pair. An example of the operational use of such extra capacity was the first generation of so-called 'pair gain' systems. This provided a two-channel analogue FDM system which enabled two subscriber's lines to carried over a single pair between the adjacent customer-premises and the serving local exchange [11]. This method of saving on the cost of copper pairs, which provides standard telephone service to the subscribers, is still used today by many network operators – although, the more recent deployments use digital pair-gain systems, based on DSL technology, described in Section 4.4.2.

The provision of multiplexed payloads over copper lines for customer-service reasons has until recently been directed at business subscribers, providing leased lines and multiple channels from their premises to the local exchange. Initially, FDM systems giving up to 12 channels (48 kHz) capacity were carried over the copper pairs. Since the 1980s, digital payloads (1.5 Mbit/s T1 in the United States and 2 Mbit/s E1 in Europe) have been deployed over copper pairs in the Access Network. Interference between pairs carrying such payloads in a single cable is minimised by using separate pairs for Go and Return directions. A further measure is the assembly of the pairs into 'balanced quads' within the cable sheath [4], such grouping reduces the mutual interference between pairs carrying the high-bit-rate signals. However, the distances the (quad) copper pairs can carry 2 Mbit/s payloads is restricted to just a few kilometres, and only a few pairs in each cable can use such signals because of the problems of interference. Thus, more robust copper systems (with longer reach and complete fills) such as transverse screen and coaxial cable are used to carry payloads up to 140 Mbit/s between the business subscriber premises and the serving exchange. To cope with long distances the transmission signal is boosted by digital repeaters located periodically along the length of the cable, housed in boxes submerged in the street – that is footway boxes – or mounted on the poles supporting the overhead or 'aerial' cable [12].

An extreme version of long-distance metallic cable transmission is that of submarine cable systems, used throughout the World to cross rivers, seas and oceans. The first generation of submarine cables used analogue FDM payloads using different sets of frequencies for the Go and Return direction of transmission over the single cable. These specialised systems are designed to withstand the rigours of being deployed under water over rough terrain (e.g. the sea bed) and generally being inaccessible for repairs. Apart from (shark resistant) armour plating on the cables the system's sub-sea analogue amplifiers use high levels of component redundancy to achieve the necessary reliability, since repairing amplifiers located on the ocean bed is a costly and difficult task! In addition to providing a low-loss transmission path for the payload, the submarine metallic cable also needs to carry electrical power from its two end stations (cable landing points) to all the amplifiers along its length [13,14]. Finally, of course, submarine transmission systems require specialist cable ships to lay the cables initially and later to recover and repair installed cables.

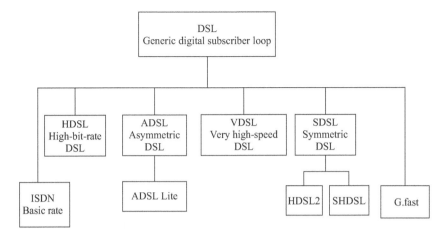

Figure 4.20   The DSL family

4.4.2   Digital subscriber line transmission systems

Advances in integrated circuit technology, particularly digital signal processors (DSPs), have enabled high-speed digital payloads to be carried over standard existing subscriber-line copper pairs originally provided for telephony only. There is a family of such DSL systems, known generically as 'xDSL', as shown in Figure 4.20. Each of the systems has the same basic architecture – that is digital send and receive transmission equipment located at each end of the copper local loop. However, there are important differences within the range. There are two main requirements for an xDSL system. The first requirement is to cope with the impairments introduced by the copper local network. The copper line is able to carry signals far in excess of those used for simple telephony – which are either constrained to 4 kHz by filters or, if unfiltered, extend to only 15 kHz – typically up to some 20 MHz over short distances. However, the attenuation introduced by the copper line rapidly increases with the higher frequencies. In addition, the use of HF signals over the copper Access Network causes interference between adjacent pairs within cables. The name for this interference is 'crosstalk', because of the way that speech from one pair is picked up and can be clearly heard by the listener on the other pair. The interference can occur at the near end to the transmitted signal – 'near-end crosstalk' (NEXT) or at the far end – 'far-end crosstalk' (FEXT). There are several techniques used by the xDSL systems to cope with the attenuation, as well as the FEXT and NEXT introduced by the copper lines.

The second requirement is for both Go and Return signals to be carried by the digital line system over the copper pair – that is 'duplex' working. Even though the copper pair can support both directions of transmission simultaneously, as described in Chapter 1, the xDSL electronic equipment at either end is inherently unidirectional because of its use of semiconductor technology. Generally, xDSL systems use a form of multiplexing to separate the Go and Return signals, for example, TDM (use of 'ping pong' alternate Go and Return bursts); FDM (use of separate frequency bands for the two directions); or use of separate Go and Return pairs (a form of space-division multiplexing, SDM). Other techniques, such as echo cancellation, are also used.

The first xDSL systems were associated with providing the so-called 'basic rate' ISDN over copper. These systems extend the IDN to the subscriber's premises allowing digital data and telephony calls over the PSTN, as introduced in Chapter 3 and further described in Chapter 6. However, the xDSL family is now considered to refer mainly to those systems that enable 'broadband' access over the local loop. Such systems include (see Figure 4.20) the following [3,15,16].

4.4.2.1   Asymmetrical digital subscriber line system

This system was designed originally to provide 'video on demand' service over a copper pair, but is now widely deployed over many PSTNs in the world to provide broadband services (mainly high-speed Internet access) to consumers and small businesses, where the deployment of optical fibre to the premises is not warranted. ADSL uses a filter at the subscriber's premises and at the exchange end of the local loop to extract the standard (4 kHz) telephony signal from the composite broadband signal; thus, the telephony service can be used simultaneously and separately from the broadband service. The latter comprises a medium speed upstream data channel and a high-speed downstream channel – hence the term 'asymmetrical DSL'.

The system copes with crosstalk and other forms of interference by using a technique known as discrete multitone (DMT) modulation, or alternatively known as orthogonal frequency division multiplexing (OFDM) [17]. This approach conveys the data signal over a large set of individual carriers, each at a precise frequency in the available range – for ADSL the frequency separation is 4.3 kHz. At the receiving end of the link the composite waveform is processed so that each modulated carrier is sampled at the appropriate frequency. The use of OFDM enables the maximum possible rate of data through-put despite the limitations of the line due to crosstalk and interference across the whole range of frequencies used. This results in ADSL systems having variable data rates which continuously differ in real time from the nominal advertised rate, as conditions on the line vary. Therefore, DSL systems require a short period of automatic measurement at the DSLAM to set the optimum transmission rate for the conditions on the line. This action, known as 'train up', occurs when initialised or whenever there is a significant change to the induced noise or crosstalk on the line.

Duplex working – that is transmission in both directions over the single copper pair – is achieved using separate bands of frequencies for up and down stream, a technique known as frequency division duplex (FDD). Figure 4.21 shows the spectral allocation of ADSL in the United Kingdom [15], indicating the frequency spreads of the low-band splitter filter and the high-band data extraction filter. With perfect conditions, that is in the absence of interference, data rates of up to some 2 Mbit/s can be carried over about 5 km of copper line, with up to 8 Mbit/s over lengths of about 2.7 km. These figures will improve with progressive upgrades to the ADSL equipment, but they are in any case highly dependent on the

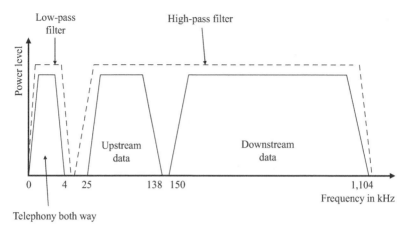

Figure 4.21   Spectral allocation for ADSL

quality of the local loop line plant within the PSTN. More details, including a block-schematic diagram showing the architecture of an ADSL line, is given in Chapter 5 and Figure 5.4.

4.4.2.2   High-bit-rate digital subscriber line system

The high-bit-rate digital subscriber line (HDSL) provides high-speed data transmission symmetrically over two or three pairs of copper cable. Rates of 2 Mbit/s symmetrical can be carried over three pairs up to distances of 4.6 km [3]. Network operators typically use HDSL to provide the local ends of digital private circuits (leased lines) for business customers, where optical fibre local ends cannot be justified or in advance of later deployment of optical fibre.

4.4.2.3   Symmetric digital subscriber line system

Symmetric digital subscriber line (SDSL) is a derivative of ADSL, but with equal (symmetric) transmission capacity upstream and downstream. This means that the NEXT interference dominates at each end and consequently the achievable speeds are much lower than for an ADSL system over a single pair in the same network.

4.4.2.4   Very high-bit-rate digital subscriber line system

The very high-bit-rate digital subscriber line (VDSL) is a derivative of ADSL which provides significantly higher data rates by restricting the use of copper line used for the broadband signal to just a few hundreds of metres. This requires the deployment of optical fibre from the exchange building to the VDSL DSLAM equipment housed in street cabinets – typically physically adjacent to the primary connection points of the Access Network (see Figure 2.8 in Chapter 2 and Chapter 5). This hybrid optical fibre-copper system provides data rates up to some 52 Mbit/s over 300 m of copper pair or some 12.96 Mbit/s over 1,350 m of copper line [16,18]. The spectrum of the VDSL signal over the copper pair goes as high as 30 MHz.

4.4.2.5   ADSL2 and VDSL2

The performance of high-speed digital signals over bundled pairs of copper wires, as in Access Network cables, is hindered by two main impairments, namely: crosstalk and power loss. Unfortunately, just increasing the signal power to compensate for the attenuation due to increasing copper length and signal frequency worsens the effect of crosstalk. Therefore, developments of the xDSL family have focussed on addressing these two impediments so as to increase the speed of broadband transmission. Both ADSL and VDSL have been enhanced – ADSL2 and VDSL2 – using the following techniques.

(a) **Vectoring**. Crosstalk in a copper pair is the induced electrical interference from adjacent pairs in the vicinity, particularly in the case of multi-pair cables. The technique of vectoring relies on being able to generate an inverse signal at the sending end which is applied at the receiving end to cancel the induced voltages due to far-end crosstalk. (Chapter 11 covers the phenomenon of crosstalk in more detail.) This requires a 'vectoring engine' at the DSLAM which computes the appropriate vectoring signal for all the pairs in a bundle terminating on that DSLAM. The effect of vectoring achieves the same bandwidth on a pair of copper wires in a bundled cable as on a single cable – which gives a significant improvement in achieved broadband performance in practical copper-pair networks [19].

(b) **Forward error correction (FEC)**. In addition to crosstalk, induced noise along the line can causes errors to the received digital signal. The effect becomes more pronounced as the speed of transmission increases since the receiver has proportionally less time to detect whether a '1' or '0' was sent. One way of keeping the error rate acceptable is to employ FEC, in which the data to be sent is grouped into blocks which are treated with a suitable error-correcting code (e.g. Reed–Solomon or Hamming codes). At the subscriber's end, any received blocks with errors are detected and, if within limits, automatically corrected. Whilst this technique allows for higher transmission rates over noisy lines, it does introduce some additional delay and an overhead of additional bits (even if there are no errors incurred on the line) which itself reduces the potential gain in speed. An alternative used in some systems is to rely on retransmission of any errored blocks – which saves on the overhead (no correction code bits required) and the repeated content is carried only when errors are incurred [20].

4.4.2.6   G.fast

The latest development in the DSL family is known as G.fast. This system is based on taking the DSLAM very much closer to the subscriber's premises so that the length of copper wires is 500 m or less. In practice, the length of copper pair can be about 20–30 m. This means that transmission is substantially over optical fibre with just a short tail of copper pair delivering the broadband signal over the last few metres to the subscriber's premises. Broadband speeds up to about 1 Gigabit/s are therefore possible – perhaps justifying the name 'G.fast?' Actually, the name is an acronym of 'fast access to subscriber terminals' and the 'G' denotes the series of ITU-T Recommendations (G.9700 and G9701) specifying the system.

The specification has several variants, but the key features are as follows [21]:

- DMT is used to modulate data onto the carrier (as in ADSL and VDSL).
- A wider set of frequencies is used over the copper pair than in the rest of the DSL family. The spectrum of G.fast extends up to 108 MHz, which necessitates the use of special waveform shaping to avoid interference with FM radio broadcasts.
- Duplex working is achieved using TDD. There are several ratios of duplex working, ranging from 50:50 symmetry to 90:10.
- FEC is used to minimise effect of impulse noise.
- An improved form of vectoring is used to eliminate crosstalk.

4.4.3   Point-to-point optical fibre

Optical fibre systems offer the network operator the highest bandwidths of all transmission systems – potentially up into the Tbit/s (1,000,000,000,000 bits/s) range. They were initially introduced as replacements for the coaxial metallic cable systems carrying PDH payloads in the trunk network, but they are now used extensively throughout the national Trunk and International networks and increasingly in the Access Networks, carrying SDH/SONET payloads.

A generic picture showing the basic architecture of an optical fibre point-to-point digital transmission system is given in Figure 4.3. The carrier modulation system uses a LED or laser diode to pulse a single colour light beam at the send end and Avalanche photo diodes or PIN diodes are used to detect the light pulses at the receive end, as described earlier in this chapter. The operating characteristics of the send and receive transducers and the 'windows' in the attenuation curves of the single mode optical fibre now used in transmission systems – as shown in Figure 4.5 – give a wide range of potential wavelengths (colours) of light, centred on 850 nm, 1,300 nm, 1,500 nm. (Note: a nanometre, nm, is 1 billionth of a metre.) Special line codes, for example, 'code-mark inversion' are used in conjunction with the payload coding (e.g. scrambling) to ensure that there are sufficient on-off transitions in the pulsed light sent down the optical fibre for successful timing recovery by the regenerators along the line [14]. Actually, the very low rate of attenuation of modern optical fibre systems means that the regenerators can be spaced at least 50 km apart – a figure that is continually improving with equipment development. This means that on many optical fibre transmission links within medium-or-small-sized countries, such as the United Kingdom, no regenerators are required along the route, saving equipment, accommodation and operational costs.

Normally, separate optical fibres within a single cable are used to provide the Go and Return directions of transmission. The alternative configuration – that is the two directions of transmission carried as separate frequency bands (FDM) over a single optical fibre incurs extra electronic equipment costs and so is used only where the number of optical fibres in a single cable is unduly restricted.

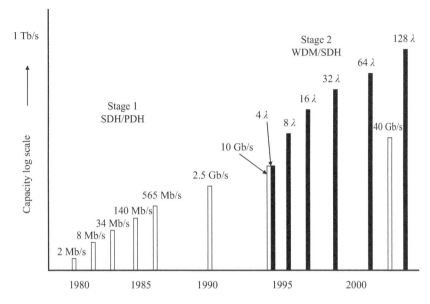

Figure 4.22   Transmission capacity per fibre

4.4.4   Dense wave-division multiplex system

Network operators have progressively introduced to their transmission networks digital line systems of increasing speed, to gain improvements in the cost-per-channel carried. Of course, the full economy of scale on transmission systems is only achieved if the transmission systems are fully loaded with traffic. Since the early 1980s the demand for transmission capacity has generally grown in line with the development of ever increasing line speeds. For example, between 1980 and 1990 the development of digital transmission systems over optical fibre has increased from 2 Mbit/s to 2.5 Gbit/s. However, this progression is based on increasing the speed of optical transmitters, receivers, and the repeater systems, all operating on a single wavelength of light over the optical fibre. Since the mid-1990s development work focused on the use of several different wavelengths of light – that is colours – as a way of increasing capacity on line systems beyond the 10 Gbit/s achieved on one wavelength in 1995. These systems, which are referred to as 'dense wave-division multiplexed' (DWDM), require very precise optical filters to separate the component colours. Figure 4.22 presents a historical view of the development of transmission capacity up to 10 Gbit/s over a single wavelength of light over an optical fibre, and the trend for increasingly higher numbers of wavelengths within DWDM systems.

Figure 4.23 illustrates the general arrangement for a DWDM system with *n* wavelengths. At the sending end *n* lasers are required, each generating a different colour (i.e. wavelength) of light. The optical fibre carries the composite DWDM signal containing the *n* wavelengths. At the receiving end of the system a filter is

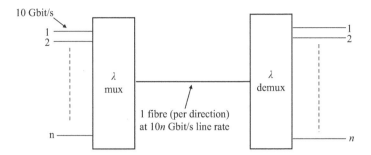

Figure 4.23   Dense wave-division multiplexing

required to separate the n wavelengths and these are passed to n individual pho-todiodes to detect the set of sent signals. For the 16-wavelength system described for TAT-14 above the potential capacity is 16×10 Gbit/s, i.e. 160 Gbit/s – which is the equivalent capacity of some 2.4 million voice channels!

4.4.5   Passive optical fibre network

So far, we have considered only point-to-point optical fibre transmission systems. Whilst this configuration is appropriate for delivering large capacities between network nodes, or to those business customer sites requiring many broadband and telephony circuits, it is not economical for delivering small payloads. Network operators have been seeking a cost-efficient method of deploying optical fibre systems in the Access Network to serve broadband services (e.g. Internet access) to residential and small business customer sites, as described in Chapter 5. There are two main cost factors that need to be addressed: the cost of deploying the many thousands of optical fibres from a local exchange to each subscriber premises, and the terminating equipment at each end needed to extract/insert the signal to/from each fibre. Passive optical fibre network (PON) systems have been developed to address these factors.

The basic premise of the PON system is that the cost of fibre and electronics is minimised by using a tree and branch configuration, whereby a single fibre from the local exchange splits several times in succession in order to serve many sub-scriber premises. The splitting is achieved by fusing bundles of optical fibres so that the light energy splits equally into each tributary. Since this is done along the fibre route without the use of electronics, it is known as a 'passive' system. The tree and branch configuration also minimises the cost of electronics since only one device per fibre spine (or 'tree trunk') is needed at the exchange end. Figure 4.24 shows the basic concept of a PON system. The optical fibre splitters are typically housed in footway boxes beneath the street. In the example shown, the first splitter is four-way, and the second set of splitters is eight-way; thus, 32 subscriber pre-mises are served by a single optical fibre spine from the exchange.

At the exchange end, optical line termination (OLT) equipment associated with each optical fibre spine assembles the Go transmission for the 32 channels into

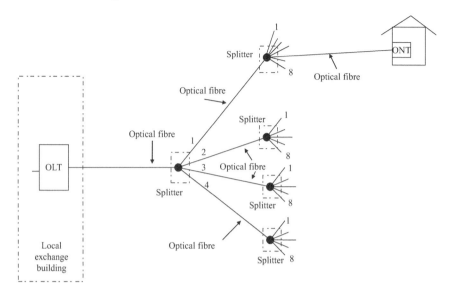

Figure 4.24   Passive optical fibre network

a time-division multiplexed signal that is broadcast over the PON fibre tree structure. Thus, at each subscriber premises a terminal unit, usually called an optical network termination (ONT) extracts the appropriate time slot from the received TDM broadcast signal. For the Return direction of transmission, each ONT is allocated a fixed time slot during which a burst of the subscriber's digital signal is sent. In effect, this is a dispersed version of a TDM multiplexed system, similar to that described in Chapter 3, but with each of the 32 tributaries located in different subscriber premises. An important variant on this configuration is the amalgamation of a set of subscriber ONTs in a single unit, known as an optical network unit (ONU), which is located at a business customer's premises where it serves the many terminations via internal wiring. With the business configuration of PON there is either just one or no stages of splitting in the network, with the necessary TDM channel extraction/injection performed by the ONU.

The PON was first developed in the BT Laboratories and trialled in 1987, with the first application designed for telephony subscribers – known as telephony over passive optical network (TPON). Since then there have been a series of PON developments defined by the full-service access network (FSAN) consortium and the ITU-T; more recently in the United States the IEEE 802.3ah standard has been defined. There are several types of PON, as follows [22,23].

APON: This broadband version uses ATM (asynchronous transfer mode – see Chapter 8) over the PON system to provide broadband service. The third-generation system provides a total capacity of 622 Mbit/s symmetrical for the maximum 32-split PON system.

BPON: Broadband PON is an expanded version of APON designed to carry video. Use is made of DWDM to add a separate wavelength for the video services

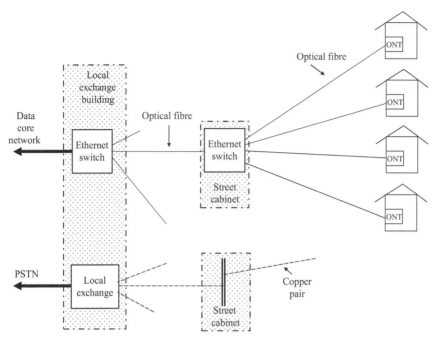

Figure 4.25   Ethernet fibre to the premises (home)

above the non-video traffic on the PON. For example, a BPON system may carry voice, data, and bidirectional video for 32 customers over a single fibre spine. Two wavelengths are used to carry voice and data at 622 Mbit/s downstream and 155 Mbit/s upstream, and a third wavelength is used to carry video up to 1 Gbit/s bidirectionally.

EPON: Ethernet PON offers the LAN packet-based protocol of Ethernet as an interface to the data networks on business customers' premises.

GPON: Gigabit/s PON is the latest high-speed PON design, offering up to 2.5 Gbit/s symmetrically.

4.4.6   Ethernet fibre to the premises

An alternative means of providing broadband delivery over optical fibre to residential or business customers' premises (i.e. FTTH, FTTP, FTTB) is through using Ethernet as the bearer system. This has the advantage of providing an exclusive fibre to each customer's premises from an Ethernet switch, which can deliver the full system bandwidth. As the schematic diagram of Figure 4.25 shows, the number of fibres carried through the Access duct network is kept to a manageable level by siting Ethernet switches in street cabinets (usually next to the copper-cable cabinets). This means that there is contention of the broadband capacity (see Chapter 8), which will restrict customers' data rates during busy periods. It should be noted that the telephony service to the customers continues to be delivered over the existing copper distribution network.

4.4.7   Dark fibre

The principle of sending a modulated light beam down optical fibre cables gave rise to the term 'dark fibre', which originally referred to unused fibres in an operator's cable, that is spare capacity. However, the increasing desire for businesses to build their own data corporate networks has given rise to the now more usual meaning of the term as the leasing of unused fibres in a network operator's cable. These can then be used by the business customers to construct their own data networks, say between two campus sites. Since the dark fibre is essentially raw and transparent, customers are free to use their own broadband transmission systems running any form of protocol. The network operators are, therefore, responsible only for the physical upkeep of the fibre.

Whilst network operators generally have several spare fibres in their cables, they may not always be able to meet customers' demand. Also, there is an understandable reluctance by the operators to release their prime infrastructure – since their business models and mode of operations are designed to provide communication services. One approach to addressing both issues is for the operators to lease individual wavelengths within a DWDM system on optical fibres rather than the fibres themselves. Again, the leased wavelength can be used by the customers to carry any protocol data transmission stream generated by their private network equipment [24].

4.4.8   Submarine cable systems

One of the biggest challenges facing network operators in planning the deployment of cable systems is crossing over large expanses of water within the country or international links to other countries. If a bridge exists across the waterway the first preference is to make use of a cable duct way slung within the bridge. In the case of large waterways or seas, submarine cable systems are needed. These were originally introduced to carry telegraph signals in the late 1890s and the first telephone submarine cables were laid in the 1920s. A big step forward was achieved with the first trans-Atlantic telephone system, TAT-1, opened in 1956 based on a single coaxial cable. As is now normal practice, international submarine cables are owned and managed by a consortium of operators and service providers. In 1986 the first international optical fibre submarine cable system was laid between the United Kingdom and Belgium, and 2 years later the TAT-8, the first trans-Atlantic optical fibre system, was laid [25]. A network of submarine cables now criss-crosses the oceans, seas and channels linking islands and continents, carrying the vast majority of the World's inter-landmass voice and data traffic. (Only a small proportion of such traffic is carried over radio or satellite systems.)

Figure 4.26 shows a simplified view of a typical multifibre submarine cable system. There are several important differences between submarine and terrestrial communication systems. The first is that once laid under the sea it is extremely difficult and expensive to gain access to the submerged cable system for repairs or upgrades. Therefore, the submerged parts of the system are designed to have an

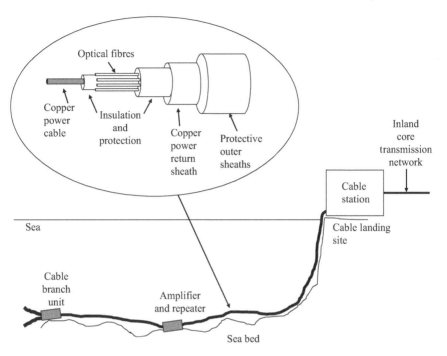

Figure 4.26   Submarine optical fibre cable system

operational life of 25 years. The second main difference is the harsh environment; the cable system and submerged electronics are designed for working at up to 8,000 m depth, experiencing some 800 atmospheres of pressure [26]. Thirdly, the underwater electronics in the optical amplifiers and repeaters need electric power, which means feeding electricity along the cable from both ends. The power is fed down the central copper cable with the return path provided by a copper sheath, as shown in Figure 4.26.

The submerged amplifier/repeaters are spaced about 50 km apart. Branching units are also used to give multiple accesses from landing sites in neighbouring countries to long-distance submarine cables (e.g. cables from the Netherlands, the United Kingdom, and France joining a trans-Atlantic cable at a branch in the English Channel). Finally, the outer layers of the cable are often armoured – to cope with shark attacks and fishing trawlers, etc., and contain various layers of protection wrappings against marine burrowing creatures. At the landing site, a cable station comprises of the submarine transmission termination, power-feeding termination, and a network-management centre for operations and maintenance of the system. The linkage to the terrestrial network is via the Core Transmission Network as shown in Figure 5.12 of Chapter 5.

The potential capacity of each new system laid continues to increase each year. For example, the trans-Atlantic telecommunications (TATs) optical fibre cable system laid in 2001, known as 'TAT-14', is capable of carrying 1,024 STM-1's across the Ocean (compared to TAT-1 which opened in 1956 with 36 telephone circuits). The system achieves this huge capacity by carrying payloads of STM-64 (10 Gbit/s) on each of 16 different wavelengths of light over the fibre pair using DWDM, as described in Section 4.4.4. It is configured as a pair of optical fibres (Go and Return) looping from Manasquan in New Jersey, the United States, to Denmark, then Holland, France, and the SW tip of the United Kingdom (Bude-Haven) and back to Tuckerton in New Jersey, the United States, totalling some 15,800 km. Transmission over current submarine cables is based on DWDM at 10 Gbit/s, 40 Gbit/s, and 100 Gbit/s, centred on wavelengths in the 1,552 nm range.

4.4.9   Line-of-sight microwave radio systems

The basic configuration of a line-of-sight (LOS) microwave radio system is shown in Figure 4.8(a). The key components are narrow-beam antennas, which can operate as precision point-to-point LOS systems, allowing reuse of the same carrier frequencies around the country without mutual interference. Frequency bands allocated by the Regulator and the ITI-R for this use by network operators reside at 4 GHz, 6 GHz, and 11 GHz – these are mainly used on systems providing trunk links between exchanges. In addition, frequency bands around 19 GHz, 23 GHz, and 29 GHz are also available, and are mainly used for shorter distance trunk links and subscriber access [27] and higher frequency bands are periodically being made available. In general, the size of the required antenna reduces with increasing carrier frequency, but the allowable spacing between antennas also reduces and the radio transmission become more susceptible to water absorption, that is rain and clouds.

The LOS microwave radio systems were initially introduced into operators' networks in the 1960s using analogue transmission, as an economical solution for the conveyance of TV programmes between the various studios, TV switching centres and the TV broadcast mast antennas around the country. Later, FDM pay-loads of up to 3,600 telephony channels were carried using FM over these radio systems between trunk exchanges.

Since the 1980s the microwave radio systems have been designed for digital transmission, initially carrying the standard PDH payloads of 34 Mbit/s and 140 Mbit/s (and multiples of 45 Mbit/s in the United States). With the advent of SDH/SONET, the microwave systems carry the standard 155 Mbit/s STM-1 digital block.

An alternative configuration for LOS microwave radio systems is point to multipoint, in which an omnidirectional central antenna radiates uniformly to any antenna/receiver in the vicinity. Such systems can be attractive in providing sub-scriber access links. Several broadband multipoint microwave radio systems have been developed to support the delivery of data services to business customer pre-mises. One such example from the United States is the local multipoint distribution service (LMDS), which operates in the 28 GHz range and offers capacities up to 155 Mbit/s per customer. The deployment of point-to-point and multipoint micro-wave radio systems is covered further in Chapter 5.

4.4.10   Earth satellite systems

The established system of telecommunication global satellites uses a configuration of three satellites equally spaced on an equatorial plane at a distance of 36,000 km above the Earth. At this height the satellites spin around the Earth at exactly the same rate as the rotational speed of the Earth, and they therefore appear stationary to antennas on the Earth's surface [28,29]. Whilst this positioning enables all areas (except the far North and South) of the World to communicate with at least one of the satellites, the high orbit does incur a transmission propagation time for the microwave radio signal of some 250 ms, even travelling at the speed of light. This means a round trip delay of about half a second. Such a delay is noticeable to two people speaking over a satellite link, resulting in some difficulty in maintaining a natural conversation. Consequently, telephone call connections are never provided over more than one satellite hop; the remaining distance is always provided over terrestrial links with their lower delay.

However, the delay issues with geostationary satellites is not a serious problem for one-way communication, such as broadcast TV (e.g. Sky) or data communication. More recently, two new satellite configurations have been introduced: medium Earth-orbit (MEO) and low Earth-orbit (LEO) systems, both of which incur less transmission distance and hence delay [29]. MEO and LEO systems comprise of a constellation of satellites set in an elliptical orbit around the World, designed to give coverage to specific areas only of the Earth's surface. Since the satellites are not in a geostationary orbit they appear to move continuously across the sky; thus, the Earth-station antennas need to be able to track and move to receive the satellite's signal. Once a satellite has moved to the limit of visibility the Earth station antenna needs to switch to a following satellite, which should have just come into view. Typically, each satellite in a LEO system is visible for about 10 min, orbiting about every 120 min at some 780 km above the Earth. These LEO systems tend to be low capacity but can be tailored to give good coverage of polar regions and incur only 10 ms propagation delay. MEO systems use orbits around 10,000 km and incur some 70–80 ms propagation delay.

Unlike the three satellites necessary for full coverage with a geostationary system, MEO systems require some 12 satellites, whilst LEO systems may require up to 200 satellites. The complexity of the antenna steering control, the overall satellite system control and the large number of satellites required means that LEO and MEO systems are not currently economical for large-scale commercial application – although they are used for specialist applications, such as communications in the remote north of Russia and Canada.

4.4.11   Wireless local area networks

There is a family of short distance microwave radio systems designed for conveyance of data services between a user's terminal – for example, a laptop computer – which is normally static and some serving node in the network. Since the configuration is really an extension of the local area network (LAN) concept, originally used in offices but now also deployed in residences, the systems are known as 'wireless LANs'. These systems have been specified in the United States within the 'IEEE 802.11' range, and have subsequently been widely adopted World-wide. They are also popularly referred to as 'Wi-Fi'. These systems also use the OFDM techniques described earlier for ADSL to cope with the multipath distortions to the radio signal (see Chapter 9). The transmission network aspects of wireless LANs and their data network aspects are covered in Chapters 5 and 8, respectively.

4.4.12   Wireless metropolitan area networks

A new generation of LOS microwave access systems has been defined which provide broadband data links between LANs or between a LAN and a central server, creating so-called wireless metropolitan area networks (MAN). The system, defined by the IEEE 802.16 standard, is popularly known as 'WiMax' [30,31]. It uses 256- or 2048-carrier frequency OFDM systems to provide maximum data throughput over the multipath conditions experienced with radio links. However, the standard allows many variations in the way that manufacturer's equipment is designed, thus giving a range of performance and cost options for users. Although WiMax is specified for use in the 10–66 GHz range of the radio spectrum, initial deployments use the unlicensed band of 5.725–5.850 GHz, and in the licensed bands of 2.5–2.69 GHz and 3.4–5.80 GHz. The data throughputs possible range from about 3–7 Mbit/s over a 5 MHz wide radio band; however, this capacity is shared among all the users in the WiMax cell. The use of WiMax to provide broadband access is considered in Chapter 5.

4.5   Summary

In this chapter, we looked at the all-important subject of transmission systems as used in telecommunications. In Section 4.2 we began with the basis mechanism of transmission and then reviewed the physical transmission media: copper cable, coaxial cable, optical fibre and radio. Section 4.3 addressed the various ways of building a digital multiplexed payload for carrying over the transmission media, covering:

- PCM and the basic 2 Mbit/s or 1.5 Mbit/s digital block – the basic building block of digital networks.
- PDH, the original digital transmission multiplex payload system;
- SONET and SDH, the synchronous digital payload system that facilitates a managed transmission network, as well as eliminating the multiplexer mountain problems of PDH.

    The wide range of transmission systems currently in use by network operators were then briefly reviewed in Section 4.4, including:

- Metallic cable
- The family of xDSL (i.e. ISDN, ADSL, VDSL, HDSL, and SDSL) and G.fast systems

- Optical fibre systems, primarily point-to-point (PDH, SDH, and Ethernet)
- Passive optical network systems, for cost-effective access applications
- DWDM, which by using many colours of light is providing ever increasing capacity over optical fibre
- Submarine cables
- Microwave radio systems
- Earth satellite systems (geostationary, LEO, and MEO)
- Wireless LANs and wireless MANs.

The deployment of digital multiplexed payloads over radio, metallic line or optical fibre systems in the Access, Core and International transmission networks is covered in the next chapter.

References

[1]   Flood, J.E. 'Transmission principles'. In Flood, J.E. and Cochrane, P. (eds.). *Transmission Systems*. Stevenage: Peter Peregrinus Ltd.; *IEE Telecommunications Series No. 27*, 1995, Chapter 2.

[2]   Dorward, R.M. 'Digital transmission principles'. In Flood, J.E. and Cochrane, P. (eds.). *Transmission Systems*. Stevenage: Peter Peregrinus Ltd.; *IEE Telecommunications Series No. 27*, 1995, Chapter 7.

[3]   Anttalainen, T. *Introduction to Telecommunications Network Engineering*. Norwood, MA: Artech House; 1998, Chapter 4.

[4]   Flood, J.E. 'Transmission media'. In Flood, J.E. and Cochrane, P. (eds.). *Transmission Systems*. Stevenage: Peter Peregrinus Ltd.; *IEE Telecommunications Series No. 27*, 1995, Chapter 3.

[5]   Dufour, I.G. 'Local lines – The way ahead', *British Telecommunications Engineering*. 1985;**4**(1):47–51.

[6]   Kingdom, D.J. 'Frequency-division multiplexing'. In Flood, J.E. and Cochrane, P. (eds.). *Transmission Systems*. Stevenage: Peter Peregrinus Ltd.; *IEE Telecommunications Series No. 27*, 1995, Chapter 5.

[7]   Redmill, F.J.and Valdar, A.R.'SPC digital telephone exchanges'. Stevenage: Peter Peregrinus Ltd.; *IEE Telecommunications Series No. 21*, 1990, Chapter 10.

[8]   Ferguson, S.P. 'Plesiochronous higher-order digital multiplexing'. In Flood, J.E. and Cochrane, P. (eds.). *Transmission Systems*. Stevenage: Peter Peregrinus Ltd.; *IEE Telecommunications Series No. 27*, 1995, Chapter 8.

[9]   Sexton, M.J. and Ferguson, S.P. 'Synchronous higher-order digital multiplexing'. In Flood, J.E. and Cochrane, P. (eds.). *Transmission Systems*. Stevenage: Peter Peregrinus Ltd.; *IEE Telecommunications Series No. 27*, 1995, Chapter 9.

[10] Wright, T.C. 'SDH multiplexing concepts and methods', *British Telecommunications Engineering*. 1991;**10**(2):108–115.

[11] Taylor, G. 'Electronics in the local network'. In Griffiths, J.M. (ed.). *Local Communications. IEE Telecommunications Series No*. 10, 1983, Chapter 13.

[12]   Flood, J.E. 'Transmission systems'. In Flood, J.E. (ed.). *Telecommunication Networks*, 2nd ed. *IEE Telecommunications Series No. 36*, 1997, Chapter 4.

[13]   Howard, P.J. and Catchpole, R.J. 'Line systems'. In Flood, J.E. and Cochrane, P. (eds.). *Transmission Systems. IEE Telecommunications Series No. 27*, 1995, Chapter 10.

[14]   Petterson, O. and Tew, A.J. 'Components for submerged repeaters and their qualification', *British Telecommunications Engineering*. 1986;**5**:93–96.

[15]   Kerrison, A. 'A solution for broadband'. In France, P. (ed.). *Local Access Networks. IEE Telecommunications Series No. 47*, 2004, Chapter 5.

[16]   Abraham, S. 'Technology for the fixed access network'. *British Telecommunications Engineering*. 1998;**17**(2):147–150.

[17]   Engels, M. *Wireless OFDM Systems: How to Make Them Work?* Norwell, MA: Kluwer Academic Publishers; 2002, Chapter 3.

[18]   Clarke, D. 'VDSL – The story so far', *Journal of the Institute of British Telecommunications Engineers*. 2001;**2**(1):21–27.

[19]   Marshall, J. and Benchaim, D. 'The rural challenge for high speed broadband', *Journal of the Institute of Telecommunications Professionals*. 2013; **7**(2):10–14.

[20]   Marshall, J. 'Super-fast broadband – The engineering of high speed over existing tracks', *Journal of the Institute of Telecommunications Professionals*. 2014;**8**(3):36–41.

[21]   Gorshe, S., Raghavan, A., Starr, T., and Galli, S. *Broadband Access: Wireline and Wireless – Alternatives for Internet Services*. Chichester, UK: John Wiley & Sons; 2014, Chapter 8.

[22]   James, K. and Fisher, S. 'Developments in optical access networks'. In France, P. (ed.). *Local Access Networks. IEE Telecommunications Series No. 47*, 2004, Chapter 9.

[23]   Low, R. 'What's next after DSL – Passive optical networking?' *Journal of The Communications Network*. 2005;**4**(1):49–52.

[24]   Chomycz, B. *Planning Fibre Optics Networks*. Philadelphia, PA: McGraw Hill; 2009, Chapter 11.

[25]   Horne, J.M. 'Optical-fibre submarine cable systems – The way forward', *Journal of the Institute of British Telecommunications Engineers*. 2001; **2**(1):77–79.

[26]   Chesnoy, J. and Jerpahagnon, J. 'Introduction to submarine fiber communications'. In Chesnoy, J. (ed.). *Undersea Fiber Communications Systems*. San Diego, CA: Academic Press (Elsevier Science); 2002, Chapter 1.

[27]   de Belin, M.J. 'Microwave radio links'. In Flood, J.E. and Cochrane, P. (eds.). *Transmission Systems. IEE Telecommunications Series No. 27*, 1995, Chapter 12.

[28]   Nouri, M. 'Satellite communication'. In Flood, J.E. and Cochrane, P. (eds.). *Transmission Systems*. Stevenage: Peter Peregrinus Ltd.; *IEE Telecommunications Series No. 27*, 1995, Chapter 13.

[29]   Schiller, J. *Mobile Communications*, 2nd ed. Boston, MA: Addison Wesley; 2003, Chapter 5.

[30] Eklund, C., Marks, R.B., Stanwood, K.L., and Wang, S. 'IEEE Standard 802.16: A technical overview of the WirelessMAN^{TM} air interface for broadband wireless access'. *IEEE Communications Magazine*, June 2002, pp. 98–107.

[31] Ghosh, A., Wolter, D.R., Andrews, J.G., and Chen, R. 'Broadband wireless access with WiMax/802.16: Current performance benchmarks and future potential', *IEEE Communications Magazine*, February 2005, pp. 129–136.

Chapter 5

Transmission networks

5.1 Introduction

Chapter 4 introduces the concept of modulating the information (voice, data, video, etc.) onto a variety of transmission systems. We now need to consider how these systems are used in telecommunication networks. As discussed in Chapter 1, all telecommunication networks, and indeed most other sorts of networks, comprise of an access portion which connects the population of users to a serving network node, and a core portion which interconnects the set of serving network nodes. These portions are transmission networks in their own right, containing links and nodes, and are usually known as the Access Network and the Core Transmission Network (or similar), respectively. (Of course, the core portion also contains switches or exchanges, data routers, control nodes, etc., which are contained in separate core switching, data routeing, control networks, etc.) The Access Network may be provided over fixed transmission links (using wires, optical fibre cable or radio systems) or over mobile radio links (using a mobile handset), whilst the Core is always provided over a fixed link. This chapter describes the structure and characteristics of the fixed Access and Core Transmission Networks; whilst Chapter 9 covers the mobile access.

5.2 Access networks

5.2.1 Scene setting

The role of the Access Network is to provide a link between each user (referred to as 'subscriber') and their serving node in the network. The public switched telephone network (PSTN) of each country provides the basis for the Access Network, with the deployment of copper cable to (practically) all parts of the country to provide basic telephony service for residential and business subscribers. However, overlaying the copper or 'local Loop' network of the PSTN is a variety of transmission systems – transverse-screen copper, optical fibre, terrestrial microwave radio, or satellite microwave radio – provided specifically to deliver many types of services in addition to telephony to business subscribers. Where possible, the near-ubiquitous physical infrastructure of the local loop – that is the duct ways, poles, and joint boxes, carrying the cables – is used to carry these overlaid transmission systems.

It should be noted that the so-called 'local-loop network', which comprises of the copper pairs connecting subscribers to the telephone exchange, is also used by several non-telephony services, such as analogue and digital-leased lines (private circuits), alarm circuits, telex and, of course, broadband (e.g. using xDSL (digital subscriber line)). Indeed, there are a few local exchanges in central London where copper pairs used for private circuits outnumber the telephony pairs terminating at the exchange.

5.2.2   The copper (local loop) access network

The structure of a copper access network is best described as a set of trees and their branches radiating from the serving exchange to all the subscribers in the catchment area. Figure 5.1 shows a stylised view of the physical architecture of a typical copper network. At the exchange, all of the copper pairs are connected to a large metal distribution frame, located on the ground floor. This frame, known as the main distribution frame (MDF), is the demarcation point in the exchange between external cabling on its input and internal cabling on its output. Jumper wires are used to link each external pair, connected to a particular subscriber, to the internal wires to the appropriate termination equipment dedicated to that subscriber (i.e. the 'line card' as described in Chapter 6) on the local switching unit in the exchange building. It is also at the MDF that jumpers are used to link the external pairs to internal wiring associated with leased line equipment, broadband terminating

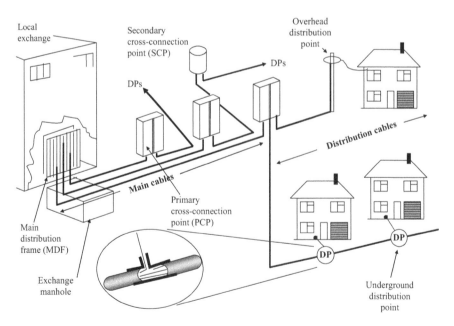

Figure 5.1   The physical architecture of a copper access network

equipment (as described later in this chapter), etc. In general, the MDF is the physical point of presence in the exchange where an exclusive electrical contact can be made to a particular copper pair for operational activity, such as testing or application of special tones, etc. Apart from its dominance functionally as the exchange end of the copper Access Network, the MDF is physically an imposing sight for a visitor to an exchange building, being a metal structure some 2.5 m high and arranged in rows of 10–15 m long, often occupying most of the ground floor and, exceptionally, covering several floors.

Whilst considering the exchange end of the copper Access Network, it is convenient to describe the other major physical entity usually on the ground floor of a serving (i.e. local) exchange – namely, the battery. As described in Chapter 1, a voltage of 50 V direct current (DC) (with slight variation between countries) is applied to each telephony copper pair to power the telephone instrument. Actually, most PSTN operators are required to provide sufficient power for several instruments on any one subscriber's line. The limit is set by the power required to ring a standard bell at the instrument – defined as ringing equivalent number (REN). In the United Kingdom, the REN is set at four, which allows any number of phones to be connected to the one line, provided the combined REN values do not exceed four.

The exchange battery is also an impressive sight within a local exchange building. It can best be described as similar to a large car battery, about 1 m high, occupying a large-sized room. The 50 V DC power from the centralised battery is carried to the switching equipment over aluminium or copper bus bars suspended from the ceiling and walls, measuring about 20 cm wide and 1.3 cm thick. The exchange battery is charged form the mains electricity on a trickle basis. During normal operation, the power to the exchange is provided by the electrical main supply with the battery connected in parallel (with its voltage, the so-called 'floating') across the charge input, creating a stabilising effect on the 50 V DC supplied to the telephones. In the event of a mains failure, the battery can provide power to the subscribers' telephones for several hours. A local generator may be used in the event of long mains power outages. This ability to enable the telephones to work during a mains electrical power outage is viewed seriously by the regulatory authorities because it allows subscribers to call the emergency services at all times.

The progressive introduction of electronic equipment within exchange buildings since the mid-1980s, serving not only digital telephony switching, but also data transmission and routeing, has increased the need for much lower voltages than the standard 50 V DC. This has been met by distributed power-equipment racks located in adjacent rooms to the relevant equipment within the exchange building. These racks, which are directly fed by the standard alternating current electrical mains (e.g. 240 V AC) include power rectifiers to convert to the required direct current voltage for the served electronic equipment, for example, 6 V DC, as well as compact stand by batteries. The power is distributed over the metal bus bars described above. Most exchange buildings tend to have a mixture of the central and distributed systems [1,2].

Immediately outside of the serving (local) exchange, all the cables from the MDF are hubbed at an overhead or, more usually, underground assembly (known as the 'exchange manhole' in the United Kingdom). Here the copper cables are carried through ducts laid under the streets radiating from the exchange building. As Figure 5.1 shows, those subscribers living close to the exchange are served directly from one of these cables. For the more distant subscribers, the exchange cables terminate on frames within street cabinets (known as primary connection points (PCPs)) where distribution cables radiate down individual streets within the PCP catchment area. For the most distant subscribers, or for the more rural areas, a further stage of connection at secondary connection points, SCP (also called 'pillars' in the United Kingdom), is used. Jumpers are run across the frames in PCPs and SCPs to create a continuous circuit (loop) between the exchange-side cable and the distribution-side cables – known as 'E-side' and 'D-side', respectively.

The final link (or 'final drop') to the subscriber's premises is made either via a drop wire from a pole in the street (referred to as 'overhead distribution') or via a lead-in pair from the cable running underground along the street. For the overhead case, the cable from the PCP or SCP runs underground to the base of the pole and then terminates at the top of the pole in a distribution box containing, typically, a 10- or 20-pair terminal block. Each drop line is then slung from a secure point on the outside of the subscriber's premises. With underground distribution, the lead-ins are run to joints made, along the cable, in footway boxes below street level, as it passes the subscriber's premises. In general, business premises and all premises in urban areas are served by underground distribution, with the suburban residential premises generally served by overhead distribution. It is interesting to note that in the United Kingdom about half of the residential subscriber final drops are provided overhead.

The arrangement of PCPs and SCPs described above allows a flexible development of a tree and branch structure, which can minimise the total cost of the copper cables to serve an exchange catchment area. Firstly, the quantity of pairs deployed is minimised by taking fewer pairs to the E-side of the PCP/SCP than on the D-side, a jumper-link between the two sides is being made only when a new subscriber is added to the exchange. Secondly, the cost is minimised by using as thin a gauge of copper wire as possible. The limit to the size of gauge is set by the required electrical characteristics of an exchange line, since the thinner the gauge, the greater the attenuation of the local loop and hence the greater the reduction in loudness of the telephone connection. (E.g. in the United Kingdom, the attenuation limit is set at 15 dB, measured at a frequency of 1,600 Hz.) The maximum allowable local-loop resistance is set by the need to for a sufficient current to operate the signalling relays in the exchange line card. Typical limits for local loop resistance are 2 kΩ. In practice, a combination of different gauges is used, with narrow gauge serving the shorter subscriber lines and a combination of narrow, medium and thicker gauges serving the longer subscriber lines. Figure 5.2 illustrates a typical arrangement of the gauges of the copper pairs and the size of the cables in the Access Network.

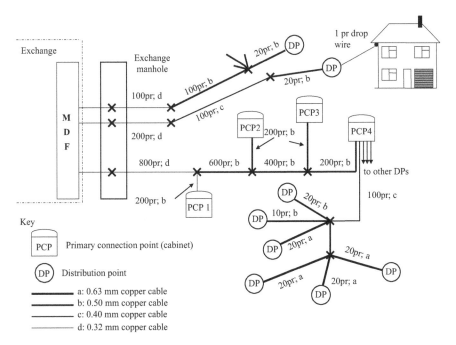

Figure 5.2 Typical local loop make-up

In rural areas, particularly where the terrain is rocky or mountainous, the E-side cables are carried overhead, slung from poles running along the road or cross-country, as appropriate. Special rugged cables are used to cope with the exposure to wind encountered in such situations and the weight of the cable over long catenary lengths between poles. Very long overhead cables also require amplification or reduction in loss using loading coils at points along the route to ensure appropriate electrical characteristics are achieved.

5.2.3 The optical fibre access network

5.2.3.1 Direct fibre to the premises

Optical fibre cables are also carried in the underground duct of the Access Networks. The optical fibre is deployed to serve the larger businesses (providing telephony as well as leased-line, data and other services) – where the link takes a large capacity directly to the terminating point in the business premises. However, direct fibre to the premises (FTTP) is also being deployed to residential and small businesses to carry broadband access, as discussed later. (Alternatively, the fibre is deployed as a hybrid optical-copper delivery method for residential and small business subscribers, i.e. xDSL, as we discuss in the next section.)

The optical fibre taken directly to the subscriber's premises terminates on a bank of multiplexing equipment which extracts the various channels and services.

These may be leased lines, data circuits, digital multiplexed speech circuits, analogue individual voice circuits, etc. Most business subscriber premises contain an equipment room for the terminating equipment (e.g. multiplexors) of the network operator, together with a smaller version of an MDF for copper wires and a DDF (digital distribution frame) for jumpering the digital channels to the various internal communication equipment, for example, PABX (private automatic branch exchange), ATM (asynchronous transfer mode) switches, IP routers, Frame Relay switches, Ethernet, and LANs (local area networks).

In the case of plesiochronous digital hierarchy (PDH) transmission systems, a typical arrangement is an 8 Mbit/s digital block carried over the optical fibre from the serving exchange to the subscriber premises. This is demultiplexed to four 2 Mbit/s digital blocks in the subscriber's equipment room, giving, say, one 2 Mbit/s ISDN (international subscriber digital network) link from the PSTN switch to the PABX and three 2 Mbit/s leased lines (or two 2 Mbit/s leased lines and one spare slot for future expansion).

Now the preferred high-capacity delivery system within the Access Network for businesses is synchronous digital hierarchy (SDH). Typically, this is at the STM-1 (synchronous transport module) (155 Mbit/s) level, delivered over the fibre from the exchange with add-drop multiplexors in the subscriber's equipment room extracting the required tributaries. In addition, Ethernet over fibre is becoming an increasingly attractive form of delivery of broadband service, as described in Chapter 4.

For PDH, SDH and Ethernet delivery, the optical fibre network is structured as a tree and branch network, similar to that of the copper Access Network. Large capacity optical fibre cable (up to 96 fibre pairs per cable) radiate from the local exchange, with smaller cables branching out at joints in boxes at appropriate points – with the final drops to customers' premises using typically four-fibre (Go and Return with spare pair) optical cables. The optical fibre network then extends from the serving local exchange, the parent trunk exchange and perhaps to other trunk exchanges and neighbouring exchanges, as appropriate – thence joining to the Core Transmission Network. For the residential subscribers, given the large numbers and moderately low traffic generation, the use of Ethernet switches at street cabinets enables a tree and branch structure, giving a net saving of fibres back to the exchange node (see Figure 4.25 in Chapter 4).

5.2.3.2 Passive optical network delivery

Passive optical networks provide an alternative to the use of a single optical fibre per customer's premises, and hence they potentially hence offer network operators an economical solution to avoiding further deployment of copper cables. The basic system is described in Chapter 4. In practice, the PON systems are installed in parts of the Access Network where there is growth in demand for standard telephone, ISDN service or medium speed leased-line services. The optical splitter assemblies are usually housed in footway boxes, since no power supply is required en route. As with the direct optical fibre systems, the active line terminating equipment

(LTE) is housed in the customer's premises. The derived tributaries are connected at small distribution frames to the internal wiring to the terminal devices within the customer's premises.

5.2.3.3   Hybrid fibre-copper telephony delivery

Here, the optical fibre terminates on street-located cabinets, usually adjacent to the PCPs, where individual channels are derived from the high capacity on the fibre by multiplexing equipment. The channels are carried by copper cable to the PCP, where they are jumpered to the appropriate copper final drops to the subscriber's premises.

Such arrangements are used by PON systems, in which the street-based multiplexors extract 48 subscribers' circuits (Go and Return Channels), which are delivered over copper pairs either as underground or overhead final drops. This use of PON, as opposed to taking the PON fibres direct to each served premises described above, is primarily to provide relief for the existing copper Access Network by delivering capacity for new telephony subscribers at the outer edge of the exchange catchment area without requiring new copper cables in the E-side.

5.2.4   Radio access network

5.2.4.1   Fixed line-of-sight microwave radio

Radio links are also an important part of many operators' Access Networks. Such links are provided on a permanent or semipermanent basis with one end terminating on a fixed radio receiver attached to the subscriber's premises and the other end termination at the serving exchange building. This use of radio is therefore part of the fixed network and is different to the way radio is deployed in a mobile network, as described in Chapter 9.

Multipoint radio systems, based on an omni-directional antenna, typically located on the serving-exchange roof, are mainly used to provide permanent access links to subscribers wanting single-line telephony. The choice of whether to use multipoint radio is based on the degree of geographical remoteness of the subscriber and practical problems with using the cheaper copper-pair option, e.g. the serving of subscribers in the mountains or across rivers. Consequently, multipoint radio has been deployed extensively since the 1970s in such areas as the Highlands and Islands of Scotland. Many of these earlier systems use analogue transmission. However, more recent deployment use digital transmission, offering a better performance quality (e.g. clarity and lack of noise) for the user.

Digital microwave point-to-point systems using two dish antennae, one on the roof of the subscriber's premises (usually an office block) and the other on the roof of the exchange, provide higher-capacity access links. A typical example is the use of a 19 GHz digital radio system to provide for a bi-directional 2 Mbit/s digital leased line from an exchange to the business customer's premises (i.e. the 'local end'). Another use of microwave radio is that of providing high-capacity distribution of satellite links, where the provision of individual satellite terminals on customers' premises is not possible. Here, the 29 GHz transmitter is located on the roof of the serving exchange building along with a satellite receive antenna. The video transmission received from the satellite is then relayed via the 29 GHz radio system to antennas on several customer's buildings on a point-to-point or multipoint basis [3].

Unfortunately, the performance of radio links, particularly microwave radio, is impaired by atmospheric conditions, for example, rain, as well as reflections from buildings. (The biggest reflection problems are caused by the tall cranes on building sites which are constantly swinging around!) Many operators, therefore, install microwave radio links initially to serve a business customer, taking advantage of the speed of provision of radio compared to running a new optical fibre link. Later, once the optical fibre route is completed, the service is swapped to the optical-fibre system and the microwave radio link recovered, ready for deployment elsewhere.

The key aspect for a network operator deploying a microwave radio system is the need for so-called 'radio-frequency planning'. Each network operator has a portion of the spectrum allocated by the national regulator for their use of specific types of radio systems. The operator, therefore, must ensure that their proposed deployment of a radio system will not interfere with any existing radio systems in the vicinity or other operators or future potential radio users in the area. They must also consider the profile of the landscape and potential points of interference or areas of severe attenuation so that the radio path provides the required level of transmission performance. Planning tools, usually computer based, are used to determine the appropriate power levels and frequencies within the allocated range to be used within the proposed fixed radio system.

5.2.4.2   Wireless LANs

In contrast to the fixed microwave radio systems above, wireless LANs (WLANs) use frequencies in the unregulated part of the spectrum. This means that non-radio-expert service providers, as well as network operators, can rapidly deploy WLAN hot spots around the country. The low power and short distance of the wireless LAN coverage means that there is little chance of interference with other radio users. Typically, Wireless LANs are provided in airport lounges, coffee shops, railway stations, etc., enabling customers to gain broadband data access from portable data terminals, such as computer lap tops. This form of radio access net-work gives the user the ability to move within the catchment area of the antenna, typically 10–20 m; the user may also log on to other Wireless locations around the country – giving a form of the so-called 'nomadic mobility'.

Figure 5.3 illustrates the arrangement [4] for a public service. The WLAN antenna is centrally located at the serving location, for example, a coffee shop, giving a catchment area determined by the power of the transmission and the characteristics of the building, since this determines the absorption of the radio signal. The antenna may be connected by a leased line over the Access Network,

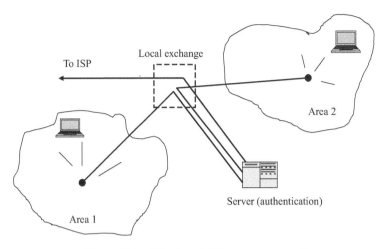

Figure 5.3   Wireless LAN access network

typically using optical fibre to the serving local exchange building and then to a server on the service provider premises, as shown in Figure 5.3. Alternatively, this backhaul link is carried on an xDSL system. The service provides authentication and logging-on functions. Access to the Internet is then provided by the serving ISP, as described in Chapter 8.

5.2.4.3   WiMax

As described in Chapter 4, the second generation of wireless MAN, standardised as IEEE 802.16 and popularly known as 'WiMax', offers the possibility of ubiquitous broadband data access. It conveys a broadband link from an exchange to an external antenna on a customer's building and thence to a LAN or other in-building network. Both unlicensed and licensed parts of the frequency spectrum may be used, giving a trade-off between freedom of use and minimal planning but with greater chances of interference versus more predictable performance but with more planning required, respectively. This radio technology can prove useful for providing broadband data coverage to many customers who cannot be reached economically by xDSL or optical fibre.

5.2.5   *Broadband access for the small business and residential market*

So far in this chapter we have considered the delivery of telephone service over the copper cables and the delivery of higher-capacity services primarily to business customers using optical fibre or radio in the Access Network. However, since the late 1990s telecommunication network operators and cable TV operators have been delivering high-speed data services, primarily fast Internet access, to residential and small business customers – the so-called 'broadband' service. Cable TV operators provide broadband service to their subscribers by adding cable modems to carry the fast Internet-access data over a spare TV channel, as described in Chapter 2 (see Figure 2.9).

Telecommunication network operators have deployed ADSL electronic equipment at each end of copper lines to convey the fast Internet access data channel in addition to the telephone service. xDSL is now the main method that incumbent network operators use to provide broadband service to the residential and small business market. This new use of the copper local network originally deployed just for the provision of telephony (i.e. the so-called 'narrowband' service) represents a significant new lease of life for the copper cables. Since it is only the incumbent operators that have the ubiquitous copper network, there is usually a regulatory requirement for this asset to be made available to alternative network operators so that they can deploy their own xDSL equipment over the copper pairs, through a process known as 'unbundling', as described in the following section.

5.2.5.1   ADSL and the unbundling of the local loop

Chapter 4 introduces ADSL equipment as one of a family of digital transmission systems designed to operate over copper pairs in the Access Network, and Chapter 2 (see Figure 2.8) describes the overall routeing of the Internet access over ADSL. Now we will take a more detailed view of the arrangement within the copper Access Network, as shown in Figure 5.4. For a network operator to provide

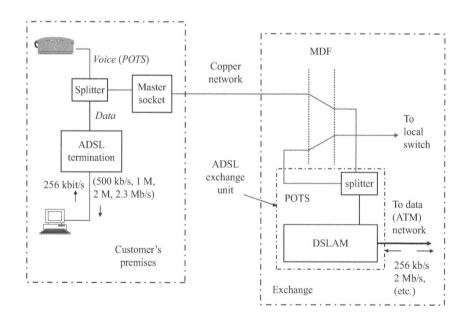

Figure 5.4   Basic ADSL architecture

broadband service ADSL terminating equipment, that is, the DSLAM, needs to have been installed at the serving local exchange. In practice, the functions of splitter and DSLAM multiplexor are incorporated within equipment capable of serving several hundred subscribers.

(Chapter 8 considers the data transport capabilities of ADSL systems.) When a subscriber requires broadband service an ADSL splitter and termination are installed at the subscriber's premises, and a jumper connection is made at the exchange MDF to link the copper pair to a port on the DSLAM equipment. The need for the network operator to send a technician to the subscriber's premises is avoided by the self-installation of small filters at the subscriber's main telephone sockets.

In many countries, particularly those in Europe and North America, the Regulator requires the incumbent PSTN operator to make parts of the Access Network available to other public network operators (PNO) so that they can provide ADSL-based broadband service to their own customers. The term 'unbundling' is used to describe this opening up of the Access Network at various points. In theory, the local loop can be unbundled at any point between the subscriber premises and the MDF – for example, by allowing the PNO to connect to the subscriber's pair at a street cabinet or primary connect point. The latter, which is known as 'sub-loop unbundling' is applicable to VDSL or G.fast where the PNO optical fibre terminates at a multiplexor/DSLAM in the street and uses the copper loop for just the final drop. Full loop unbundling is, however, applicable to ADSL delivery. There is a final unbundling option, in which the incumbent network operator shares the copper pair with a PNO; the incumbent providing telephony (i.e. plain old telephone service (POTS)) over the bottom 4 kHz of bandwidth (also known as 'baseband'), whilst the PNO uses the higher frequencies for the broadband service. The line-sharing unbundled option requires both operators to agree careful operational procedures for the maintenance of the copper local loop, as well as the provision and maintenance of the telephony and broadband services [5].

However, the most common approach is the use of full unbundling, in which the demarcation between the two operators is at the MDF in the serving local exchange. Here, a pair of copper jumper wires is extended from the subscriber's line termination on the MDF to a 'hand-over distribution frame' (HDF) associated with the PNO's DSLAM equipment, which is typically located in a segregated area of the exchange. The incumbent network operator allocates a protected and segregated area of the exchange to various PNOs providing unbundled services. The latter pay rental charges to the network operator for both the subscriber pairs and the floor space in the exchange, use of power, standby arrangements, etc. The alternative arrangement is for the PNO to extend a tie cable from the incumbent's MDF to the remotely-located HDF and DSLAM housed in their own accommodation.

The performance of ADSL was progressively upgraded by advances in the underlying electronics (i.e. discrete multitone or OFDM) technology. However, it is important to appreciate that the rate at which data can be carried over a copper pair is decreased by the presence of:

(a) Crosstalk due to interference from signals (e.g. ADSL, VDSL, HDSL) on other pairs in the same cable, and

(b) Impulsive noise outside the control of the network operator (e.g. due to electrical induction from power cables or from other operator's cables in the vicinity).

Both effects increase with distance and with the proportion of pairs in a cable carrying DSL systems (known as 'the fill').

The first-generation ADSL is capable of about 8 Mbit/s downstream over pairs up to some 2.5 km in the absence of interference, but nearer 1.5 km on cables with realistic amounts of interference from other pairs in the same cable. The second-generation equipment (ADSL2) provides an increased maximum rate of about 12 Mbit/s (due to the improved frame structure). However, the increased rates will be available only on short lines, and with only a modest improvement over first-generation ADSL over longer lines. A further development, 'ADSL2+', increases the rate to about 24 Mbit/s using a wider spectrum over the line – that is 2.2 MHz instead of the 1.1 MHz used by ADSL and ADSL2. But again, due to the rapid attenuation of the higher frequencies with distance, in practice the higher data rates will be restricted to the shorter lines. Thus, overall ADSL2 and ADSL2+ will provide higher downstream data rates on only a relatively small proportion of lines. Therefore, most xDSL deployments are now based on the hybrid optical fibre-copper systems such as VDSL and G.fast, in order to better meet the subscriber's need for higher broadband speeds.

5.2.5.2   Broadband options for the access network

There is a wide range of transmission systems that a telecommunication network operator or cable TV network operator can use to provide broadband access to customers, as illustrated in Figure 5.5. As described earlier in this chapter, this range covers:

- Copper pair cable:
  ○ ADSL
  ○ ADSL2
  ○ ADSL2+
  ○ HDSL
- Hybrid optical fibre and copper pair cable:
  ○ VDSL (FTTK)
  ○ G.fast
- Hybrid optical fibre and coaxial cable (HFC)
  ○ Cable modem
- Optical fibre:
  ○ Direct fibre (FTTH, FTTO)
  ○ PON
  ○ Ethernet

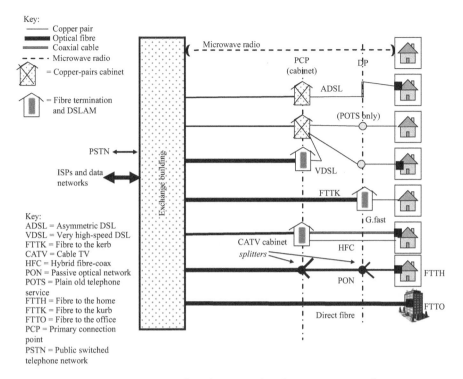

Figure 5.5   Broadband options for the access network

- Microwave radio (generically known as BRA):
  - ○ Wi-Fi (IEEE 802.11)
  - ○ WiMax (IEEE 802.16)
  - ○ LOS microwave radio
  - ○ Point-to-point and point-to-multipoint radio.

All of these systems have differing characteristics in terms of bandwidth delivered, technology used, operational and distance constraints, and – of course – cost.

In general, the unit cost of transmission systems deployed in an Access Network depends on the density of subscribers served. This is because the cost of the electronics located in an exchange or street cabinet must be apportioned across the number of subscribers served within the equipment's catchment area. Against the cost dimension, the network operator needs to offset the potential revenues from the services delivered.

The copper pair cable is still the main basis of PSTNs, even though the technology and design dates from the early 1900s. Although optimised for voice, the copper network has proved remarkably versatile in its ability to carry ever greater bandwidths using additional electronic systems, such as xDSL, and so continuing its utility in today's 'Internet age'. Nevertheless, the high levels of costly manual interventions involved in the provision of service, rearrangements, and maintenance of the copper network has been a driver for network operators to seek replacements with lower operational expenditure (opex). Unfortunately, all the potential alternatives – for example, direct optical fibre, fibre to the kerb, PON, multipoint radio, etc. – have proved to be uneconomical as wholesale replacements of the copper network. This is due to the difficulty in achieving opex savings in practice with such conversions, and the absence of sufficient extra revenues from services over the higher-speed optical fibre systems. The latter is often referred to as the lack of a 'golden bullet' – that is a profitable application or service that can be supported only by the new technology.

The continual downward pressure on prices, and the relatively low-value people place on extra bandwidth, may prolong the search for a business justification for the replacement of copper. Furthermore, the growth in the unbundling of the copper network, which creates a dependency of an increasing number of alternative network operators on copper, is likely to exacerbate the position. This problem has been tackled in some countries by their governments providing subsidies for the move to a politically attractive high-speed fibre broadband Access Network. In the absence of such subsidies or a suitable golden bullet, for the near future the Access Networks are likely to continue to have a large component of copper pairs, together with a mix of optical fibre and radio systems.

Finally, it is important to note that even if the commercial conditions were attractive, the sheer scale of the fixed line Access Network means that it would take several years (probably at least 10) to install fibre as a replacement for all the copper-pair cables. The general view by incumbents is therefore that deployment of hybrid systems like G.fast will be the fastest way of getting high-speed broadband access to the majority of subscribers. However, there will be many initiatives by government and specialised broadband network providers to supplement the hybrid approach by the focussed deployment of optical fibre broadband systems where possible.

5.2.6   The future of fixed line telephony

Of course, the incumbent network operator will still need to provide fixed line telephone service whilst the broadband systems are deployed. As Figure 5.5 shows, with the hybrid fibre-copper or fibre-coaxial systems the telephony service continues to be delivered over the existing copper pairs. However, the full fibre systems (i.e. FTTH/O) do not provide a means of carrying a telephone channels; fixed line telephony to those fibred premises, therefore, needs to remain on the copper pairs!

There are three future scenarios for telephony from premises served by optical fibre, as follows:

(i)   The use of the copper network for telephony continues indefinitely;
(ii)   The copper line is ceased and the customers rely on the use of a mobile network for their telephony service;
(iii) The telephony service is provided over the broadband network (i.e. VoIP over broadband, as described in Chapter 8).

Already, there are households in the United Kingdom where no fixed-line telephony is taken, the users relying solely on mobile services – so the likelihood of scenario (ii) will probably increase. In addition to scenario (ii), broadband customers are expected to progressively increase their adoption of scenario (iii), using either voice service from OTT providers (see Chapter 8) or taking a VoIP service form the network operator. The gradual replacement of traditional fixed-line telephones by IP-phones, associated with scenario (iii) will enable network operators to move more easily to NGN replacements for the PSTN, as discussed in Chapter 11.

5.2.7   Planning and operational issues

Planning and operating the fixed Access Network represents the major annual cost of any PSTN operator. The copper Access Network involves substantial amounts of manual intervention in the provision of service to new customers, maintenance, and the installation of new or replacement capacity. Consequently, by far the largest number of an operator's technicians work in the Access Network, incurring the highest contribution to the current account cost.

Network operators constantly endeavour to reduce the amount of manpower involved in the routine actions which are performed many thousands of times a day, such as the provision of a new telephone line, or the repair of a fault on the line. Such reductions are enabled by improvements to the processes being followed (i.e. 'process engineering') and using computer support systems to provide line plant inventory, track fault histories, etc. Computer support systems can also help the dispatch supervisor in the deployment of technicians to an area, taking into account the location of the various jobs that need to be done that day, so minimising the amount of travelling time spent going to subscriber premises.

5.2.7.1   Provision of a new subscriber's line

In a copper Access Network, the provision of a new subscriber's line involves a visit to the subscriber's premises and relevant parts of the external network as well as actions within the exchange building, the service-management centre (SMC), and within the administration offices of the network operator. Briefly, the actions may be summarised as follows:

- Order taken by sales administrator (over the telephone, by e-mail or fax), who can check availability of line plant to the street address and allocate a telephone number. Both actions are usually supported by a computerised order-handling system.
- The order is received from the sales administrator by the technician dispatcher, who schedules an external line plant technician job. In parallel, a job is set up for the internal exchange work and for the SMC.
- At the SMC a technician at a control console enables the allocated line card in the local concentrator switch of the serving exchange (see Chapter 6) against the given telephone number. Dial tone is then detectable on the wires connecting the line card to the MDF.
- A technician adds a jumper wire from the exchange side of the MDF to the allocated pair in the external cable to the serving street cabinet (i.e. the PCP).

- The external technician then visits the street cabinet and installs a jumper wire to link the scheduled E-side (the line going to the exchange) and D-side (the line going to the subscriber) copper pairs. If there is more than one flexibility point involved (e.g. an SCP) then this needs to be jumpered also. The technician should be able to detect 50 V DC and the dial tone on the completed loop to the exchange. The technician makes the final drop connection by either running an overhead drop wire or an underground lead in to the subscriber's premises, as described above. The exchange line is terminated in the subscriber's premises at an appropriate block terminal – the network termination and test point (NTTP) (see Box 5.1). This is the point from where the subscriber is able to run internal wiring to their own equipment, such as cordless telephone systems, answer machines. At this stage of the process, the subscriber should then be able to verify that telephone service has been provided.
- Successful completion of line provision is then noted in the administrative system so that directory information is recorded and the billing process initiated.

Box 5.1   Network termination

The key feature of an Access Network is the user-to-network interface presented at the subscriber's premises. The actual interface depends on the telecommunication service being provided by the network operator and the type of user's terminal. Because of the influence on terminals and service – and hence the competitive opportunities and constraints – this interface is defined by the national telecommunication Regulator. In the United Kingdom, a network termination and test point (NTTP) is identified for all operators [6] usually referred to as the NTP.

As Figure 5.6(a) shows, the NTP is a reference point located inside the network terminating equipment (NTE) on the user's premises. The NTE terminates the line transmission system and may also include a safety barrier between the subscriber's equipment – which may be powered by mains electricity – and the Access line. This provides protection for the operator's technicians working on the copper lines as well as protecting the users from any dangerous induced voltages on the copper line.

The NTE is deemed to be the edge of the network, and hence the point where the operator's network service ends, as shown in Figure 5.6(b). However, the user's application and the actual service seen by the users really extend between the ends of the apparatus or terminals at both ends. Very often, users will attach several terminals directly and indirectly to the NTE.

The NTP is also a major test point for maintenance activities, such as fault location and diagnosis. Operators arrange for a loop to be placed across the line at the NTP to check continuity within the access link, so that a fault can be located to the network side or to the user's terminal side. Often, this looping within the NTE can be initiated under remote control from a maintenance centre in the network.

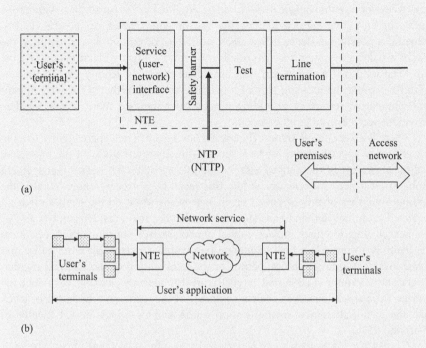

Figure 5.6   Network termination. (a) The network terminating equipment functions. (b) The extent of network service

A corresponding process is required for the installation of ADSL to this basic copper pair, as mentioned above. In addition, there are processes for the installation of all the broadband access systems listed in Section 5.2.5.2. For business premises with sufficient demand for telephony, corporate voice networks, private circuits, VPN, Centrex and data services (as described in Chapter 2) the network operator will deploy SDH transmission payload over direct optical fibre. Chapter 11 looks at the architectural aspects of the management of operations and maintenance within a telecommunications network.

5.2.7.2   Spare plant

Most network operators aim to be able to provide a new subscriber line within a matter of a few days after the order is received. Clearly, for this to be achieved there needs to be a stock of spare pairs throughout the copper Access Network sufficient to meet the demand for new service. If there are not any spare pairs in the serving D-side cable or E-side cable connecting the PCP to the exchange then either new capacity needs to be added to the network or some alternative means found to provide service. The installation of new capacity will incur an appreciable delay (e.g. weeks or months) before service can be provided to the subscriber. Alternative means of providing service include the use of other cables off another PCP – sometimes possible in dense urban areas, or the use of radio links. Sometimes, the use of a pair-gain system (as described in Chapter 4), using electronics to derive two local loops off a single pair, can enable service to be readily given to a new subscriber using the existing pair of a neighbouring subscriber. However, in general, alternative means of providing a subscriber line are more expensive than providing service from pairs in stock.

There is an economic trade-off between having sufficient spare copper pairs in the ground to meet demand for new service directly from stock on the one hand, and the sunk cost of the unused asset of the idle capacity. This asset management problem for a network operator is like that faced by a grocery shop: What is the optimum number of tins of baked beans that needs to be on the shelf so that all expected customer demand can be met – not having tins available means loss of business, having too many ties up capital unnecessarily? For most PSTN operators the inability to provide a new telephone line within a reasonable time will either cause loss of business to a competitor or the intervention of the national regular. Operators determine an internal target figure for meeting demand for telephone service from stock, based on their assessment of this economic trade-off of stock size, the cost of alternative methods of provision, and loss of business, typically set at around 95%.

Copper Access Network planners therefore use the 95% (say) figure with their forecasts of future demand for new telephone lines to dimension the cable network. Finally, it is worth emphasising that the highly granular nature of the Access Network means that the planning decisions are critical – capacity is needed down the actual streets to serve potential subscribers, having plenty of spare capacity in cables running down other streets will not enable this demand to be met.

The planning and operations of Access Networks is covered in more detail in the companion book, References [7] and [8], respectively.

5.3   Core transmission networks

5.3.1   Scene setting

Unlike the Access Network, which may link some tens of thousands of subscribers to a single serving local exchange, the Core Transmission Network provides links between relatively small numbers of network nodes, typically spread across the whole country. These transmission nodes are points in the national network where bundles of circuits, serving telephony, leased lines, data services, etc., are extracted from or entered onto the required transmission links. Economies of scale are achieved by multiplexing onto as few, but large, transmission systems as

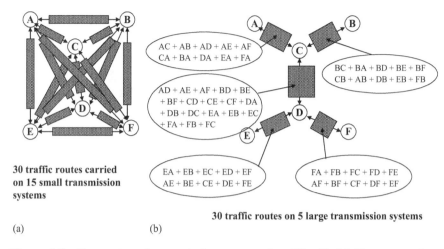

*Figure 5.7 Economies of transmission aggregation [Ward]. (a) No aggregation.
(b) With aggregation*

possible – since the unit costs of transmission systems decreases with system size. Thus, in many cases it is economical to unpack and repack transmission links at intermediate points within a network to achieve optimum loading of transmission links. This principle is illustrated in Figure 5.7, where 30 unidirectional traffic routes require 15 small transmission systems when carried directly (Figure 5.7(a)), whereas the same set of traffic routes could be carried over just five large transmission systems using aggregation (Figure 5.7(b)). Whilst this leads to economies of scale in the transmission cost, it does also incur the need for packing and unpacking at the C and D nodes. This activity involves demultiplexing and multiplexing together with a manual or automatic means of jumpering (i.e. cross-connection) between them – a process known as 'transmission-flexibility'. The nodes at which this flexibility is provided are known generally as 'Core Transmission Network stations' (CTS), and called 'transmission repeater stations' (TRS) in the United Kingdom.

The CTS are usually located at each of the trunk and international exchanges and many of the larger local exchanges. However, in addition there may be some CTSs located in buildings which contain only transmission equipment. Figure 5.8 illustrates the concept with an example of five CTSs (A to E), which are supporting the core transmission between a set of trunk telephone switching units, data nodes and private circuit nodes. Exchange building A contains the CTS-A which serve the co-located trunk switching unit, data unit and private circuit node, and a similar situation exists at exchange buildings B, D, and E. However, the CTS providing transmission flexibility for core transmission links between A, B, D and E is not located in an exchange building, but rather at a transmission-only building at an appropriate transmission network nodal point.

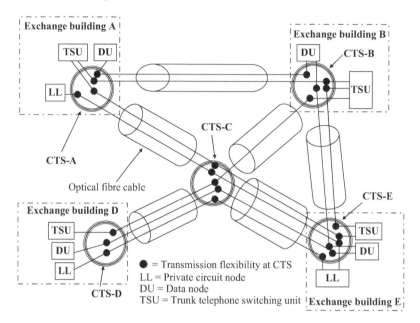

Figure 5.8   Transmission network configuration

5.3.2   PDH network

Chapter 4 and Figure 4.13 introduced the concept of the 'multiplexor mountain' associated with Core Transmission Networks using PDH equipment. The practical arrangement at a CTS is shown in Figure 5.9, in which a 2 Mbit/s block is extracted from the incoming 140 Mbit/s PDH transmission system on the left and inserted into the outgoing 140 Mbit/s PDH transmission system on the right (and vice versa for the return direction of transmission). The CTS is composed of racks of transmission terminal equipment (not shown in Figure 5.9 for clarity) and three stages of multiplexors: 140/34, 34/8 and 8/2 – each connected to a common DDF, or to one of several DDFs. In this example, the incoming 140 Mbit/s system terminates on 140/34 Mbit/s mux. Number 1, from which a jumper wire is run from output Number 1 to the input of 34/8 Mbit/s mux. Number 1, and so on via the central DDF to the output multiplexor mountain through to 140/34 Mbit/s mux Number 11.

Figure 5.9 also shows the extraction at the DDF of 2 Mbit/s blocks from the input 140 Mbit/s transmission system (via 34/8 Mbit/s muxs Numbers 1 and 4, and 8/2 Mbit/s muxs Numbers 1 and 13) to the co-sited telephone switching unit.

Transmission flexibility is also possible at any of the intermediate transmission rates of 34 Mbit/s and 8 Mbit/s through appropriate jumpering at the DDFs. Such a facility would be used, for example, to route a 34 Mbit/s digital private circuit through the CTS.

The jumper wires on the DDFs are coaxial type cables, similar to those used to connect the domestic TV to its aerial. Each of the jumperings have to be made manually when the routes are set up. Not only does this represent a current account

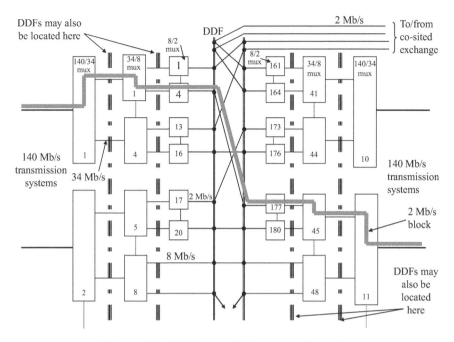

Figure 5.9   PDH transmission network station

cost for the network operator, but the periodic intervention on the DDFs to add or remove jumper wires also introduces disturbance to the established jumpers, often resulting in damage and faults. Electronic replacements for the jumper wires and DDFs, using digital cross connect equipment, known as digital cross-connect (DXC) in the United States, are used by some network operators. However, for cost reasons most network operators tend to use DXC to provide the transmission flexibility only for the high-value international transmission systems at the larger international gateway CTSs.

5.3.3   SDH network

An SDH CTS uses a combination of add-drop multiplexors (ADMs) and DXC equipment (see Chapter 4) to provide the necessary transmission flexibility, as shown in Figure 5.10. This automated system contrasts with the more complex and manually intensive arrangement for PDH shown in Figure 5.9. The configuration of the ADM and the DXC are managed through a computer-based controller, which may be co-sited or remotely located. As described in Chapter 4, use of SDH equipment enables a fully managed transmission network to be created. The SDH configuration controller allows the network operator to manage the configuration of the CTS flexibility points, through planning and assignment processes, as well as reconfigurations in real time to compensate for transmission link breakdowns. SDH also allows the end-to-end performance of an SDH routeing through several CTSs

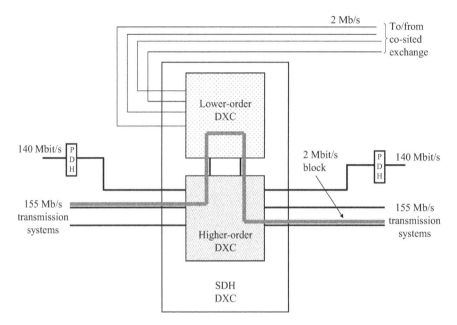

Figure 5.10   SDH core transmission station – using DXCs

to be monitored automatically – a particularly attractive feature for digital-leased-line customers.

An SDH CTS comprises DXCs on which the high-speed transmission links terminate. For the example of a 2 Mbit/s block extraction from the incoming 155 Mbit/s link the DXC needs to be able to identify and manipulate the appropriate 2 Mbit/s tributary from the incoming SDH link and pass it to the outgoing link. This requires the DXC to have a 2 Mbit/s 'granularity' – that is to be able to cross-connect at the 2 Mbit/s level. Typically, a DXC will need a range of granularities to cope with the PDH rates still used in the network (i.e. 2 Mbit/s, 8 Mbit/s, 34 Mbit/s), as well as the SDH line rates (i.e. 155 Mbit/s, 622 Mbit/s, 2.5 Gbit/s) to enable a practical set of transmission flexibility levels. This is illustrated in Figure 5.10, in which the DXC is divided into a higher-order DXC switch-block handling the SDH transmission rates and a lower-order DXC switch-block handling the 2 Mbit/s and other PDH rates. Any legacy PDH transmission systems still in the network terminate on special PDH terminal interfaces to the SDH DXC system, as shown in Figure 5.10 [9].

At smaller transmission nodes ADMs only are used to extract digital transmission blocks (at the PDH rates of 2 Mbit/s, 8 Mbit/s and 34 Mbit/s) for the co-sited telephone switching unit, private circuit, and data units within the exchange building associated with the CTS, as shown in Figure 5.11.

In order to maximise their management capability, SDH networks are usually structured in a set of hierarchical levels. Figure 5.12 illustrates a typical SDH network structure using DXCs and ADMs. A useful analogy which can help in

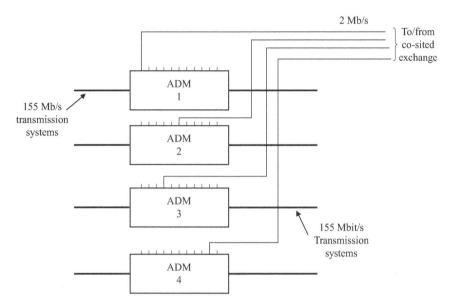

Figure 5.11   SDH core transmission network station – using ADMs

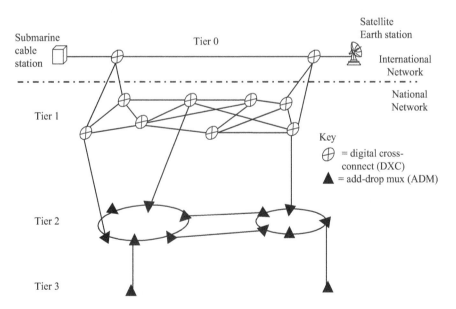

Figure 5.12   SDH transmission network structure

visualising this structure is to consider it as an upside-down set of trees, with the trunks at the top level of the national SDH network and the succession of branches reaching down into the lower transmission levels. (Unfortunately, the tier designations increase the lower the level!) At the top level (Tier 1) of the national network is a mesh of high-capacity SDH transmission links between flexibility nodes (CTS) of DXCs. This forms the inner portion, or backbone of the Core Transmission Network and links the major trunk exchanges, as well as leased lines and data nodes. Hanging off this level at Tier 2 is a set of SDH rings linking ADMs within a region of the country, serving smaller trunk exchanges, local exchanges and other nodes. Tier 2 nodes serve the Access Network, with links to large business customers (Access SDH), small local exchanges or other customer service aggregation points. Above the Tier 1 of the National Network is the International portion of the Core Transmission Network, the Tier 0, which links to the network's transmission gateways to the transmission networks of other countries – via submarine cable landing stations, microwave radio stations or satellite Earth stations [10].

Of course, the technology used in both the Access and Core Transmission Networks continues to develop in terms of cost per circuit or unit bandwidth carried and management and service features. Consequently, new types of transmission systems will be introduced periodically to the network. For example, DWDM, described in Chapter 4, has been added to the existing SDH over optical fibre major transmission links within the Core Network to increase the payload capacity. Whilst adding DWDM on a link-by-link basis gives proportionate capacity increases, the widespread availability of multiple wavelengths over the optical fibre network offers the attractive possibility of managing transmission routeings across the network on a wavelength basis. Wavelengths can be added and extracted to/ from optical fibre networks using wavelength add-drop multiplexors (WADM) and interconnected by optical cross connects (OXC) to create the so-called 'λ-routeing' ('lambda routeing'), thus creating the so-called 'managed optical platform', in which a connection-pattern of different wavelengths of light support the SDH payloads carried over the optical fibre network [11]. This may give rise to new network structures and architectures, as discussed in the section on NGN in Chapter 11.

5.3.4   Carrier Ethernet backhaul links

Ethernet, which earlier in this chapter and in Chapter 4 has been described as a LAN or broadband Access technology, has been developed into a version suitable for WAN applications. The version is known as Carrier Ethernet (CE), indicating that it has a long-distance transmission capability. It is also capable of supporting several different LANS on the same system – creating virtual LANs (VLAN), similar in concept to the VPN application of MPLS described in Chapter 8. CE circuits may be transported over SDH/SONET, PDH or TDM transmission systems or they may have their Layer 1 mapped directly on to an optical fibre transmission system. In addition to supporting VLANs, CE offers network performance monitoring capabilities on a transmission-link basis, so making it suitable for long-distance transmission of IP-based data traffic throughout the network.

An important application of CE is the provision of IP-based backhaul circuits for 4G (LTE) mobile networks, linking the cell sites to their eNodeBs, as described in Chapter 9. This is now the preferred transmission network arrangement because all ports within the LTE mobile network are Ethernet LAN-based, reflecting the fact that this 4G mobile system is all-IP – thus, carrying users' data and voice as composite data traffic.

5.3.5   Transmission network resilience

In practice, from time to time there will be events that cause a transmission link to fail. Typical examples of such events include the physical severing of the link resulting from storm damage of an overhead cable, or a digger slicing through an underground optical fibre, alternatively the terminal equipment in the CTS might become faulty. There are other occasions where planned work by the operator, for example, adding new capacity or rearranging the use of transmission links, will necessitate an arranged outage on one or several links. Whatever the cause, the effect is the withdrawal of connectivity across the Core Transmission Network, affecting perhaps several thousand channels of capacity between two nodes. The ability of the network to cope with this disruption is often referred to as 'resilience'.

The inherent resilience of the network can be enhanced by remedial action within the Core Transmission Network, as described in this section, and at the switched traffic level, as described in Chapter 6 – whilst the overall contribution to network performance and quality of service as perceived by the customers is considered in Chapter 11. All methods of increasing resilience rely on the addition of some extra network capacity. Thus, a resilient network is one in which its optimum structure and capacity is augmented by some degree of redundancy. The appropriate extent of this over provision is, of course, determined by a cost-benefit balance set by the value of the traffic being carried over the network which would otherwise be lost.

Figure 5.13 illustrates the various levels of resilience possible in a tele-communications network. At the top is the switched network, where a set of calls (i.e. 'traffic') between switching units at exchanges A and B could automatically be routed via the switching unit at Exchange C in the event of a failure on traffic route A-B. This alternative routeing of the calls between exchanges A and B would incur throwing extra load on the traffic routes A-C and C-B, as well as the cost of switching the extra calls at C. Although this use of alternative routeing gives a good increase to the overall network resilience, its use is limited by the ability of the alternative routes, A-C, C-B, which are already carrying their normal load of calls, to handle the offloaded traffic from A-B without incurring unacceptable overload and congestion. Automatic alternative routeing (AAR) is considered further in Chapter 6. The remainder of this section will consider the resilience measures that can be taken in the four levels of the Core Transmission Network below the traffic switched level shown in Figure 5.13.

The various measures described below fall into two camps: those that enable instantaneous full replacement of the lost capacity, so ensuring service with little or no perceived interruption, and those that mitigate the loss to an acceptable degraded level of service. At the transmission-circuit level of the Core Transmission Network

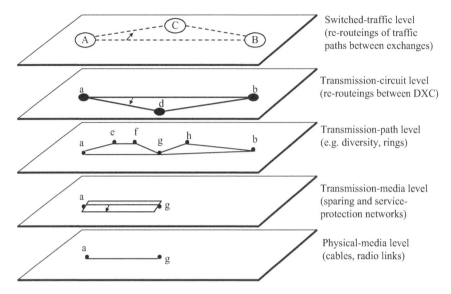

Figure 5.13   Levels of network resilience

(Figure 5.13) the traffic route between exchanges at A and B can be provided by a direct *a–b* transmission circuit route or, in the event of failure, routed via a transit CTS *d*. Each transmission route can itself be protected at the transmission path level using physical diversity or the use of rings. Each of the diverse routes, or links of the rings can be further protected at the transmission-media level by the automatic switching to standby capacity using sparing or service-protection networks. Finally, at the physical media level the resilience can be enhanced using physically-protected line plant.

5.3.5.1   Transmission-circuit level: switching at digital cross-connection units

Apart from facilitating the deployment of capacity the use of digital cross connection units at the top tier of an SDH Core Transmission Network (see Figure 5.12) enables the circuits to be re-routed in the event of link failures. This form of resilience relies on appropriate spare link capacity being available, as well as the ability to detect link failures and decide on the suitable re-routeings. The latter may be handled manually or through a fully automated control system, which is usually complex and expensive. Obviously, this form of resilience is available only at the Tier 1 of the Core Transmission hierarchy and it is, therefore, limited to protecting just the links (which also tend to be large capacity) between the major nodes.

5.3.5.2   Transmission path level: diversity

This simple, but highly effective, method of increasing resilience is through the spreading of the link capacity across two, three or sometimes four parallel separate

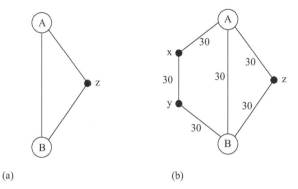

Figure 5.14 Transmission-path resilience: diverse routeing. (a) Physical diversity of two. (b) Physical diversity of three

transmission routeings, as shown in Figure 5.14. As an example, we will consider a typical three-way diverse routeing of, say, 90 circuits capacity on the traffic route between exchanges A and B (see Figure 5.14(b)). This means that 30 circuits are carried over each of the three parallel paths (engineering routeings) – Number 1 (A-B direct), Number 2 (A-z-B), and Number 3 (A-x-y-B). Each of these 'routeings' is chosen to give as much physical diversity as possible – that is, using separate cables and CTSs following different paths (roads and towns) between exchanges A and B.

In the event of failure of any one of the diverse routeings only one-third, i.e. 30 circuits, of the traffic route A-B capacity would be lost. If telephony service was being carried, this loss would be perceived by the users as an interruption to the calls in progress over the effected link and a general increase in the chance of call congestion on the whole traffic route due to its effective reduction to 60 circuits. This is significantly more acceptable than the total failure of all calls between exchanges A and B that would otherwise have occurred (in the absence of resilience at the switched traffic level). The cost to the operator of increasing resilience by diversity results from the use of three smaller sets of transmission links, with their inherently higher unit costs, rather than a single large more cost-efficient link.

5.3.5.3 Transmission path level: SDH and SONET rings

The SDH and SONET transmission systems have the valuable facility of a supervisory channel which enables them to work in a managed transmission network, as described in Chapter 4. Resilience at the transmission-path level can be provided by the use of spare capacity between add-drop multiplexors (ADM) when associated with SDH or SONET carried over optical fibre deployed in a ring structure. These systems are referred to as self-healing rings (SHRs). There are several types of SHR used in current networks, for example:

- Multiplex section – shared protection ring (MS-SPRING)
- Multiplex section – dedicated protection ring (MS-DPRING)

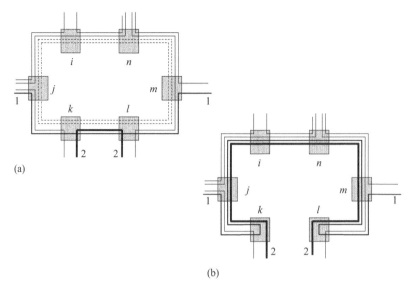

(a)

(b)

Figure 5.15   Transmission-path resilience: SPRINGS. (a) Ring with vacant standby paths. (b) Break between k–1; Circuits 1 and 2 routed over standby paths around RING

These may be based on two or four optical fibre configurations.

Figure 5.15(a) shows a simple example of a two-fibre MS-SPRING comprising six ADMs, labelled *i* to *n*. The spare capacity within the optical fibres joining the ADMs is shown as dotted lines, and the working capacity between the various ADMs is shown as full lines. Attention is drawn to circuit 1, entering at ADM *j*, routeing via ADMs *k* and *l*, and leaving at ADM *m*; and circuit 2 entering at ADM *k* and leaving at ADM *l*. (Note, only one fibre carrying one direction of transmission is shown for simplicity; a corresponding fibre is required for the return direction.) In the event of a break in link *k–l*, ADMs *k* and *l* initiate a switch of circuit 1 onto the spare capacity back to ADM *j*, and hence via *i*, *n*, *m*, and switch at *l* and back out at *m* – thus restoring the *j–m* connectivity, as shown in Figure 5.15 (b). Similarly, the circuit 2 connectivity is restored by switching within ADM *k* onto spare capacity which routes via ADMs *j, i, n, m* and switches out at ADM *l*.

5.3.5.4   Transmission-media level: use of spare links and service protection network (SPN)

An individual transmission link can be protected by running a duplicate standby link in parallel. There are three main configurations for the use of spare links, as shown in Figure 5.16. The simplest but most robust and expensive arrangement is the so-called 'hot standby', or '1+1 Protection', in which the live traffic is run over both the worker link and standby duplicate link simultaneously (see Figure 5.16(a)). Both line system terminal equipment (LTE) continuously monitor

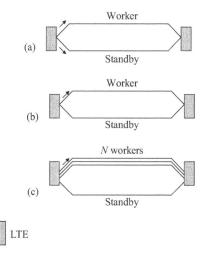

Figure 5.16 Transmission-media resilience: Use of spare links: (a) 1 + 1 protection, (b) 1:1 protection (automatic protection systems (APS)) and (c) 1:N protection (APS)

the signal from both links, and in the event of failure of the worker the output from the standby link takes precedence.

A more complex, but generally cheaper arrangement is the 1:1 protection system shown in Figure 5.16(b) – one of several automatic protection systems (APS). Here, the standby link is normally idle and the traffic is switched at the LTEs from the worker link in the event of its failure. This has the advantage of allowing the standby link to be used by the network operator to carry other, but lower priority, traffic whilst it is in standby mode, on the basis that it will be lost whenever the standby link needs to take over from the worker.

Finally, Figure 5.16(c) shows the more usual '1:N Protection' scheme for transmission networks, in which 'N' worker links are protected by a single standby link. Whilst this is cheapest of all three schemes in terms of the least level of link capacity redundancy, it does suffer from the limitation of being only able to handle the traffic from one failed worker link at a time. Again, the standby link can be used to carry low-priority traffic when in standby mode.

An extension of this concept of the use of spare links is used in SPN – in which a full network of standby links extending across the country is deployed. In the event of failure transmission capacity is switched over to the SPN, but unlike the single link 1:N spare arrangement, several links across the SPN may be used to carry the offloaded traffic. An SPN therefore needs not only a set of spare transmission links, but also an automatic control and switch system to detect and respond to link failures rapidly. The expense and complexity of an SPN means that they are used to protect only the major transmission links between the most important nodes in a network.

5.3.5.5   Physical media level

At the individual physical media level, resilience can be increased using extra strengthening in the construction of the optical fibre cables, copper cables, etc. Two typical examples of such measures are the use of armour-plated cable in submarine cables – providing protection against shark attack and fishing trawlers – and strengthened aerial cable for use in exposed, windy terrain, such as hills and mountains.

5.4   Summary

This chapter considered how the various transmission systems described in Chapter 4 are used to construct transmission networks. In Section 5.2 we addressed the Access Network, covering the ubiquitous copper local loop network, as well as the use optical fibre and microwave radio. We noted that a key aspect is the ability of the Access Network to carry increasingly higher capacities to the residential and small business users – the so-called 'broadband access' – as well as for the larger business customers. Some of the factors influencing the network operators' choice of broadband technology were discussed. The importance of the planning, operational and regulatory issues for the Access Network was also described.

In Section 5.3 we considered the structure of the Core Transmission Network and the role of the Core Transmission Stations at the transmission nodes, looking at:

- PDH networks
- SDH networks

Finally, the various ways in which the resilience of the Core Transmission Network can be increased were identified.

A simple example of dimensioning a Core Transmission Network in association with the switched telephony network is given in Chapter 6. Subsequent chapters describe intelligence, signalling, data and mobile networks.

References

[1]   Parsons, J.R. 'The development of power and building engineering services in telecommunications', *British Telecommunications Engineering*, 1995;**14**:71–80.

[2]   Parsons, J., Akerlund, J., Riddleberger, C., *et al.* 'Powering the internet: Data communications equipment in telecommunications facilities: The need for a DC powering option', *British Telecommunications Engineering*, 1998;**17**:185–197.

[3]   Scott, R.P. and Cooke, P.J. 'A 29 GHz radio system for video transmission', *British Telecommunications Engineering*, 1991;**10**:141–146.

[4]   Begley, M. and Sago, A. 'Wireless LANS'. In France P. (ed.), *Local Access Network Technologies, IEE Telecommunications Series No. 47*. London: Institution of Electrical Engineers; 2004, Chapter 12.

[5]   Buckley, J. 'Telecommunications regulation', *IEE Telecommunications Series No. 50*. London: Institution of Electrical Engineers; 2003, Chapter 8.

[6]   Valdar, A.R. 'Provision of telecommunications service in a competitive and deregulated environment', *British Telecommunications Engineering*, 1989; **8**(3):150–155.

[7]   Valdar, A.R. and Morfett, I. 'Understanding telecommunications business', *IET Telecommunications Series No. 60*. Stevenage: Peter Peregrinus Ltd.; 2015, Chapter 6.

[8]   Valdar, A.R. and Morfett, I. 'Understanding telecommunications business', *IET Telecommunications Series No. 60*. Stevenage: Peter Peregrinus Ltd.; 2015, Chapter 9.

[9]   Balcer, W.R. 'Equipment for SDH networks', *British Telecommunications Engineering*, 1991;**10**:126–130.

[10]   Gallagher, R.M. 'Managing SDH network flexibility', *British Telecommunications Engineering*, 1991;**10**:131–134.

[11]   Hawker, I. 'Evolution of the BT UK core transport network', *British Telecommunications Engineering*, 1999;**17**:298–300.

Chapter 6

Circuit-switching systems and networks

6.1 Introduction

Having considered in Chapters 1 and 2 how a call is conveyed across one or more networks, this chapter is all about how the exchanges – or more precisely, the switching units within the exchanges – in a public-switched telephone network (PSTN) actually work. Section 6.2 considers the basic components of circuit-switching systems, often referred to as 'voice switches'. For clarity, this chapter describes mainly the arrangement for the fixed-network exchanges (also called 'wireline exchanges'); however, mobile exchanges use the same digital circuit-switching technology but with additional elements providing mobility management, as described in Chapter 9.

In Chapter 5 we looked at how a Core Transmission Network carries a required capacity between two network nodes. But, how does a network operator determine what capacity is required to be carried? For example, are 90 circuits between Leicester and Coventry sufficient to ensure the expected volume of calls between these exchanges can be carried satisfactorily, or could the operator save money by providing just 60 circuits? Section 6.3 of this chapter considers how the size of a traffic route between two exchanges is determined, and how the flow of traffic through a network of many exchanges is routed to give maximum utilisation of the capacity while achieving the appropriate level of resilience (at the switching level). Finally, we consider a simple example of how a network is dimensioned at the switching level, and how this impacts the dimensioning of the Core Transmission Network.

6.2 Circuit-switching systems

6.2.1 Introduction

Circuit-switching systems form the nodal functions of the PSTN and they are thus optimised for voice, although they do successfully carry non-voice (i.e. data) traffic also. There are two basic types of switching unit: those that terminate and switch subscribers' lines – 'local exchanges' (or 'Class 5 exchanges' as they are known in the United States and Canada), and those that terminate and switch trunk and international lines only, with no subscriber lines. Examples of the latter include

Box 6.1   Exchange designations in North America

The terminology used in the United States and Canada for the exchange units in the PSTN differ from those used in the United Kingdom. They are summarised below:

Class	Name	Equivalent UK designation
1	Regional centre	Trunk exchange
2	Sectional centre	Trunk exchange
3	Primary centre	Trunk exchange
4	Toll centre	Junction tandem, wide-area tandem
5	End office	Local exchange

The ten Class 1 centres in the United States and two in Canada are fully interconnected (i.e. top of the routeing hierarchy). There are some 70 Class 2, 230 Class 3, 1,300 Class 4 and around 19,000 Class 5 exchanges in North America.

'trunk exchanges' also known as 'Main Switching Units', 'junction tandems', 'Toll switches', and 'Classes 1, 2, 3 or 4' in North America; as well as the international exchanges. (Box 6.1 describes the North American designations for exchanges in a PSTN.)

The switching of voice calls in 2G, 2.5G and 3G mobile networks is also performed within digital circuit-switched Mobile Switching Centres (MSC). Calls on 4G (LTE) mobile networks may also 'drop back' to an MSC for switching, as described in Chapter 9.

As is explained below, it is the switching of subscribers' lines that incurs the major complexity and cost, and so we will start by considering the local exchange units and later note the cut-down version that constitutes the trunk, junction tandem, international, and mobile exchanges.

6.2.2   Subscriber switching (local) units

A simplified generalised local switching unit is shown schematically in Figure 6.1. This representation combines the functional diagrams of Figure 1.7 of Chapter 1 with Figures 3.2(a) and (b) of Chapter 3. At the left side of the unit subscriber's lines terminate on line cards forming the front-end of a subscribers' Concentrator switch. (These line cards incorporate the off-hook detector, ring-current feed, and power feed shown in Figure 1.7.) A call initiated by Subscriber A is connected across the Concentrator to one of the chosen free outlets on the internal link to the Route switch. If the call is to another subscriber on that exchange – known as an 'own-exchange call' – it is connected through the Route switch to a chosen free

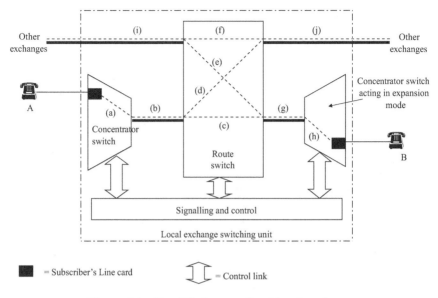

Figure 6.1   Simplified generalised local exchange

outlet to the called subscriber's (i.e. B's) Concentrator switch. The call can be completed by connecting through this Concentrator switch to the line card of subscriber B, and hence to the terminating local line. This path, shown as *a–b–c–g–h* in Figure 6.1, is established exclusively for this call until subscriber A or B clears down – a characteristic of circuit switching. The signalling and control functions necessary to set up the calls across the switches are considered in Chapter 7. There are two key points concerning Figure 6.1 to note at this stage.

(i)   Although the call is initiated by A and so the traffic flow is deemed to go from A to B via the three switches along the connection path described above, transmission paths within the switches are of course required to flow from A to B ('Go' direction) and from B to A ('Return' direction) to enable a conversation, corresponding to the 4-wire circuit described in Chapter 3 (see Figure 3.14).

(ii) With the direction of traffic flow shown for path *(a)* subscriber A's Concentrator switch is concentrating calls from many lines onto fewer outlets (a concentration ratio of 10:1 being typical). The path *(c)* through the Route switch is not concentrated, since the latter has an equal number of inlets and outlets; the switch is therefore performing the function of interconnecting subscriber A's Concentrator switch to subscriber B's Concentrator switch. However, on the exit side of the exchange the traffic flow over path *(g)* is passing through subscriber B's Concentrator switch over path *(h)* from concentrated side to non-concentrated side, since it has, say, ten times more outlets than inlets. The call path is therefore experiencing 'expansion' through this switch.

Figure 6.1 also shows that external (traffic) routes from other exchanges terminate on the Route switch. This enables calls from these exchanges to be connected via the Route switch to subscriber B via its Concentrator switch (albeit, acting in expansion mode), as shown by traffic path i–e–g–h.

Calls from subscriber A that need to be routed via other exchanges are also switched across the Route switch (separate 'Go' and 'Return' transmission paths) to the appropriate outgoing traffic route, shown as path a–b–d–j in Figure 6.1.

Finally, the switching unit also could act as a tandem by providing connectivity ('Go' and 'Return') between traffic routes terminating on the Route switch, as shown by path i–f–j in Figure 6.1.

We now need to redraw the simplified configuration of Figure 6.1 to reflect the fact that Subscribers A and B may actually be on the same concentrator. This leads to the concept of folding the switch configuration so that all subscribers' lines are on the left-hand side and all external traffic routes are on the right-hand side of the Route switch. The connections through the Route switch, thus, may need to loop back, as shown in Figure 6.2 for path a–b–c–g for calls between subscribers A and B on two co-sited Concentrator switches, and path a–b–k–l–m for calls between subscribers A and C on the same Concentrator switch. We now have a generalised circuit-switching unit, comprising several Concentrator switch units and a Route switch. Thus, the common, but somewhat trivial, labelling of a telephone exchange as a 'switch' hides the fact that in practice it will contain several switches. Indeed, as we shall see later in this chapter, the Concentrator and Route switches themselves contain several stages of switches. Confusion can be avoided by using the

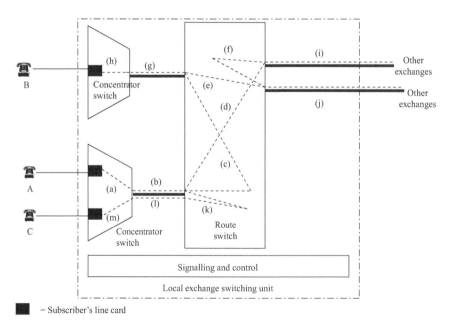

= Subscriber's line card

Figure 6.2   Generalised local exchange (folded)

term 'switch-blocks' to denote an assembly of switching stages forming the role of concentration or route switching.

It is a simple matter now to consider the generalised block schematic diagram for a trunk, junction tandem, international, or mobile exchange, since these comprise of a single large Route switch-block only – there being no subscriber lines terminate on these exchanges. We will look at these in a little more detail in the description of digital switching later in this chapter.

6.2.3 Digital telephone switching systems

Chapter 3 introduced the concept of converting analogue speech into a digital format for the purposes of switching and transmission across the PSTN. This conversion gives capital and current account (operational) cost savings for the PSTN, as well as giving the telephone users a significantly improved quality of the sound. In fact, the vast majority of the telephone exchanges in the World, both fixed and mobile, are digital. They are also computer controlled – using a technique known as 'stored-program control' (SPC) [1]. In this section, we will look at how a call connection is established across a digital switch-block within a fixed network or an MSC exchange; we consider the additional functions involved in a mobile switching system in Chapter 9, after we examine the data-network-based alternatives to switching voice calls in Chapter 8. The signalling and control functions for all types of switching systems are considered in Chapter 7.

However, before we start to examine the process of digital switching it may be helpful to review the two standard forms of digital multiplexed systems. Chapter 3 introduced the concept of slicing a single channel of analogue speech into a series of samples taken 8,000 times a second, each of which is converted to an 8-bit binary number. This stream of 64,000 bits/s (64 kbit/s) is then multiplexed with the 64 kbit/s streams of other speech channels. The process is formally known as pulse-code modulation (PCM). There are two standard formats for PCM systems: the 32-channel international standard used throughout Europe and many parts of the world as well as on all international links, and the 24-channel North American standard, as shown in detail in Figures 4.9 and 4.10 of Chapter 4, respectively, and summarised for this chapter in Figure 6.3. Both systems contain their channels within a frame size of 125 μs (i.e. 125 millionths of a second), being the reciprocal of the universally agreed sampling rate of 8,000 times a second. However, there are differences in the algorithm used by the systems to convert the speech samples to digital format (the encoding laws). In addition, there are the following differences in frame format which are important to consider in our treatment of digital switching.

- The 30-channel PCM system has its frame divided equally into 32 time slots of 3.9 μs duration, each containing 8 bits, making a total frame size of 256 bits. Time slot 0 (TS0) is used only for PCM system management – primarily to carry the frame-alignment signal, which is used by the receiving terminal to determine the start of the frame. Time slots 1–15 carry speech channels 1–15, respectively, and time slots 17–31 carry speech channels 16–30, respectively.

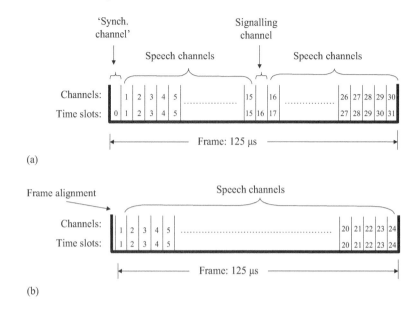

Figure 6.3   Summary of PCM frame formats. (a) 30-channel international standard. (b) 24-channel North American standard

Time slot 16 (TS16) is normally not used for speech, instead it is used as a bearer for signalling between exchange control systems – mainly, the international standard signalling system number 7 (SSNo.7), as described in Chapter 7. The total rate of the 30-channel PCM system is 2,048 kbit/s, usually written as '2 Mbit/s'.

- The 24-channel PCM system has its frame divided into 24 time slots of 8-bits each, and a single bit at the front used to carry the frame alignment signal, making a total frame size of 193 bits. If SSNo.7 signalling needs to be carried this is placed in one of the time slots, typically TS 24, which means that such PCM systems can support only 23 speech channels. The total rate of the 24-channel PCM system is 1,544 kbit/s, usually written as '1.5 Mbit/s'.

For simplicity, the rest of this chapter will consider digital switching using just the example of the 30-channel PCM format.

Figure 6.4 illustrates a simple block schematic diagram of the main functional entities in a digital local exchange; this can be seen as a development of the generalised local exchange of Figure 6.2. Each subscriber copper line is terminated on the main distribution frame (MDF) (see also Figure 5.1 in Chapter 5) from where an internal copper pair cable is run within the exchange building to the appropriate subscriber line card on the Concentrator switch-block. As described later, the line card performs all the necessary subscriber-line termination functions as well as converting the speech from the analogue signal on the subscriber's line to a 64 kbit/s digital stream. The outputs from 30 of these line cards are multiplexed together to form a standard 30-channel PCM frame, with a line rate of 2 Mbit/s.

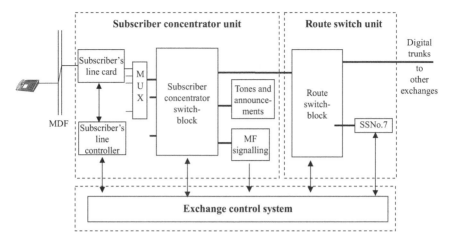

Figure 6.4   SPC digital exchange: functional overview

Each multiplexor is connected at the 2 Mbit/s level to the input of the Concentrator switch-block. From the other side of this switch-block 2 Mbit/s links are taken to the co-located Route switch-block for connection to the 2 Mbit/s digital trunk links to other exchanges. As described in Chapter 5, the 2 Mbit/s output from the switching units are multiplexed together with other links onto either plesiochronous digital hierarchy (PDH) or synchronous digital hierarchy (SDH) higher-speed transmission links at the serving CTS, which is usually in the same exchange building as the switch-blocks. The Route switch-block is also connected with 2 Mbit/s links to the exchange's SSNo.7 signalling equipment, for signalling to the control systems of distant exchanges.

As also shown in Figure 6.4, each subscriber's line card is monitored by a controller which is used by the exchange control system to manage the setting up and clearing of calls, while the signalling from subscribers' telephones is received by the MF equipment, both systems are described in Chapter 7. The tone equipment supplies supervisory tones, for example, 'network busy', as well as recorded announcements, such as 'Sorry, we are unable to connect your call'.

Before we start to look inside the subscriber Concentrator switch-block and Route switch-block, some attention should be paid to the importance of the sub-scriber line card. This piece of equipment, physically about the size of a man's wallet, performs the functions of terminating a copper line on the switching system. First and foremost of these functions is the provision of an interface between the high voltages and currents present on the external copper pair and the relatively low voltages of the semiconductor electronic equipment of the digital switch-blocks. Thus, the line card includes equipment that cannot be constructed using semi-conductor technology – this makes it relatively expensive. The key point is that a line card must be provided for each subscriber's line, irrespective of the number of calls made. In fact, the capital cost of all the line cards on a typical local exchange is about 70% of the total exchange cost!

It is important to note that all costs of a PSTN are allocated on a per-line basis for the network between the local-loop termination at the subscriber's premises and the line card, while all network costs from the output of the line card and through all the switching and Core transmission networks is allocated on a per-call basis. The United Kingdom regulator identifies the former as the 'Line Business' and the latter as the 'Calls Business'. The way that these different sets of costs are recovered by the telephone price structure is briefly addressed in Chapter 10.

In fact, the issues associated with the subscriber's line card are fundamental to telecommunication networks, all forms of user terminations on any type of network (data, voice, fixed, mobile) incur the cost and complexity which must be borne on a per-user access basis. This principle will also be addressed in Chapters 8 and 9, looking at data and mobile networks, respectively.

The set of functions required by a copper-line termination and provided by a subscriber's line card are best remembered by applying the acronym 'BORSCHT', as follows:

- **B**attery: Application of the 50 V direct current (DC) power supply to the subscriber's line (as described in Chapter 1).
- **O**verload: Protection of the delicate semiconductor equipment of the digital switch-blocks from any induced voltages on the external copper line – it is particularly important to provide protection against lightning.
- **R**inging current: Provision of an interrupted electrical signal, with an appropriate cadence, of about 75 V DC and 200 mA is required to ring a subscribers' set of telephone instruments.
- **S**upervision: That is, the detection of the subscriber going off-hook to signify call initiation, and detection of off-hook conditions in the presence of ringing current to signify call answer.
- **C**odec: The analogue-to-digital (A/D) conversion in the Go direction and the digital-to-analogue conversion in the Return direction – A/D coding and decoding (hence the term 'codec').
- **H**ybrid: Two-to-four wire conversion (as described in Chapter 1) from the local loop to the Go and Return format of the switches and Core Transmission Network.
- **T**est: The application of electrical continuity testing at the endpoint of the local loop. This is usually applied automatically under remote control from an exchange maintenance console.

Figure 6.5 illustrates the items on a typical subscriber's line card and shows how these correspond to the BORSCHT functions listed above [2].

Now to consider how the digital switch-blocks work, Figure 6.6 presents a simple example of a digital switch-block (either Concentrator or Route) with five 2 Mbit/s PCM highways at the inlets and five 2 Mbit/s PCM highways at the outlets. It is necessary for all the PCM systems to synchronised and aligned with each other so that the timing of all their timeslots coincide. Thus, for example, all the PCM systems pass the 8 bits relating to their time slot 1's (TS1) simultaneously. Two call connections are due to be set up. First a call arrives on the channel in time

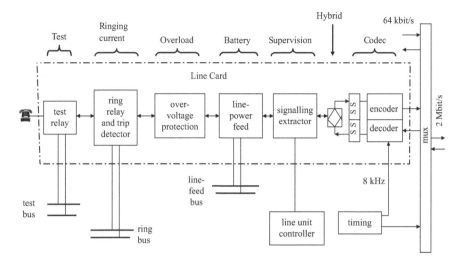

Figure 6.5   The components of a subscriber's line card

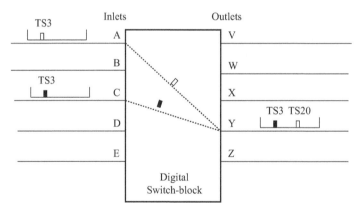

Figure 6.6   Digital switch-blocks

slot 3 (TS3) of highway C destined for highway Y, and a connection is set up across the switch-block to exit onto TS3 of highway Y. The stream of 8 bits now flows every time TS3 occurs, that is, every 125 μs, from inlet C to outlet Y across the switch-block. However, a problem occurs if another call were to arrive in the channel occupying TS3 of highway A also Destined for highway Y, since the TS3 on the outlet Y is already occupied with the call from highway C. The only solution is to transfer the call carried on TS3 of highway A to a vacant timeslot channel on the required outlet highway Y – for example, TS20, as shown in Figure 6.6.

The conclusion to the above reasoning is that two forms of switching are required in a practical digital switch-block. The first is the switching of a timeslot's contents from an input 2 Mbit/s highway to the same number (i.e. coincident) timeslot on an outgoing highway, a function known as 'space switching'. The second is the switching of the contents of a time slot of an inlet highway to a different value (i.e. non-coincident) time slot on the outlet highway, a function known as 'time switching'. All digital Concentrator and Route switch-blocks comprise both space (S) and time (T) switches. Typical configurations for a Concentrator switch-block are T–S and T–S–T and for a Route switch-block are T–S–T and T–S–S–T.

Digital time switching. The process of time switching is achieved using computer storage to delay the contents of an incoming timeslot until the moment when the required output time slot arrives. A simple example is given in Figure 6.7, in which we assume the input highway has a digital stream with just four time slots (as opposed to some 512 time slots in practical switches). The required connection pattern, as determined by the exchange-control system is as shown – that is, input TS1 to output TS4, and similarly TS2 to TS3, TS3 to TS1, TS4 to TS2. This connectivity is achieved by using a simple electronic counter to set the write address for the contents of each time slot, and the 'Connection Memory' (CM) to hold the read address. Thus, the contents of incoming channel during TS1 1 are 'written' (i.e. transferred) into location 1 of a computer store, known as the 'Speech Memory' (SM); the contents of the channel during TS2 are written into location 2, and so on. The SM has as many locations, or cells, as there are time slots on the inlet highway. After the contents of the channels in the four timeslots have been passed sequentially into the SM the counter continues with another round, and the

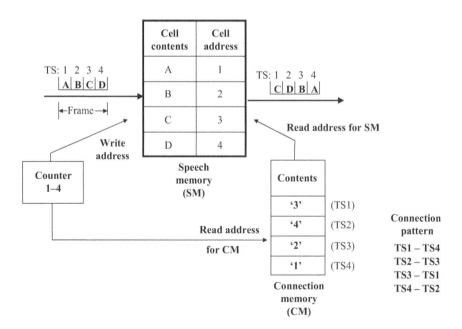

Figure 6.7   The principle of time switching using storage

next set of speech samples in the time slots are written into the respective cells in the SM. This continues for the lifetime of the switch-block (assuming no system failures)!

The CM also has four locations, the contents of which are 'read' (i.e. transferred out) under the control of the counter (cycling 1 to 4). The contents of each location in the CM indicate the read address of the cell in the SM that must be read during that time slot. The exchange control system places the appropriate read addresses in the respective locations of the CM to achieve the desired output sequence from the time switch. The shifts in time for the four channels arriving in time slots 1 to 4 is illustrated in Figure 6.7. The connection pattern will continue cyclically 8,000 per second until there is a need to cease one of the calls or establish a new one, when the CM contents will be altered by the exchange-control system appropriately.

Digital space switching. Digital space switching is achieved through the time-division multiplexing of a simple matrix array, with the functional equivalent of digital logic AND gates at the cross-points. As usual, the basic requirement is for all the highways and the matrix array to be synchronised and aligned, so that all corresponding time slots precisely coincide. The operation of the Space switch relies on the scheduled closure of the specific cross-points during the period of each time slot so that the contents can be shifted between input and output highways. Figure 6.8 illustrates a simple example, again with just four timeslots on the input and output highways. Each column of the matrix is controlled by a CM, the contents of which contain the address of the cross-point that should be closed during the time that it is being read. Thus, each of the CMs is read simultaneously during

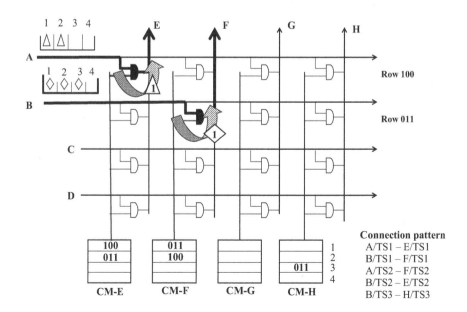

Figure 6.8 Digital space switching

each time slot period under the read-address control of a counter, cycling 1 to 4. When TS1 occurs, the address read from the first location of CM-E is *100*; this appears on the control bus for the column, which is decoded by the cross-point on row 100 causing it to close; the contents of TS1 on input highway A are then transferred to TS1 on output highway E. At the same time, the address *011* is read from the CM-F onto the control bus causing the cross-point on row 110 to close, allowing the contents of TS1 of input highway B to be transferred to TS1 of output highway F. During the instance of TS2 the second locations of all the CMs are read and the addresses of the appropriate cross-points used to establish the required connections across the matrix: contents of TS2 on highway A to TS2 on highway F; similarly, TS2/B to TS2/E. During TS3 only one cross-point opens giving TS3/B to TS3/H connectivity, and there are no connections required during TS4. This connection pattern across the matrix continues 8,000 times a second until the contents of one of the connection memories is altered by the exchange control system to enable another call to be established or an existing one to clear down, and the new pattern then continues.

Digital switch-blocks. Practical digital switch-blocks use space and time switches constructed from high-speed semiconductor technology using a large degree of integrated circuitry. Maximum efficiency is gained by further time-division multiplexing the 2 Mbit/s highways terminating on the switch-block up to typically some 512 time slots per 125 μs frame (i.e. still with a new frame every 8,000 times per second). An example of a typical T–S–T switch-block is shown in Figure 6.9. Here, a call connection is to be established between TS10 of the high-way A2 entering input Time switch A2 and TS45 of output highway C1 leaving output Time switch C1. The exchange control system determines that the next

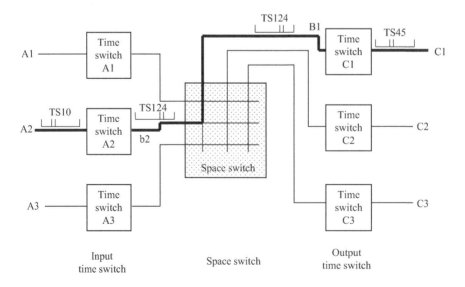

Figure 6.9   T–S–T switch-block

freely available pathway through the Space switch is via time slot 124. Thus, Time switch A2 is set up to shift the contents of TS10 to TS124 on the highway b2. The Space switch provides connectivity during TS124 between highway b2 and highway B1 and into the output Time switch C1. The latter performs time switching between TS124 and TS45 onto its outgoing highway C2. This pattern of connectivity continues until the contents of the connection memories are changed [3].

6.2.4   *Private automatic branch exchanges*

The concept of private automatic branch exchanges (PABX or PBX) is introduced in Chapter 2 (see Box 2.1). These privately owned exchange systems are essentially the same as the public network equivalents, using digital switch-blocks, as described above. However, there are many different designs of digital PBX resulting from the wide range of sizes from just a few extensions (i.e. internal telephone lines) to tens of thousands of extensions, depending on the extent of the customer's business. Generally, the PABXs are connected to the serving public local exchange over one or more 2 Mbit/s (or 1.5 Mbit/s in the United States) digital links. Small PBX systems are based on T–S switch-blocks and the larger systems based on T–S–T switch-blocks.

6.2.5   *Digital exchange structures*

We are now able to consider the key structural features of digital exchanges. Figure 6.10 shows a block schematic diagram of a digital local exchange serving

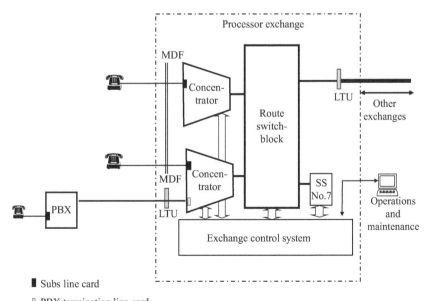

Figure 6.10   Local exchange unit structure

both subscriber copper lines and PBXs. The line card is provided by the Concentrator switch-block in the Local Exchange in the case of subscribers served by copper local lines, and by the PBX switch-block for all its extensions. The 2 Mbit/s digital links (carried over optical fibre or high-speed transmission over copper, e.g. high bit-rate digital subscriber line (HDSL)) from PBXs are terminated on a digital line termination unit (LTU) at the exchange building, close to the MDF, from where a digital link terminates on a special PABX line card on the Concentrator switch-block. This special line card simply needs to provide a timing interface between the digital line and the switch-block; it does not need to perform all the BORSCHT functions, since these are provided by the line card in the Concentrator switch-block of the PBX. A special signalling link between the control systems of the PBX and the Local exchange is carried over the 2 Mbit/s digital link (in TS16), enabling the PBX to establish and receive calls to and from the PSTN, as described in Chapter 7.

Digital public exchanges are designed to meet a high level of system reliability to ensure continuity of telephone service to the users – typical objectives being no breaks in service more often than once every 20 years! This means that all the key components of the exchange of the exchange contain redundancy and employ automatic cut-over to standby units in the event of failure. In particular, the digital Concentrator and Route switch-blocks are fully duplicated, operating on a worker-standby arrangement [3] (see also Chapter 5).

The maintenance and operations (O&M) systems for the local exchange have a man–machine interface (MMI) with technicians via a computer console, which may be within the exchange building or remotely located, as shown in Figure 6.10.

Another element in the reliability of the exchange system is that of synchronisation. As will be apparent from the description of digital switching, it is necessary for all the digital line systems terminating on the exchange switch-blocks, as well as the switch-blocks themselves to be operating at exactly the same speed – that is, the number of bits per second. This means that in all exchanges in the PSTN either the digital clocks generating the bit streams need to be precisely accurate and stable, or some mechanism is required to synchronise the speed of all the clocks. The actual arrangements for synchronising may differ between national networks, but they usually involve the dissemination the timing from a central atomic clock through a mesh of synchronisation links to control equipment in each exchange [4,5]. More recently, network synchronisation systems have been introduced that use reception of GPS (global positioning satellite) timing pulses at each network node. It should be noted that network synchronisation is used by all digital components in the various networks (not just the PSTN) of all network operators in a country. Usually, the incumbent network operator provides the definitive timing source which is taken by all the other network operators (fixed voice, data and mobile) as well as subscribers operating private networks and PBXs.

In practice, digital Concentrator switch-blocks are designed to terminate some 2,000 subscriber lines. Therefore, in an exchange configuration as shown in

Figure 6.10 there may be several Concentrator switch-blocks co-located within the exchange building, each connected to the single large Route switch-block. Typical capacities of such exchanges range from about 5,000 to 50,000 lines. When serving a catchment area of fewer than about 4,000 lines a digital local exchange based on the configuration of Figure 6.10 becomes uneconomical, with unacceptably high per-line cost due to the dominance of the fixed costs of the exchange-control system. There are two main solutions to this problem: the use of special cost-reduced very small-exchange systems, or the use of remote concentrator units (RCUs), as described below.

Cost-reduced small exchange systems. These very small systems, which typically range in capacity from some 10 to 600 lines, have a cut-down exchange-control system providing a reduced set of service features for the users and a severe restriction in their traffic carrying capacity. These very small systems tend to be deployed in remote rural areas, for example, British Telecommunication (BT) has deployed some 500 units (designated 'UXD5') in the Highlands and Islands of Scotland, parts of Wales and Northern Ireland [6].

Remote concentrator units. An economical solution for serving some 100 to about 4,000 lines is achieved by locating just a subscribers Concentration switch-block at the focal point of the copper network within a small town, and linking this to the Route switch-block and the exchange-control systems at a distant parent local exchange. The latter is often called the 'Processor Exchange', because that is where large exchange control system, comprising of a multi-processor cluster is located. With this configuration, the fixed cost of the exchange-control system is spread over the subscriber lines connected to all the co-located and remote Concentrator switch-blocks served. The RCU has a small local control system to support the Concentrator switch-block and line-card functions, but this has to be supported by the main control system at the parent exchange, as shown in Figure 6.11. In fact, the RCU is dependent on the parent for its operation. All calls on the RCU, including those to subscribers on the same Concentrator switch-block, must be switched to the Route switch-block at the parent (processor) exchange, from where it routes to the destination Concentrator switch-block – this may be at the parent exchange, another RCU off that exchange, the originating RCU, or taken via a trunk route to another exchange to complete the call. In the case of an own-RCU call the to-and-fro routeing via the parent exchange is known as 'tromboning' for obvious reasons.

If the digital link between the RCU and the parent exchange fails the RCU will be isolated. In such conditions a restricted form of own-RCU calls can usually be provided by the local controller, but without many of the facilities including charging being available. Normally, several parallel PCM digital links are provided from the RCUs to their parent exchanges, with as much spatial separation in the line plant routeing as possible (see the Section 5.3.5 on resilience in Chapter 5), to minimise the risk of RCU isolation. In the United Kingdom, the RCU-to-parent exchange distances are typically some 5–20 miles, although in some other countries far greater distances are involved.

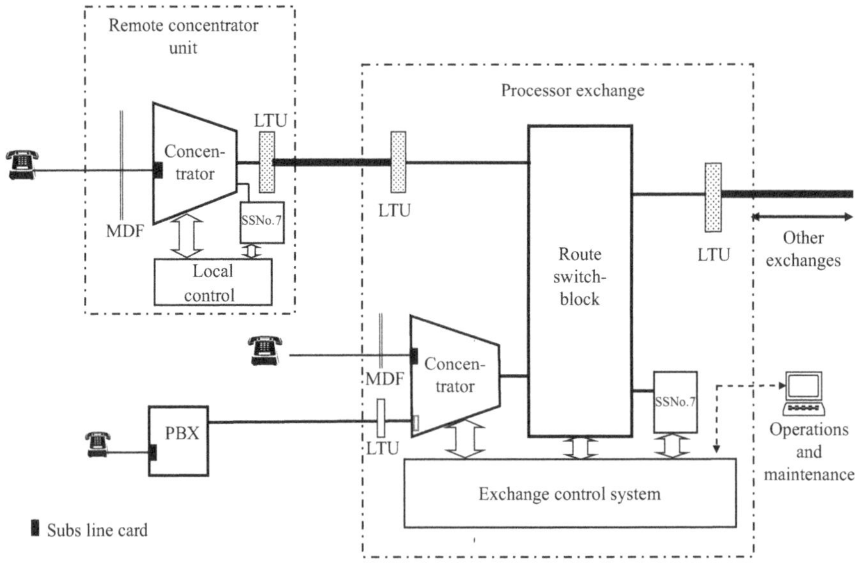

Figure 6.11   Local exchange structure with remote concentrator units

6.2.6   Integrated services digital network exchanges

Following the successful introduction of digital switching for voice in the 1980s, the next logical step of evolution was that of the integrated services digital network (ISDN). The concept of the ISDN is that the network of digital telephony exchanges is enhanced to allow the switching of data calls. This requires the following additions to the digital PSTN:

(i)   The introduction of a digital transmission system onto the subscriber line to extend the digital path from the exchange to the subscriber's data terminals at their premises. (The ISDN digital line system is covered briefly in Chapter 4.)

(ii)   The provision of enhanced signalling between subscriber and the exchange (see Chapter 7).

(iii) The provision of enhanced signalling between exchanges (see Chapter 7).

(iv)   The provision of extra control software and SS7 signalling to enable the more complex form of data calls to be handled (see Chapter 7).

Two forms of internationally standardised ISDN systems are available:

Basic rate: Two 64 kbit/s channels (called the 'B' channels), together with a 16 kbit/s signalling channel ('D' channel). This is designated '2B + D'. The user rate is 144 kbit/s (i.e. 2×64 kbit/s + 16 kbit/s) and is normally carried over a single copper pair.

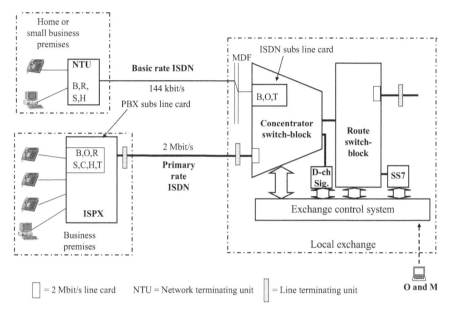

Figure 6.12   ISDN lines on a digital local exchange

Primary rate: 30 × 64 kit/s (B) channels, together with a 64 kbit/s signalling channel (D). This is designated '30B + D'. The user rate is 2 Mbit/s and this is carried over an optical fibre, microwave radio link or possibly HDSL over three parallel copper pairs.

In practice, most PSTN exchanges today are combined ISDN and telephony systems. When a subscriber wishes to have basic-rate ISDN service the existing copper line is shifted from the analogue line card to an ISDN line card on the same Concentrator switch-block (requiring a re-jumpering at the MDF). A network ter-minating unit (NTU) is installed at the subscriber's premises to terminate the digital line system and present a set of sockets for the subscriber's data and telephony terminals. There is usually a terminal adaptor associated with the NTU to provide a conversion of the analogue telephone to digital so that it can use the ISDN line. In the case of Primary-rate ISDN service the termination at the subscriber con-centrator is at the 2 Mbit/s level, requiring an appropriate digital line termination – as provided for digital PBX. Figure 6.12 shows the arrangement for ISDN, with an indication of how the BORSCHT functions are spread between the ISDN line card and the NTU on the subscriber's premises, instead of being solely in the line card as for analogue subscriber lines.

Non-data calls from ISDN subscribers are routed through the switch network as for telephony with each B channel being treated separately. Data calls are also normally handled on a single B-channel basis, with the required extra features indicated through the call set-up across the network through the ISDN signalling system. ISDN lines are given standard telephony numbers and are charged at the same rate as telephony. In addition, some network operators provide the ability for basic-rate ISDN subscribers to use the two B channels as a composite 128 kbit/s channel; the ISDN exchange then routes two parallel calls through the network simultaneously to the same destination ISDN line. (However, these 128 kbit/s services have not proved to be a commercial success.)

6.2.7   Non-subscriber digital switching units

As discussed earlier in this chapter, the structure of the digital switching units on which subscribers' lines do not terminate is centred on just the Route switch-block, since there are no Concentrator units required. The two main examples of this situation are MSC and fixed network tandem units (i.e. trunk, junction tandem, and international exchanges).

MSC. A simplified view of the structure of an MSC is given in Figure 6.13, in which the voice traffic of 2G (and 2.5G) and 3G is switched within the digital-TST-based Route switch-block. As described in Chapter 9, maximum use is made of the available spectrum for the mobile systems by encoding speech at lower bit rates, typically 13 kbit/s. Therefore, the transmission links between the base transceiver station (BTS or Node B for 2G, 3G, respectively) and the base station controllers (BSC or RNC) uses sub-multiplexing of the low rate channels onto the 64 kbit/s timeslots. Each of the subchannels is rate adapted at the BSC or RNC nodes so that the speech circuits occupy a standard 64 kbit/s channel and therefore are able to be switched through the Route switch-block in the normal way. (The reverse happens for transmission in the return direction.) Figure 6.13 also shows that in addition to the SS7 signalling sender receiver (S/R), there are several units providing various aspects of mobility management, some of which may be collocated with the MSC (e.g. VLR) and some of which will be remotely located at other nodes within the

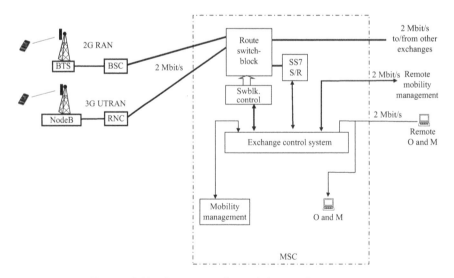

Figure 6.13   Structure of a mobile switching centre

mobile network (e.g. HLR). The signalling and switch-block control arrangements are described in Chapter 7, and the functions of the mobility-management systems are described in Chapter 9.

Trunk, junction tandem and international exchanges. Figure 6.14 shows the structure of a digital trunk exchange within a typical operational building. The exchange primarily provides 64 kbit/s switched connections between incoming and outgoing traffic routes to/from other telephony exchanges. The links to the other exchanges (e.g. dependent local exchanges, other trunk exchanges, MSCs and interconnection to exchanges of other operators) all terminate on the digital switch-block at the 2 Mbit/s level (i.e. 30 speech channels). The Figure 6.14 also shows the O&M systems collocated with the trunk unit, although they may also be remotely located as well.

Looking now at the whole operational building, the external transmission systems (mainly optical fibre cables) terminate within the CTS, as described in

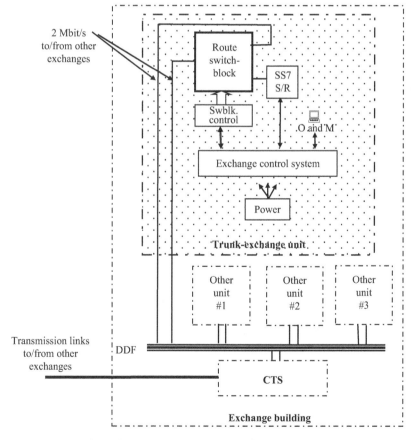

*Figure 6.14   Trunk-exchange structure within a typical operational
            exchange building*

Chapter 5. Appropriate connections from the CTS are made at the DDF to the various telecom units in the building. Examples of such other units within the building include mobile exchanges (i.e. MSC); data switches; leased-line DXCs; and other PSTN units like junction tandems. For completeness, Figure 6.14 also shows the power system in the trunk exchange structure, which comprises of a central master unit and separate distribution systems for each shelf within the equipment racks. Similar arrangements for power supply apply to each of the other telecom units in the building.

6.3   Network dimensioning

6.3.1   The concept of switched traffic

Most people are comfortable with the concept of traffic, say, as a flow of vehicles down a motorway or aircraft taking off and landing on an airport runway. Likewise, telephone traffic is the set of calls being carried over the line between a user and the serving exchange or over the link between two telephone exchanges (i.e. a 'traffic route'). Just as there are vehicles of various lengths on the motorway, and planes requiring long and short take-off runs, telephone calls also vary in duration. The need is to be able to dimension the number of channels on the traffic route between two exchanges (equivalent to the number of lanes on the Motorway) so that, at the expected level of calls during the busiest time of day, there is a sufficiently low probability that any new calls arriving will encounter a lack of free capacity on the route, that is, congestion, – similar to all lanes on the motorway being so tightly packed that no more cars can join it. With the existing circuit-switched telephone exchanges, any calls meeting congestion will be 'lost', with the user experiencing equipment-busy tone leading to the call attempt being abandoned. However, there are switching systems in which calls are delayed rather than lost when meeting congestion – so-called 'delay systems'. This alternative form of telephone switching using packet technology is considered in Chapter 8.

Network operators use dimensioning techniques to determine the size of traffic routes between exchanges – the number of channels or circuits required to avoid congestion – as well as the capacity of the exchange equipment that handles the calls. Both of these dimensioning activities involves an implicit economic trade-off between the cost penalty of over-providing exchange switching capacity or circuits on a traffic route on the one hand, and the impairment to the quality of service offered to the user due to under-provision on the other hand. This is a common dilemma for any service industry.

We have already used the analogy of the motorway to describe the dimensioning of the traffic routes. Another analogy may help in considering the dimensioning issues for the exchange equipment. For example, consider the staffing levels required in a public post office in which several attendants serve simultaneously at the counter. By arranging for a single queuing system arriving customers can be directed to the next available attendant. Obviously, the serving time for the customers will vary depending on the nature of their transactions and so there will be the chance that all attendants are busy when the next customer arrives in the post office. This chance can be reduced by increasing the number of attendants at the counter. The dimensioning challenge for the post office management is to determine how many attendants should be provided: trading their cost against the disadvantages of having dissatisfied customers having to wait to be served. For the post office situation, a short wait in the queue is usually deemed acceptable, whereas the loss-system nature of a telephone exchange means that a lack of available servers will cause a call to fail (be lost), with a perceived degradation of quality of service for the customer and a potential loss of revenue if the repeat attempt is made via another network operator.

6.3.2 Call distribution

For a network operator to dimension a telephone exchange a forward view of the likely amount of calls on that exchange is required. While the instantaneous demand from individual telephone users will vary with circumstances, the total population of lines on a telephone exchange will tend to follow a pattern of peaks and troughs of numbers of calls from day-to-day which is fairly stable, and over a course of time predictable growth (or decline) forecasts of traffic can be made. A typical daily pattern of telephone demand on a local exchange is shown in Figure 6.14, arranged in hourly segments. This shows that during the typical weekday the demand for calls increases from a low level at night from around 8 a.m. to a peak around 11 a.m., followed by a slight drop in early afternoon and a second lower peak in mid-afternoon – driven by activity at the work place. There may also be a domestically driven evening peak around 8 p.m. A different pattern applies at weekends and public holidays with most calls being generated by residences rather than from the workplace. For those exchanges serving seaside or holiday resorts there is usually a distinctive seasonal pattern, for example, with much higher levels of demand in the summer months.

6.3.3 Traffic flow

As Figure 6.15 shows, the required exchange capacity is set by the 1-hour period of highest demand, traditionally known as the 'Busy Hour'. The network operator needs to ensure that there is sufficient capacity in the switch-blocks to cope with the current demand, with an appropriate allowance for exceptional increases. In addition, the operator arranges for the exchange capacity to be augmented periodically to cope with the predicted growth in demand over the following 12–18 months. This future-proofing also ensures that there is spare capacity in the exchange to deal with any unusual daily fluctuations. The extent of this spare capacity is obviously, an economic trade-off, as described earlier.

So, now to look at the process of dimensioning an exchange. Figure 6.16 illustrates the concept of traffic occupancy, taking the example of a system of four channels – these might be channels within a switch, as described above, or channels in an outgoing route to another exchange. The incidents of calls being carried over each channel is plotted over time, indicating the different durations or holding

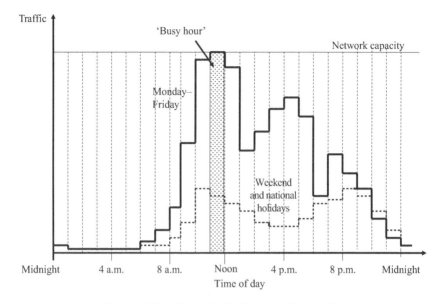

Figure 6.15   Typical telephony traffic profile

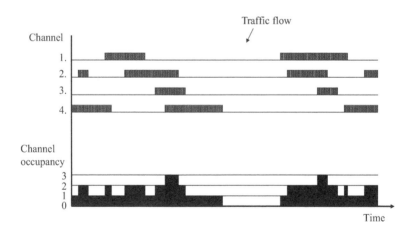

Figure 6.16   Traffic and occupancy

times of the calls. At any one time the number of simultaneous calls on the system varies from zero to three, as shown in the lower plot in Figure 6.16. The summation of all the channel occupancies over time constitutes the telephone traffic intensity, which is measured in units called 'Erlangs' (named after A. K. Erlang, the Danish pioneer of traffic theory) [4]. An Erlang is defined at the average number of simultaneous calls on a system during a 1-hour period. (An alternative unit is used in North America which measures traffic intensity in units of '100 call seconds

Table 6.1   Loading of circuits with GOS

Number of circuits	Erlang capacity with grade of service				
	1:1,000	**1:500**	**1:200**	**1:100**	**1:50**
1	0.001	0.002	0.005	0.01	0.02
5	0.76	0.90	1.13	1.36	1.66
10	3.09	3.43	3.96	4.46	5.08
20	10.11	10.79	11.86	12.8	14.0
50	32.51	33.88	35.98	37.9	40.3
100	75.24	77.47	80.91	84.1	87.6

per hour' – CCS – that is, equal to 1/36th of an Erlang.) In general, the traffic intensity of flow A, measured in Erlangs, is given by:

$$A = C\,h$$

where C is the average number of calls during the hour and h is the average holding time of those calls. By definition, a traffic flow of one Erlang is equivalent to a circuit being occupied for a complete hour.

As described in the introduction above, there will be occasions when a system serving telephone traffic will be fully occupied and so temporarily unable to service any more calls – a state known as call congestion. Any calls arriving at that time will be lost (the subscriber must abandon the call). The probability of this occurring is defined as the 'Grade of Service (GOS)', given the symbol 'B' in traffic dimensioning formulas. Telephone exchanges are typically dimensioned to a GOS of one lost call in 200 – that is, during the busy hour there is a 0.5% chance that a call attempt will meet congestion (outside the busy periods the chances of congestion are negligible). It should be emphasised that even this low probability of call failure will only be experienced when the capacity of the exchange is just at the design level; as explained above, usually spare capacity has been provided in anticipation of growth and so the actual GOS will at all other times be even better than the design limit.

The name of Erlang is also associated with a formula for determining the number of circuits required to carry a specified level of telephone call intensity at a specified grade of service – the famous 'Erlang's Lost Call' or 'Erlang's B' Formula [7,8]. Table 6.1 shows the variation in the traffic that can be carried on a given number of circuits at various grades of service, calculated using Erlang's B formula. There are several points to note. The first is that the amount of traffic that can be carried by a set of circuits increases as the grade of service decreases, showing the trade-off between capacity carried and the chance of congestion or loss of traffic. The second point is that the larger the group of circuits the proportionately higher the amount of traffic carried by each circuit – that is, the circuit loading. For example, with a GOS of 1-in-100 a group of 20 circuits carries 12.8 Erlangs (making 0.64 E/circuit) compared to a group of 100 circuits which carries 84.1 Erlangs (making 0.84 E/circuit), an improvement of circuit loading from 64% to 84%.

It is apparent from Table 6.1 that the traffic characteristics of a group of circuits or servers are nonlinear, with improved efficiency with size. In general, therefore, network economies can be achieved by carrying and switching telephone traffic through fewer but larger exchanges with the fewest and largest size routes between exchanges as possible. This process was already identified when the role of Concentrator switch-blocks was described at the beginning of this chapter, as the traffic concentration of the lightly loaded subscriber lines (at, say, 0.05 E per circuit) onto the smaller set of more highly loaded outlet trunks (at 0.5 E per circuit), assuming a 10:1 concentration ratio.

6.3.4   Traffic routeing

We are now able to consider how traffic is handled by the exchange. Firstly, of course, the subscribers on the exchange will 'originate' traffic – that is, initiate calls, and they will 'terminate' traffic – that is, receive calls. This is shown in Figure 6.17 with the direction of the arrows indicating the traffic flow (the conveyance of speech is of course in both directions). Thus, the subscribers originate a total of 1,500 Erlangs (1.5 kE), of which 1.2 kE is own exchange and 0.3 kE is outgoing trunk traffic; the subscribers also terminate some 1.4 kE, of traffic of which 0.2 kE is from other exchanges. Although in practice the traffic distribution from the various exchanges differs, typically some 80% of originated traffic is own-exchange, and as this example shows the originating and terminating traffic is not usually symmetrical. In this simple example the total switched traffic is $1.2 + 0.3 + 0.2 = 1.7$ kE. The 40,000 subscribers originate 1.5 kE and terminate 1.4 kE in the busy hour, giving an average both-way calling rate of 0.0725 E per line. Typical both-way calling rates are around 0.06 E per line for residential subscribers, and 0.18 E per line for business subscribers [9].

Figure 6.17 shows the full exchange configuration for the above example. The Route switch-block needs to be dimensioned to carry the 1.7 kE of total switched traffic as well as provide terminations for all the trunk routes (total traffic of 0.3 kE outgoing and 0.2 kE incoming). Since an RCU can typically terminate a maximum of 2,000 lines, some 23 or 24 units (depending on the fill factor) would be required to support the 40,000 subscribers. These Concentrators will be either co-located with the Route switch-block in the exchange or located at several remote small

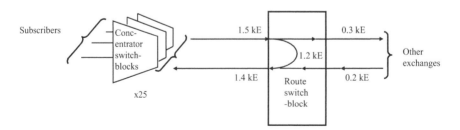

Figure 6.17   Exchange dimensioning

exchange buildings, depending on the disposition of the subscriber lines in the exchange catchment area.

Chapter 1 described a typical PSTN and introduced Figure 1.2 as a simplified diagram of BT's telephone network, which is reproduced as Figure 6.18 for convenience. A PSTN is arranged as a hierarchy of exchanges, with local exchanges connected to a parent trunk exchange, and in the case of the United Kingdom each trunk exchange is fully interconnected. This enables calls from subscribers on any exchange to be routed to subscribers on any other exchange via a maximum of four interexchange links. Calls between two local exchanges on the same parent trunk unit would normally be tandem switched at that trunk unit. However, where the level of traffic between two exchanges becomes sufficiently high it becomes economical to establish a direct route. Such routes are referred to as 'optional' or 'auxiliary', whereas the hierarchical routes (e.g. local exchange to parent trunk exchange) are mandatory.

An example of an optional route is given in Figure 6.19, in which there is 25 E of traffic between Exchange A and the parent trunk exchange B, and 30 E between Exchanges B and C (assume all traffic quantities are total both-way) and the planning question is whether the 5 E of direct traffic between A and C should transit-switch at exchange B or go directly. This is known as the direct versus tandem (*d/t*) routeing decision. The cost comparison comprises: the cost of the transmission of 5 E of traffic between A and B (in practice this may actually be carried over line plant A to B and B to C rather than over a new A to C link) versus the cost of 5 E extra on to the existing 25 E route A–B, plus the marginal cost of switching 5 E extra at B, plus

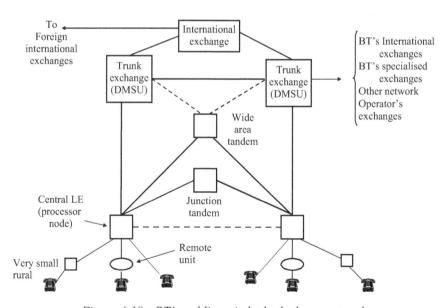

Figure 6.18   BT's public switched telephone network

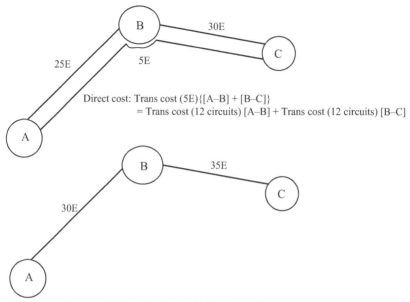

Direct cost: Trans cost (5E){[A–B] + [B–C]}
= Trans cost (12 circuits) [A–B] + Trans cost (12 circuits) [B–C]

Transit cost: Trans cost (30E–25E) [A–B] + Switching cost (5E) + Trans cost (35E–30E) [B–C]
= Trans cost (6 circuits) [A–B] + Switching cost (5E) + Trans cost (5 circuits) [B–C]

Figure 6.19   Example of direct versus tandem routeing comparison

the cost of 5 E extra on the existing 30 E route B–C. However, because of the nonlinear characteristic of traffic loading with route size, the additional 5E on link A–B incurs just six extra circuits, and on link B–C just five extra circuits compared to the need for 12 circuits to carry the new 5 E route. The cost of a direct route would not be warranted in this case. Thus, if the $d/t > 1$ it is cheaper to route via the tandem; with the $d/t < 1$ the direct route is justified. The d/t ratio does not normally approach one until the traffic between two exchanges is around 10–12 E, depending on the actual cost of switching and transmission equipment deployed in the network.

In the case of the BT PSTN shown in Figure 6.18, the potential for optional routes is shown with a dotted line. Junction tandems are used to carry traffic between the local exchanges in large metropolitan areas such as London and Birmingham; direct optional routes are only used where the level of co-terminal traffic is above the d/t threshold. Actually, the addition of these optional routes improves the resilience of the switched traffic network by imposing a mesh structure on top of the basic hierarchical star structure.

Furthermore, the appropriate use of automatic alternative routeing (AAR) enhances both the resilience and the overall cost efficiency of the switched traffic network. This facility uses the ability of the processor control system of a digital exchange to select several alternative routeings to a destination exchange on a call-by-call basis. Figure 6.20 shows a part of a network of trunk exchanges A, B, C, D

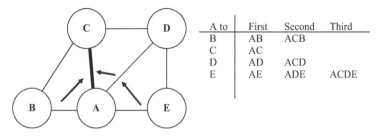

A to	First	Second	Third
B	AB	ACB	
C	AC		
D	AD	ACD	
E	AE	ADE	ACDE

Figure 6.20   Automatic alternative routeing

and E, in which the first choice and alternative routeings from A are shown in the table. For example, the first-choice routeing of A to B is over A–B, but in the event of this route being full (congested) the second choice is to route A–C–B (i.e. tandem switching at C). Similarly, for the choices of alternative routeing are shown for the other routes from A, with the case of A–E having a third-choice routeing via A–C–D–E – that is, incurring two sets of tandem switching. Clearly, this process introduces increased resilience by virtue of the quick fall back from first choice to second or third choice routeings, without the users being aware of the changes. However, it is also used to decrease overall costs by allowing the under-dimensioning of all but the final choice of route. Thus, in the example of Figure 6.20 only route A–C is fully provided (according to Erlang's B formula), all other routes are dimensioned to a lower level, known as 'high usage' – the occasional peaks of traffic experiencing congestion on these routes being switched to the alternative fully provided route. Typically, optional routes are dimensioned on a high-usage basis, while mandatory routes are fully provided.

6.3.5   Exchange capacity planning

A network operator's planners have the task of ensuring that there is sufficient equipment in each of the exchanges to meet the traffic demand on a daily basis. To this end measurements of the traffic demand are made periodically at each exchange. Forecasts of future demand are made, based usually on historical data, and the planners are then able to determine when each exchange will 'exhaust' – that is when the forecast future demand equals the installed capacity. The aim is for new capacity to be added to an exchange sufficiently in advance of exhaustion to avoid any congestion on the one hand and not to have unnecessary amounts of spare equipment sitting idly in the exchange on the other hand. Figure 6.21 illustrates this exchange capacity management process. The demand curve is shown, being the actual measured demand up to current date and forecast forward in time. Exchange equipment is supplied in appropriate capacity increments, in terms of extra switch-block (Concentrator and Route) capacity, trunk LTUs, signalling system units and exchange control system units. Each increment will give capacity sufficient to meet growth in demand for a period of several months – this is known as the 'design period'. There will also be constraints on the capacity increments due to module sizes of the equipment.

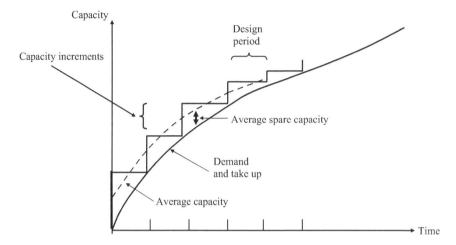

Figure 6.21   Exchange capacity design periods

Design periods are an economic trade-off between the cost of planning and installation versus the burden of undue spare capacity. Typical design periods for exchanges are around 18 months to 2 years. Figure 6.21 also shows the plot of average spare capacity; again, this needs to be kept at a level which meets the economic trade-offs, as well as giving a useful buffer for unforeseen sudden increases in demand. It is interesting to note that the alternative approach of 'just-in-time' provision would require the demand curve to be tracked closely by the installed capacity resulting in only minimal amounts of spare capacity, but this requires a rapid delivery and installation process, which is generally impractical for the telecommunication network environment.

6.4   Summary

In Section 6.2 we considered how an exchange (or more precisely the switching equipment in the exchange building) switches telephone calls. First, the generic local exchange structure was described, comprising switch-blocks to concentrate the traffic from subscriber lines (subscriber concentrator units), and a remote or co-located Route switch-block. The mechanism of digital time and space switching – the heart of modern digital exchanges – was explained. We introduced the acronym BORSCHT to help identify all the functions necessary to support a subscriber's line at an exchange.

Section 6.3 then looked at the network aspects of switching: the concept of traffic flow, how it should be routed, and the economic and resilience aspects of switched network structures. Finally, the planning of switch capacity was briefly examined.

A simple example of the stages of network planning of a PSTN, bringing together all the aspects covered in this chapter and the Core Transmission Network planning covered in the previous chapter was given in Box 6.2. The network planning for fixed and mobile networks is covered in more detail in the companion book [10].

Box 6.2   A simple example of the stages of planning a PSTN

As a simple example of the stages of planning a PSTN we will consider the small hypothetical island with population distribution as shown in Figure 6.22.

- Stage 1 is the determination of the number of exchanges to serve the island and their catchment areas, also shown in Figure 6.22.

Figure 6.22   Planning example; Stage 1; Customer distribution [Ward]

- Stage 2 (Figure 6.23) is the identification of the exchange locations, which need to be as close as possible to the centre of gravity of these catchment areas so as to serve all potential customers with minimum total subscriber-line costs.

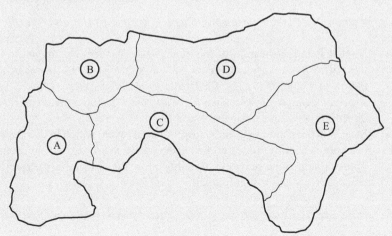

Figure 6.23   Planning example; Stage 2; Exchange location [Ward]

• Stage 3 is the estimation of the traffic flows between each of the five
  exchanges, in both directions, as set out in the matrix. In practice these
  flows would be determined from measurement of existing calling dis-
  tributions, or in the absence of data use of the gravitational formula
  (traffic between $(a - b) = $ [population of a × population of b]/the square
  of the distance apart) may be appropriate (see Figure 6.24).

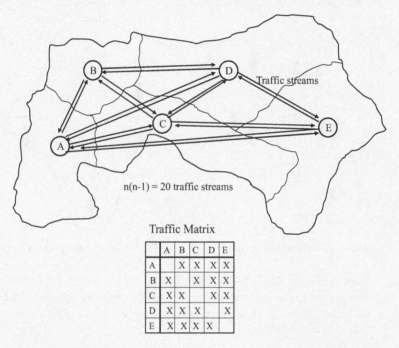

Traffic streams

$n(n-1) = 20$ traffic streams

Traffic Matrix

	A	B	C	D	E
A		X	X	X	X
B	X		X	X	X
C	X	X		X	X
D	X	X	X		X
E	X	X	X	X	

Figure 6.24   Planning example; Stage 3; Traffic distribution [Ward]

• Stage 4 is the application of the direct versus tandem calculation to all
  the traffic flows, resulting in the 'd' or 't' decision, as shown in the
  matrix of Figure 6.25. The final pattern of traffic routes is then a star
  structure centred on exchange C, with direct traffic routes (in either
  direction) between exchanges A and B.
• Stage 5 is the planning of the transmission network, which in this simple
  example is a cable route following the single main road on the island.
  Transmission cable networks usually follow the road infrastructure for
  practical reason (see Figure 6.26).

- Stage 6 is then the important process of mapping the traffic network to the physical cable infrastructure – the latter are known as the 'engineering routes'. The resulting mapping is shown in Figure 6.27.

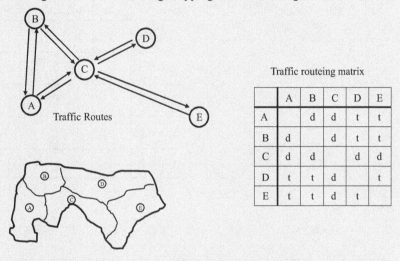

Traffic routeing matrix

	A	B	C	D	E
A		d	d	t	t
B	d		d	t	t
C	d	d		d	d
D	t	t	d		t
E	t	t	d	t	

Figure 6.25   Planning example; Stage 4; Traffic routeing [Ward]

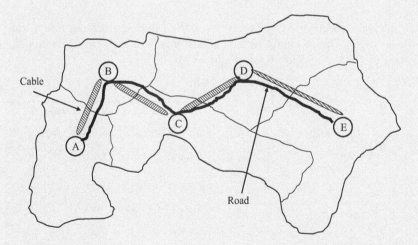

Figure 6.26   Planning example; Stage 5; Physical cable routes [Ward]

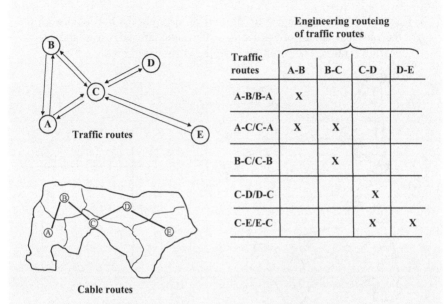

Traffic routes	Engineering routeing of traffic routes			
	A-B	B-C	C-D	D-E
A-B/B-A	X			
A-C/C-A	X	X		
B-C/C-B		X		
C-D/D-C			X	
C-E/E-C			X	X

Figure 6.27   Planning example; Stage 6; Traffic route to engineering route matrix [Ward]

Finally, in Stage 7 the dimensions of the cable, engineering routes, can be determined by summing all the traffic routes over each transmission link, using the matrix of Stage 6 (see Figure 6.28).

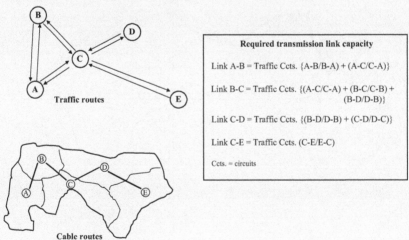

Required transmission link capacity
Link A-B = Traffic Ccts. {A-B/B-A) + (A-C/C-A)}
Link B-C = Traffic Ccts. {(A-C/C-A) + (B-C/C-B) + (B-D/D-B)}
Link C-D = Traffic Ccts. {(B-D/D-B) + (C-D/D-C)}
Link C-E = Traffic Ccts. (C-E/E-C)
Ccts. = circuits

Figure 6.28   Planning example; Stage 7; Transmission (engineering route) dimensioning [Ward]

References

[1] Redmill, F.J. and Valdar, A.R. 'SPC digital telephone exchanges', *IEE Telecommunications Series No. 21*. Stevenage: Peter Peregrinus Ltd.; 1995, Chapter 11.

[2] ibid, Chapter 7.

[3] ibid, Chapter 6.

[4] ibid, Chapter 10.

[5] Bregni, S. *Synchronization of Digital Telecommunications Networks*, Chichester, UK: John Wiley & Sons; 2002, Chapter 4.

[6] Fitter, M.A. 'The introduction of UXD5 small digital local exchanges', *British Telecommunications Engineering*, 1985;**4**(1):27–29.

[7] Flood, J.E. *Telecommunications Switching, Traffic and Networks*, Cranbury, NJ: Pearson Education; 1999, Chapter 4.

[8] Bear, D. 'Principles of telecommunications traffic engineering', *IEE Telecommunications Series No. 2*. Stevenage: Peter Peregrinus Ltd.; 1980, Chapter 7.

[9] Smith, S.F. *Telephony and Telegraphy A: An Introduction to Telephone Exchange and Telegraph Instruments and Exchanges*, London: Oxford University Press; 1969, Chapter 7.

[10] Valdar, A.R. 'Understanding telecommunications business', *IET Telecommunications Series No. 60*, 2016, Chapter 6.

Chapter 7

Signalling and control

7.1 Introduction

The concept of a telephone call being routed through a succession of exchanges was first introduced in Chapter 1, extended to the interconnection of several different operators' networks in Chapter 2, and considered from the switching systems' viewpoint in Chapter 6. We are now in a position to examine how the control of such call routeings across the networks (public switched telephone network (PSTN), mobile and others) is achieved. To this end, this chapter first considers the mechanism of signalling between nodes across the networks, and then the principal features of call switching control are introduced. In addition, we consider the broader aspects of control covering multimedia and more complex services.

7.2 Signalling

7.2.1 An overview of signalling

Signalling may be defined generically as the extension of control information from one network node or user to another over one or more links. It is employed in two distinct domains: between the user and the serving node (e.g. subscriber concentrator switch) – known as 'user' or 'subscriber' signalling; and between network nodes (e.g. exchanges) – known as 'inter-exchange' or 'inter-nodal' signalling. The latter may be between exchanges within one network or extend across network boundaries to other nodes in either the home country or via the international network to a foreign country. This is illustrated in Figure 7.1, which shows the domains of subscriber and inter-exchange signalling in both fixed and mobile networks.

Network operators create a clear demarcation within the exchanges between subscriber and inter-exchange signalling in order to ensure network integrity. It is important that the control that an individual user can have on the network – seizing of links, routeing of calls, etc. – is carefully constrained to avoid a rogue subscriber deliberately or inadvertently monopolising key parts of the network. (This is a feature that engineers are now trying to introduce into the Internet!)

Signalling in either domain has two purposes. The first is that of 'supervisory' or 'line' information, necessary to manage the link resources for the call. A simple

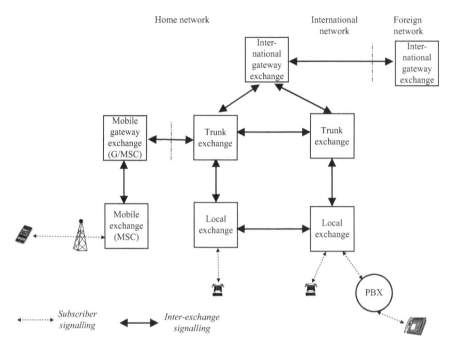

Figure 7.1   The domains of signalling

example is the 'off-hook' signal in a fixed network (a loop established in the absence of ringing current) which signifies that the subscriber wishes to initiate a call. A corresponding example for inter-exchange signalling is the indication to the distant exchange that a circuit has been seized by the calling-end exchange as part of the call set-up sequence, as described later. The second form of information sent over signalling links is that of 'address' – for example, the telephone number of the called subscriber.

Over the local copper line, subscriber supervisory signalling relies on the presence or absence of loop conditions. Since the earliest days of automatic exchanges rotating dials on the telephone injected a series of breaks in the loop at the rate of ten per second to convey the address information (e.g. three breaks indicated digit three) – a system known as 'loop disconnect' (LD) signalling. The resulting electrical waveform on the line during signalling is a train of pulses indicated by each turn of the dial (referred to as '10 PPS (pulses per second) signalling'). Most dial telephones have been replaced by the tone signalling system, as described below, but the subscriber even today is said to 'dial' a number (even if most users have never experienced a dial!). The modern tone telephone has a set of 12 push buttons used to indicate the address signalling, each of which selects the appropriate pair of tones from six frequencies – 'multi-frequency (MF) signalling' – giving rise to the characteristic musical feedback to the user. The signals are detected at the far end by a set of six filters, each tuned to one of the frequencies. In addition to the digits *0–9* the MF phones provide the 'star' and 'hash' keys for extra address information. The latter facility is useful in allowing

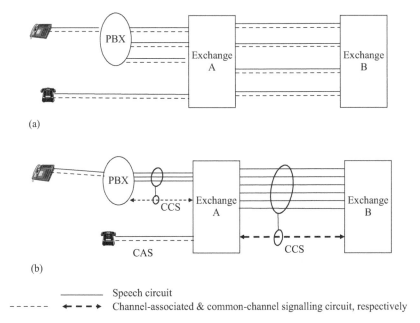

Figure 7.2 Relationships between voice channels and their signalling channels. (a) Channel-associated signalling (CAS). (b) Common-channel signalling (CCS)

users to send additional information after the call has been established, for example, the sending of PIN (process information network) codes to authorise entrance to conference calls.

A further important distinction with signalling is the relationship between the signalling channels and the corresponding speech channels. The simplest arrangement is 'channel-associated signalling (CAS)', in which the signalling and speech traffic share the same transmission path on a permanent and exclusive basis, as shown in Figure 7.2(a). The standard telephone subscriber local loop uses CAS, with both the line and address signalling carried over the subscriber's pair. Whilst CAS systems benefit from having a simple relationship between signalling and the speech channels, they do suffer from the restriction of sharing the transmission path, which can cause difficulties in signalling during the conversation phase. Also, because of the one-to-one relationship with the number of speech channels, the cost of the signalling system needs to be kept low.

The alternative arrangement is to have a single common separate channel which carries the signalling for a set of speech circuits, as shown in Figure 7.2(b). This single channel can carry the signalling at any time, including during the conversation phase, and it can be feature-rich and complex, since its costs are being shared across several speech channels. Subscriber signalling is channel-associated within the mobile networks and fixed Access Network for customers served by copper cable; whereas, common-channel signalling systems are used for the more complex needs of high-speed and data services, such as ISDN and PBXs [1].

Historically, a wide range of channel-associated systems were developed for inter-exchange signalling, each appropriate to the transmission media used at the time, for example, direct current (DC), alternating current (AC) and MF signalling systems [2]. However, today, nearly all inter-nodal signalling is common-channel, using the ITU-T (e.g. CCITT) Common Channel Signalling System Number 7 (CCSS No.7) chosen because of the wide range of messages possible and the high speed of working, as described in the following sections. (See Appendix 1 for information on ITU-T and CCITT.)

7.2.2 Applications of common-channel signalling systems

Common-channel signalling (CCS) systems provide data messaging between the computers of distant exchange-control systems. For this reason these systems are also called 'inter-processor signalling systems' and, as we shall see later, the set of data messages carried is intimately linked to the service being provided, for example, call set-up process for telephony. CCS is also used between PBXs and their serving public local exchange, as well as on ISDN local access links (e.g. ITU system Q931). A special form of CCS is used between PBXs which form a private or corporate network – for example, DPNSS (digital private network signalling system) – which enables calls to be established using private numbering schemes and special charging arrangements. However, in addition to these various telephony call signalling applications, the data message ability of CCS also enables exchange-control systems to make enquiry transactions with remote data bases to help handle complex calls. This extra application of CCS is referred to as 'non-circuit related (NCR)' signalling because, unlike the applications described earlier, there are not any speech circuits between the exchange and the control data base. Figure 7.3

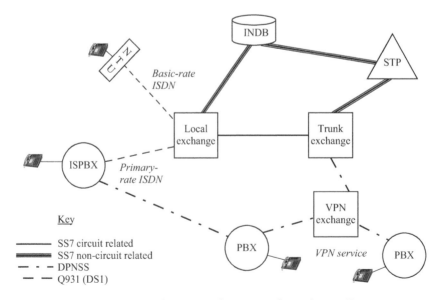

Figure 7.3   Applications of common channel signalling

illustrates all these different applications of common-channel signalling. We will now concentrate on the widely deployed CCS standard ITU SS No.7 system, which is used for both circuit-related and NCR applications.

7.2.3 ITU common-channel signalling system number 7

This universally applied international standard CSS system is called variously: CCSS No.7, SS7, ITU-T7, CCITT No.7, or C7. For convenience, in this book we use the term 'SS7'.

It is important to appreciate that the SS7 signalling links between the various nodes (as shown in Figure 7.3) really constitute a separate telecommunications network overlaying the PSTN, mobile MSC networks and the other specialised networks that it serves. Whilst the signalling nodes are physically part of the switching systems, as described in Chapter 6, and the signalling links are carried over the Core Transmission Network, as described in Chapter 5, they are functionally part of a stand-alone network, which is managed and dimensioned separately. Each SS7 node is identified by an address, known as the signalling point code. We may therefore view the basic function of SS7 as the reliable delivery of signalling messages between the various SS7 nodes. These messages can be routed directly between nodes or sent via one or more intermediate nodes, similar to the direct or tandem routeing of telephone calls. Indeed, there are special SS7 nodes which act as tandem points for SS7 messages, known as signal transfer points (STPs), which will be considered later.

Before we look at the system itself, it is instructive to consider a simple example of the flow of SS7 messages involved in setting up a telephone call through the PSTN. Figure 7.4 sets out a simple scenario with a call from subscriber

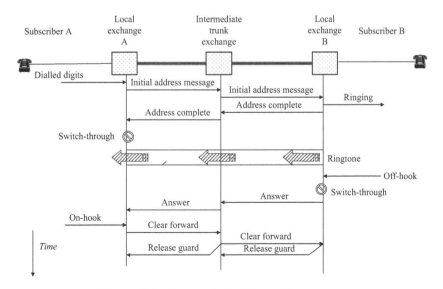

Figure 7.4 A typical telephone call sequence

A on local exchange A which is set up via an intermediate trunk exchange to the terminating local exchange B serving called-subscriber B. As time progresses down the page of Figure 7.4 the following actions take place.

(i)   On receipt of the dialled digits from subscriber A the control system of exchange A determines that the call needs to be sent to the trunk exchange for completion and so selects a free channel (time slot), say channel 17, on the traffic route to the trunk exchange.

(ii)   Exchange A then sends the first of the SS7 messages to the trunk exchange. This message indicates that a call is to be set-up for channel 17 and gives the phone number or 'address' of the called subscriber B. For obvious reasons, this is known as the 'initial address message' (IAM). The message is sent via the SS7 system using the relevant source and destination point codes of the nodes at exchange A and the trunk exchange, respectively.

(iii)   The control system of the trunk exchange then determines that the call is to one of its local exchanges and seeks a free channel on the traffic route, say channel 25, to exchange B. The control system then requests an IAM to be sent to exchange B, which contains the same destination address as the message from exchange A, but with the related channel number of 25 (rather than 17). The message is sent via the SS7 system using the source and destination addresses for the nodes at the trunk exchange and exchange B, respectively.

(iv)   On receiving this message, the control system at exchange B then checks whether the line of subscriber B is free. If it is free, ringing current is sent to the line. A reply SS7 message is then initiated by exchange B, which indicates that a correct telephone number has been received and that ringing current is being applied to subscriber B's line. This 'address complete' message is sent from Exchange B to the trunk exchange, referring to channel 25 and using the appropriate point codes.

(v)   The control system at the trunk exchange recognises that the message from exchange B relates to a particular call in progress, and sends an 'address complete' message to exchange A.

(vi)   On receipt of this response, the control system of exchange A sets up the 'switch through', that is, a continuous speech path is now established through the switch-blocks from the line of subscriber A to channel 17 on the outgoing traffic route to the trunk exchange. A continuous path then exists from subscriber A's telephone all the way through the network to exchange B. The ringing tone generated by exchange B can now be heard by the calling subscriber A. (In general, the ringing tone heard by the caller originates from the destination local exchange, which is why the style of ringing tones may vary when calls are made to different networks.)

(vii)   As soon as the control system at exchange B detects that subscriber B has answered the phone by going 'off-hook', ringing current is ceased and the switch-through exchange B's switch-blocks is established between channel 25 from the trunk exchange and subscriber B's line. The speech path has thus been extended to subscriber B.

(viii)   A return 'answer' message initiated by exchange B, indicating successful call set-up, is sent to the trunk exchange, which in turn sends a similar message to exchange A. The call now moves to the conversation phase and the control system at exchange A starts to log the call details in a 'call record' – part of local computer storage within the exchange control system. (Call records are periodically dumped off the storage at each local exchange and transferred to off-line billing centres for processing.)

 (ix)   The call-clear down sequence begins as soon as the caller replaces their handset. (If the called subscriber clears first the call connection clear-down will start once the caller clears or after a few minutes time out, whichever is the sooner.) A 'clear-forward' message is sent from exchange A to the trunk exchange, which in turn sends a clear forward message to exchange B. The speech path is then cleared down across the switch-blocks of all three exchanges. The charging for the call ceases and the call record is completed by exchange A's control system.

  (x)   Finally, return 'release guard' messages are sent from exchange B, and the trunk exchange to free up the speech channels 25 and 17 on the traffic routes, respectively.

In this typical call scenario of Figure 7.4, ten SS7 messages are sent between the three exchanges. Obviously, if there are more exchanges involved in a call the number of SS7 messages required increases. If any of the messages are corrupted during transmission or lost, a retransmission of the message is requested and made. This link-by-link progression of SS7 messages to set up and clear-down telephone calls is also applied within mobile networks, although the sequence is more complex and additional types of messages are required. We can now appreciate that the SS7 system comprises a mechanism ensuring the safe delivery of the messages, and a set of message types each related to particular services (e.g. ISDN and mobile).

Figure 7.5 gives a simplified representation of the delivery mechanism and the messages. First, we will consider the delivery mechanism. The confusingly termed 'signal unit' (SU) is a structured data packet, that is, an assembly of bits (1's and 0's) which acts as a conveyor of the signalling messages. A so-called 'flag', which is a special binary pattern (i.e. *01111110*), indicates the front end of the SU and there is another flag at the end. Between the flags within the SU there are specified areas, or fields, allocated to specific functions, although for simplicity only the main fields are shown. The first is the check sum, a 16-bit binary number which results from the multiplication of the contents of the SU by a special error-correcting polynomial binary code. Examination of this checksum at the distant end of the SS7 signalling link enables corruption or errors to be detected. The receiving end sender/receiver may then either directly correct the contents of the received SU or, if it is too severely corrupted, request a re-transmission of the SU. Next is the signalling information field, that is, the actual message to be sent. This field may vary in length up to a maximum of 272 bytes, depending on the nature of the signalling information being sent. The next field is used to indicate which of the portfolio of messages is being used. Since the size may vary according to

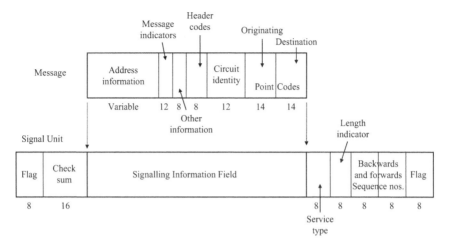

Figure 7.5   SS7 message and signal unit format

the amount of signalling information being sent, the next field indicates the length of the SU. The final two fields are used to carry the forward and backward sequence numbers. These are used to count and identify the sequence the SUs sent in each direction so that retransmissions of specific SUs can be requested and later placed in the correct sequence upon receipt.

Each type of message will have a different format depending on the nature of the signalling information to be sent [3]. However, as an example we will look at the format of the IAM, described above, as shown in Figure 7.5. The first two fields contain the destination and source points codes, each of which is the 14-bit address of the relevant SS7 node. The next field indicates the circuit number (or channel) to which the message relates, for example, 17 or 25 in the example above. Header codes in the next field indicate the type of message, in this case an IAM, so that the content can be appropriately interpreted. The final field contains the called and, if required, the calling telephone numbers.

We are now in a position to consider Figure 7.6, a block-schematic diagram showing the interworking of the SS7 system and the exchange systems (covered in Chapter 6). This shows trunk exchanges A and B connected by a traffic route comprising two 2 Mbit/s digital streams, each with 30 speech channels. The SS7 link terminates in both exchanges on a sender/receiver (S/R) system linked to the exchange-control systems and the digital Route switch-blocks. Exchange A's control system initiates a message, say an IAM, by sending an instruction to the SS7 signalling control subsystem through the control link, passing all relevant information, including: dialled number, circuit or channel identity, and the point code of the destination SS7 node (i.e. the SS7 S/R at exchange B). The formatted message is then passed to the signalling termination subsystem, where it joins a queue with other messages waiting to be sent. This subsystem then creates a signalling unit to carry the message by the addition of the SU sequence numbers,

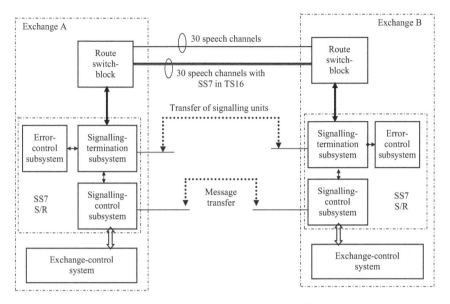

Figure 7.6   Digital exchanges and the SS7 system

length indicator, flags, etc., and the generated check sum from the error-control subsystem. As soon as a vacant slot appears on the signalling link the SU is transmitted via the time slot 16 (TS16) in one of the 30-channel 2 Mbit/s systems, as shown in Figure 7.6. The transmission rate of SS7 signalling over the network is therefore 64 kbit/s.

At exchange B the time slot 16 contents from the 2 Mbit/s link are permanently switched through to the SS7 signalling-termination subsystem. On receipt of the SU the error-control subsystem checks for errors and corrects or seeks a retransmission if necessary. The SU is then decomposed and the message part transferred via a queue to the signalling-control subsystem for decoding. The relevant information is then passed to exchange B's control system. A corresponding procedure occurs for the SS7 messages in the reverse (backward) direction. The call set up and clear down then progresses following the sequence described in the above example.

In theory, the SS7 system has sufficient capacity to handle the signalling for some 4,096 traffic channels, based on the circuit identity field of 12 bits! However, network operators view a dependency of so many calls on a single signalling link as far too vulnerable, and significantly lower capacities are used in practice. For example, the typical configuration for a remote concentrator in the PSTN is for the mandatory route to its parent exchange to be made of up to eight 2 Mbit/s (30-channel) systems, with their physical routeing spread over at least two separate cable routes following divergent paths (a diversity factor of two). The SS7 signalling link is carried over TS16s of two of the separated 2 Mbit/s streams, the TS16s of the remaining six 2 Mbit/s streams are kept empty. The signalling load is

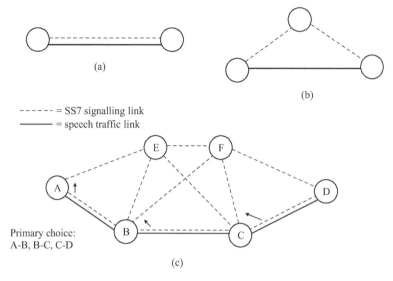

Figure 7.7   Modes of association of SS7 links with their traffic circuits. (a) Fully associated. (b) Quasi associated. (c) Alternative routeing and non-associated working

shared across the two TS16 paths, with the ability to send all signalling units over either one in the event of a cable failure.

The resilience of the SS7 signalling network is further enhanced by automatically sending signal units over alternative routes in the event of congestion or failure on the first choice SS7 route, similar in concept to the automatic alternate routeing (AAR) for telephone traffic described in Chapter 6. In this way the routeing of the signalling links can become separated from the traffic routes that they serve, as illustrated in Figure 7.7. The intermediate SS7 nodes may be located at telephone exchanges, or provided as standalone STP providing a tandem capacity exclusively for SS7 messages. The use of STPs can, in general, provide economies in terms of number of SS7 signalling routes and, as we shall see later in this chapter, more efficient routeing of signalling messages to network databases.

The SS7 has a flexible architecture which can be enhanced by the addition of new portfolios of messages as new types of service are introduced by network operators. Figure 7.8 presents a simplified architectural view, which is based on a common message transfer part (MTP), i.e., generating and transferring the signal units, and a series of user or application parts, i.e., the message portfolios. The set of messages required for telephony call control is contained in the telephone user part (TUP), as used in the example above. The messages for the more complex call control for ISDN are contained in the ISDN user part (ISUP). (Other user parts have been specified in anticipation of new services, e.g. data (DUP), broadband ISDN (B-ISUP) which have yet to be adopted.)

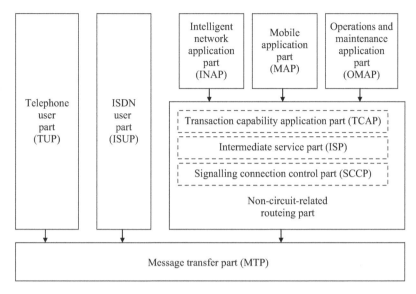

Figure 7.8   Simplified architectural view of SS7

There are also a set of message portfolios in so-called application parts, namely: intelligent networks (INAP), mobile (MAP), and operations and maintenance (OMAP), which have to be supported by what might be called the NCR routeing part, since these are not directly associated with traffic circuits. This part sets up a logical routeing for the messages to follow through the SS7 network – in effect, virtual paths – and supports the message flow, taking account of the requirements of the application or service involved. Details of this NCR part, which contains the TCAP, ISP and SCCP, can be found in [1,4].

A typical example of the use of the INAP set of SS7 messages is the enquiry by an exchange of a remote data base in order to complete a complex call. The call is initially routed from the local exchange to a trunk exchange using SS7 TUP or ISUP set of messages. Here, the exchange-control system determines that a translation of the dialled digits to an appropriate destination number is required from a remote data base – part of an intelligent network (IN) – in order to complete the call. First, an INAP message seeking a translation of the dialled number is sent over the SS7 signalling links, directly or via several SS7 links, using the NCR routeing part to steer the enquiry to the remote data base. The response to this inquiry, advising the new destination number, is then sent as an INAP message back over the SS7 network routeing to the trunk exchange control. Now the call can be completed through the PSTN using the SS7 TUP or ISUP messages in the normal way. This NCR form of SS7 signalling (using INAP messages) is now widely deployed in support of the advanced voice services provided by the various INs, as described later in this chapter.

7.2.4   *ITU-T H323 and session initiation protocol*

Although SS7 is widely deployed in nearly all PSTN, mobile and INs around the world, there is a new breed of signalling systems being developed to support the use of IP networks to provide voice calls (voice over IP) – an essential part of the so-called next-generation networks (NGNs). We will briefly consider two of these new signalling systems. The first, H323 is the signalling system specified by the ITU primarily for the setting up of calls via multimedia networks, for example a video conference call. This system uses the transport capability of the IP packet network to convey the messages (equivalent to the message transmission part of SS7) and a series of message sets to set up variable bandwidth and other features. Standard telephone numbering is used to identify subscribers [5]. It also uses the ITU-T Q931 protocol, shown in Figure 7.3, to manage the telephony aspects.

More recently a simpler system has been specified by the IETF (Internet engineering task force) specifically for the voice-over-IP networks (for more information on the IETF, see Appendix 1). The session initiation protocol (SIP) uses URL addresses (i.e. like an e-mail address) rather than telephone numbers to represent the users, and the protocol is based on a series of text-based messages. Like H323, SIP uses the transport capabilities of the IP network to convey the signalling messages between nodes. At its simplest level, a session is set up by the interchange between two computers (with telephones attached), an interchange of messages ensures compatibility between the two terminals and acceptance of the 'call'. The SIP signalling process is simply a series of requests and acknowledgements between software agents – known as 'user agents' – acting on behalf of a user or inanimate computer equipment (automatons) [6].

A simple example is shown in Figure 7.9(a), which shows the SIP message flows involved in establishing and clearing down a multimedia call, say voice and text, between users associated with computer terminals A and B. The user agent for A (resident in the software of A's computer) generates an 'invite' message which is conveyed to B over the appropriate data network – usually an IP network, as described in Chapter 8. User agent B responds positively with an 'OK' message, to which user agent A responds with an 'acknowledge' message. The media session thus initiated; the session itself comprises a flow of data (IP) packets between the A and B computers. This session is ended by either party sending a 'Bye' SIP message, which is responded to by the other party sending an 'OK' message, as shown.

The SIP messages comprise of two parts: a header, containing the logic of the message; and a body containing the identities of the parties (actually, the user agents) involved in the signalling, as shown generically in Figure 7.9(b). Special SIP names are given to identify the user agents (although for convenience these are made as similar to e-mail names as possible) – the 'caller ID' and 'called ID'. There is also space in the body for the ID of any intermediate nodes involved in routeing or interrogating the SIP message to be included in the body of the SIP message (shown as 'via proxy' in Figure 7.9(b)). Since the message will be one of several relating to a call, a 'call ID' is included in the body of the message, together with a

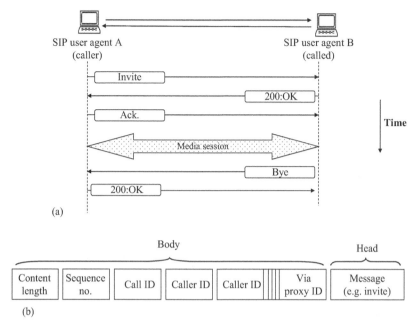

Figure 7.9   The SIP concept. (a) SIP message sequence. (b) SIP message format

sequence number to be used should a retransmission of the whole message be requested due to faulty receipt of the original.

The SIP is a simple signalling system which depends on IP and several supporting protocols for its transmission through the network. However, it is versatile and readily extendible to incorporate new service features, covering voice, data, video, etc. As we will see later in this chapter, it plays an important part in supporting call control in IP-based voice services (covered generally in Chapter 8). The move to incorporate voice and data services over a common IP platform has led to the idea of a next-generation network (NGN) architecture (see Chapter 11 for fixed NGN and Chapter 9 for 4G mobile).

7.3   Control

7.3.1   Introduction

The term 'control' when applied to telecommunication networks generally relates to the function of applying the network service when requested by the user, for example, the setting up, monitoring and clearing down of a telephone call (call control). However, the function of control also extends into the provision of service (e.g. adding a new subscriber's line to a local exchange), maintenance, exchange traffic and quality of service (QoS) management – all generally covered by the umbrella term of operations and management (O&M). In this section we will concentrate on the first aspect of service control, whilst Network QoS management and operations is addressed in Chapter 11 and covered in more detail in the companion book Reference [7].

Historically, the control of a service is provided by the operator within the network, with the user's terminals being relatively simple (dumb terminals) – which is the case for circuit-switched networks. In network-based control, the service functionality as well as steering the transmission of the users' traffic (i.e. 'transport') is provided by nodes in the network. However, with packet networks, particularly the Internet, the control of the service is usually external to the network (at the edge), provided either by the user's terminal or by third-party service provider's functional nodes. In this case it is the network that is considered 'dumb' by undertaking only the transport of the packets, whilst the service-functionality control is with the so-called 'over-the-top' service provider [8].

We will develop our view of service control in this section by first considering telephony call control for fixed and mobile networks, beginning with the most complex case, that is, local exchange. Later, we look at centralised intelligence systems for circuit-switched and packet-based networks.

7.3.2   Exchange-control systems

Having seen the close association of subscriber and inter-exchange signalling with the set up of a telephone call, it is now appropriate for us to consider the call-control system within an exchange. The block schematic diagrams of a local exchange shown in Chapter 6 (Figures 6.4, 6.10, and 6.11) show a single entity labelled Exchange-control system. In fact, this system typically comprises multi-processor clusters and a set of regional processors, as shown in Figure 7.10. The processor clusters comprise a number of central processing units (CPUs) sharing via a common bus the resources of a memory system. The total call-handling capacity of the control system depends on the number of CPUs within each cluster and the number of clusters. Normally, the control load of the originating calls of the set of subscriber lines, as well as the terminating calls received from other exchanges is shared by all clusters ('load sharing' mode). In the event of a cluster failure, a reduced capacity is provided by the remaining clusters. This load sharing approach is also applied between the CPUs within each cluster.

The supervisory line conditions detected by the subscriber line cards (e.g. 'off-hook' and loop-disconnect dial pulses) are gathered by a line controller, which periodically scans all the line cards on the Concentrator switch-block. The outputs of the line controllers associated with each of the co-located subscriber Concentrators are gathered by a regional processor which formats the off- and on-hook indications and any dialled digits into a series of computer compatible messages for analysis by the cluster controller. This regional processor also assembles the dialled digits by the tone phones as detected by the MF sender/receiver. All of these messages from the regional processor indicate the appropriate subscriber's line number so that the messages can be associated with the relevant call set up or clear down currently being handled by the processor cluster. Similarly, messages to and

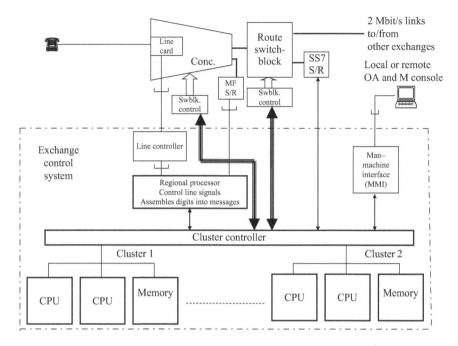

Figure 7.10   Exchange-control system in a local exchange

from the SS7 sender/receiver are passed by the cluster controller to the cluster dealing with the call.

The clusters comprise several processors (CPUs) and associated memory, containing a copy of the call-control program and relevant exchange connection data. The cluster controller shares the call control work across all the clusters. Call control is achieved by following a logical sequence that leads to appropriate actions within the switch-blocks, based on the digits received from the calling subscriber, or distant exchange in the case on an incoming call. When the controlling cluster determines that a connection is required across one of the switch-blocks a message is sent via the cluster controller to the appropriate switch-block control – in the form of a request to 'connect a particular inlet timeslot to a particular outlet time-slot'. The switch-block control is then able to set up the connection by examining the current occupancy details as held in a local store (known as 'map in memory'). A routeing table held in the cluster memories holds all the first choice and sub-sequent routeing details for all traffic routes from the exchange. These cluster memories also save all the call records for later off-line processing of the subscriber bills, as well as recording traffic-usage statistics for use by the network operators to dimension the exchange, as described in Chapter 6. The control system time shares its attention between all the active calls by processing the set-ups and clear-downs in stages, progressing each call to the next stage and then turning its attention to the other calls during the relatively long times of waiting for digits to arrive or connections to be set up [9].

Actually, the exchange-control systems provide extremely powerful and reliable real-time processing. Typically, they are capable of handling up to 1,000,000 busy hour call attempts (known as 'BHCA'), operating on a truly continuous basis ('24 × 7'). The call control and maintenance programs for these systems are complex and lengthy. To make the point, engineers have estimated that a printed version of the source program of a typical digital SPC exchange would fill about ten miles of computer print-out paper! (Indeed, some particularly feature-rich switching systems are known as '12-mile exchanges'.) The software is updated periodically by the switching system manufacturer to incorporate new features for customer services and to correct any bugs on earlier versions. Network operators typically change the software on each of their processor and trunk exchanges once every twelve to eighteen months – amazingly, without any break in service! This feat is achieved through the partitioning of the processor clusters into working and off-line status; changing the software on the latter; and then swapping the status so that the software on the remaining processor clusters can be changed. These change-overs are usually performed during the quiet periods of exchange usage, for example, in the early hours and at weekends. Obviously, this standard of reliability is significantly greater than that generally experienced on office and home computers.

A set of operations, administration and maintenance (OA&M) consoles are provided locally within the exchange building and remotely at one or more OA&M centres, supported by a man–machine interface (MMI) system attached to the cluster controller, as shown in Figure 7.10. These consoles are used by the operator's technicians and administrators to, variously:

- Set-up each new subscriber's line (service provision),
- Amend the facilities settings for individual subscribers,
- Undertake fault clearance for a subscriber's line,
- Initiate traffic monitoring of the exchange,
- Change the routeing table for the exchange,
- Manage changes to the control software,
- Change the equipment status and configuration during software upgrades and installation of new capacity,
- Manage the recording and off-loading of billing information.

The key challenge for network operators in managing the exchange call-control systems is the assembly of the correct data and software versions for each exchange – for use in initiating or upgrading exchanges. This is usually referred to as 'data build'. The set of software versions deployed at each exchange-control system in the network needs to be recorded at the central operations centre, so that upgrades and software patching can be managed accurately. As far as the data is concerned, each control system needs to be kept up to date on the latest routeing details, including any changes of codes, subscriber number blocks, or parenting arrangements of local exchanges, etc. A further major area of changes of routeing details results from the addition or withdrawal of connections to other network operators, or with changes to their points of interconnection [8,10].

We now consider the telephone call-control arrangements for subscribers on a mobile network. As described in detail in Chapter 9, there are currently three generations of mobile systems being operated: 2G, 3G and 4G, with their routeing arrangements for voice calls being shown in Figure 9.23. Digital circuit-switched exchanges known as mobile switching centres (MSC) are used for telephony calls in the 2G and 3G systems. In addition, calls on the 4G (long-term evolution (LTE)) all-IP networks must be dropped back to collocated 2G or 3G networks so they may be switched at the MSC (circuit-switched fall back (CSFB) procedure), unless the IMS control system has been deployed, as described later in this chapter and Chapter 9.

Therefore, Figure 7.11(a) shows the exchange-control arrangement for 2G and 3G mobile networks. In both systems, the telephony traffic from subscribers is concentrated, namely: within the cellular networks by the allocation of connections via the wireless channels for just the duration of the calls; and within fixed networks through the use of a concentrator switch in the local exchange. The call set-ups are initiated once the subscriber presses the send key on their handset, when the called number address information is passed over the supervisory channel to the base transceiver station (BTS) or Node B for the 2G and 3G systems, respectively. The calls are then held by the control system at these nodes (not shown in

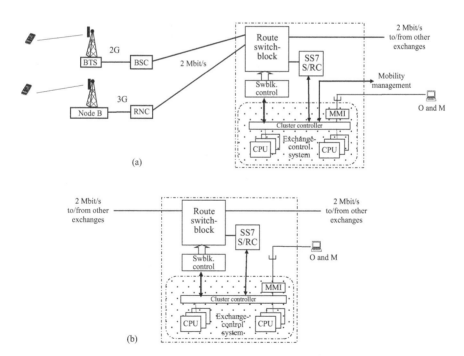

Figure 7.11   Exchange-control systems in other types of circuit-switched exchanges. (a) Mobile exchange (MSC). (b) Trunk, junction tandem or international exchange

Figure 7.11(a)) whilst signalling progresses to their controllers – namely: base station controller (BSC) or radio network controller (RNC) using the Q931 CCS system. Signalling between these cell controllers and the MSC is over SS7 carried in the usual way within TS16 of the traffic 2 Mbit/s systems, as shown in Figure 7.11(a) [11]. The MSC call control architecture is similar to that of the digital local exchange minus the systems dealing with subscribers' lines, since the telephony features are the same for fixed and mobile networks. However, of course, there are several additional control entities in the MSC providing the mobility management, as described in Chapter 9.

The simplest architecture for the exchange-control systems is with the transit-type of exchanges, namely: junction tandems, trunk exchanges, and international exchanges. These exchanges merely provide call connections between incoming and outgoing channels from/to other exchanges – there is no direct interaction with the subscribers. This is illustrated in Figure 7.11(b). Consequently, junction tandem and trunk exchanges are significantly cheaper than local and mobile exchanges; but this is not so much the case with international exchanges due to the complexity of charging and O&M support of international services.

7.3.3   Intelligent network

In the consideration of inter-nodal signalling above we recognised that call services range from simple telephony to the more-complex calls requiring some form of translation of the dialled number, or other additional processing.

Whilst the software in the exchange-control system is designed to provide the basic telephony service, it may not be able to provide the additional capabilities necessary for the service features needed by a network operator. One option is for a network operator to commission the switching system manufacturer to develop the additional program code to provide the extra features, for inclusion in the next upgrade to the exchange-control software. However, this may be expensive due to the need to include the upgrade on all exchanges despite the initial uncertainty of the amount of take-up or usage that customers will make of the new service. There will be other situations where the complexity of the new features cannot practically be provided on standard exchange-control systems. The rise in the need for ever more service features led to the development of the IN concept during the 1980s and 1990s.

The IN concept is based on the managed access from the PSTN exchanges to a centralised data base where the complex and expensive control software and any associated service data is concentrated. The standard architecture of an IN is shown in Figure 7.12, in which the IN is shown as an overlay to the PSTN. When a call is received by the PSTN that is recognised from the number dialled as an 'IN call', it is routed to an IN gateway exchange – known as a 'service switching point (SSP)'. Normally, the major trunk exchanges in the PSTN take on this additional role of being an IN SSP. The call is held at the SSP, with the normal call-control sequence interrupted (or 'triggered') so that an interrogation can be made of the remote

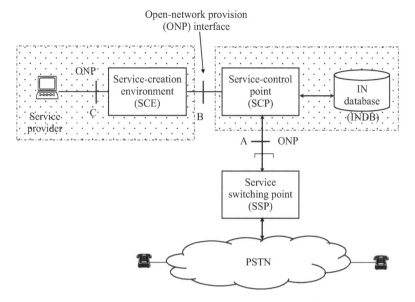

Figure 7.12   The intelligent network concept

IN service control point (SCP) and its IN data base (INDB). Typically, there are two or three SCPs located within the network serving the whole country through several hundred SSPs. The interrogation of the SCP is achieved by using NCR signalling over the SS7 network, perhaps using STPs to route the messages, as described earlier in this chapter. On receiving the response to the interrogation the SSP is able to complete the call using the new destination number and perhaps other control information, as advised by the SCP.

The other key element of the IN architecture shown in Figure 7.12 is the so-called service-creation environment (SCE), a development facility associated with the IN where new services can be constructed using software-based features as basic building blocks. This facility can act as the network operator's test bed for new services or extra features on existing services, and can also tailor specific services to the particular needs of large business customers. Finally, the standard architecture identifies open-network provision (ONP) interfaces at key points, with a view to allowing different manufacturer's equipment to be used throughout the IN. For example, the interface at point A is designed to allow different manufacturer's exchanges to operate as an SSP working to a standard SCP. Similarly, the interface B is designed to allow different manufacturer's service creation environments to work to one or more different manufacturers' SCPs [12,13].

Actually, there is no such thing as an 'IN service', since all services could be realised using resident exchange-control systems, perhaps with additional control hardware, or using an IN solution. The choice is essentially one of economics.

In general, exchange-control-based solutions are used for the following categories of services:

- Where service features are required on every call, for example, basic call set-ups with standard charging arrangements.
- Mass usage calls, for example, 'ring back when free'.
- Low-value calls (i.e. very low charges or free), for example, '1471' in the United Kingdom.

However, the following categories of services are appropriate to an IN-based solution:

- Conditional number translation, for example, 'Freefone' calls, where the destination number depends on one or more conditions (time of day, geographical area of the caller, instantaneous loading on call centre attendants, staffing arrangements at call centres, etc.).
- Complex charging arrangements, for example, 'Freefone' calls, where the charges are fully or partly paid by the recipient or other third party rather than wholly by the caller.
- Customer-specific management information, for example, where usage and distribution information on calls from their clients to a business customer are assembled and notified automatically to the customer.
- Corporate numbering schemes, for example, for corporate virtual private networks (VPN).
- Multioperator or complex number translation, for example, number portability of geographical or non-geographical telephone numbers.
- Mobile calls, for example, the determination of the location of telephone handsets in real time.
- Calls that require a network-wide view in order to determine how they need to be routed, for example, distributing calls to a dispersed set of manual call centres around the country depending on the queues at each of the centres.

As a general rule, new services are developed and provided on an IN initially. Within a few years, the volume of the customer usage and take up may increase to such a point that it becomes cheaper to arrange for the necessary enhancements to the control software and hardware at the exchanges, so that the service can be off-loaded from the IN and handled instead by the PSTN. Thus, a network operator will use both the PSTN and the IN to provide a range of services for both residential and business customers.

Figure 7.13 illustrates a practical arrangement of a set of INs, as might be used by an incumbent national network operator in conjunction with a PSTN. In this example, the 80 trunk exchanges of the PSTN act as the SSPs for all three IN networks. The NCR SS7 signalling messages from these SSPs are routed via STPs to the appropriate IN SCPs according to the nature of the service. The Main IN SCP provides several large-volume services, each based on a transaction server and associated data base. Resilience is assured through the triplication of the SCPs, each unit being located at separate buildings across the country. Signalling traffic

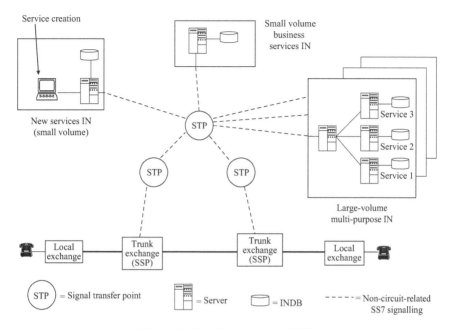

Figure 7.13 A typical set of INs

to these SCPs is spread evenly, although the three SCPs are periodically kept in step so that in the event of any single SCP failure the other two can share the work load. Relatively small volume but specialised business services are provided by a dedicated IN SCP (an example is customised automatic call distribution (ACD)). There may also be a new-services IN, comprising of a small-scale development facility with a powerful service-creation environment facility. New services are launched on this SCP, where they remain until they grow in volume to a level where it is economical to transfer them to the large volume IN. Alternatively, the services might instead be transferred (i.e. migrated) to the exchange-control systems on dedicated specialised exchanges or local and trunk exchanges within the PSTN.

7.3.4 Future network intelligence

Despite the successful implementations of INs in most countries during the 1990s enabling the provision of a range of advanced voice call services, as described above, not all of the attributes originally hoped for were achieved. In particular, the implementation of truly open interfaces (ONP) has proved difficult to achieve. Also, the extent of service creation has been limited to just those functions associated with call-control software – hence omitting the non-exchange aspects of service management. However, the biggest change in the way that intelligence is introduced to modern networks is commercial. The IN architecture is centred on a single network operator having the control of a service creation and its later

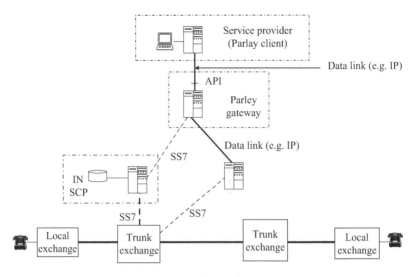

Figure 7.14   The Parlay concept

execution, a construct that does not fit well with the growing multioperator open environment that is being progressively introduced in many countries since the year 2000. There are now several ways in which intelligence based on a computer server can be made available in a network to provide the extra service capability – the IN approach may now be seen as just one technique.

One of the significant changes in the way that intelligence is managed in the network meeting the new commercial position is defined in the approach of the Parlay Consortium [14]. Figure 7.14 gives a simplified view of the Parlay architecture as associated with a fixed network operator. The key component is a Parlay gateway, provided by the operator, which offers a set of software functions that can be used by one or more service providers to create a service that can run on the network. The set of software functions is made available through an application programmer's interface (API). The Parlay consortium has defined the API in such a way that service providers can develop services on a variety of different networks: PSTN, IN, mobile, IP data networks, etc., in a commercially attractive way within the confines of maintaining the integrity and resilience of the networks used.

The Parlay concept was formalised into a set of standards by ETSI, Parlay Group and the 3GPP consortia (see Appendix 1) known as 'Parlay X'. This was then replaced by a standard prepared for mobile networks by the GSM Association (GSMA) known as 'OneAPI', which provides a simplified API platform for service providers to access network functions [15].

7.3.5   IP multimedia subsystem

The evolution of network control has now embraced the multimedia data networking nature of telecommunication networks, especially the widespread use of IP as the transport mechanism (see Chapter 8). This has led to the specification of a

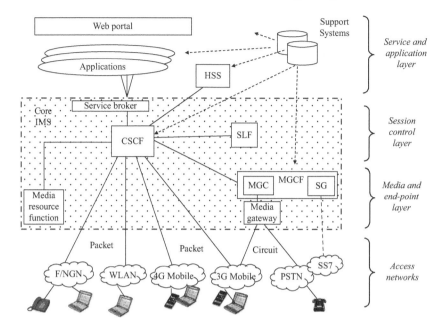

Figure 7.15   IP multimedia subsystem

new architecture for control, known as the 'IP multimedia subsystem (IMS)'. The use of the term 'subsystem' is because the IMS is only part of the infrastructure required for the job of service control within a network; the other components being the IP transport network, SIP signalling network, etc. Initially the specification of the IMS architecture assumed that it would apply just to mobile networks (especially 3G and 4G) as they became increasingly IP based, but then the move to make similar changes to the fixed network (i.e. NGN as described in Chapter 11) meant that IMS can potentially act as the universal control architecture for all types of networks. It should be noticed that the IMS provides service control intelligence from within the network, which as described in Section 7.3.1 is not normally the case for IP-based networks. Although IMS is designed to provide new multimedia type services, the biggest driver for its implementation by operators is to provide the all-IP 4G/LTE mobile network with telephony capability, that is, voice over LTE (VOLTE), as described in Chapter 9.

In principle, the subsystem comprises of a set of servers linked by SIP signalling underpinned by IP transport, which provide the control for a wide range of applications and services to users. In the full IMS architecture there are a set of 3GPP-defined interfaces between each of the components [16]. Figure 7.15 shows a simplified representation of the standard IMS architecture. Functionally the components fall into three groups: (1) the media and endpoint layer, which interfaces to the range of access networks being served; (2) the session control layer; and (3) the service and application layer. The first two layers are considered to form the IMS core. Since the range of possible networks being served includes both packet- and circuit-switched networks the IMS includes gateways to convert any SS7 signalling and PCM transport accesses to/from SIP and IP, respectively. We will now review the various components as shown in Figure 7.15.

CSCF. The heart of the IMS is the call/session control function (CSCF), which is a SIP-server orchestrating the sequence of events during the control of a data session or voice call. Actually, since the IP-based fixed and mobile access networks may be quite distant the CSCF functions are usually dispersed, with a remote presence – known as a proxy CSCF (P-CSCF) – at one or more access networks in the data session. The role of the P-CSCF is to act as the IMS port for the SIP messages to/from the access network (e.g. LTE network), providing authentication and message formatting. A particular type of P-CSCF is the interrogating CSCF (I-CSCF), which is located at the edge of the IMS within the home network and provides the additional function of SIP signalling directly to the home-subscriber server (HSS) and the subscriber location function (SLF), as will be described later. The serving CSCF (S-CSCF) is located within the IMS home network, as shown in Figure 7.15, where it acts as the central node controlling data sessions [16,17].

HSS and SLF. The home subscriber server (HSS) is a central database holding the information relating to the users (subscribers) of the service, which is actually part of the control system of the access mobile network being served by the IMS (see Chapter 9, especially Figure 9.18). The HSS holds semipermanent and real-time information on each registered subscriber covering current location, authentication details, class of service details, and the current S-CSCF controlling the call/session. In the case where there are several HSSs necessary to manage a large population of subscribers, the subscriber locator function (SLF) data base indicates which HSS holds the information for a particular subscriber.

MGCF. The media-gateway control function (MGCF) provides the interworking between PCM circuit-switched networks using SS7 signalling and the IP/SIP-based IMS. The actual conversion between PCM and IP for the traffic channel is provided by a media gateway (MG) under the supervision of the media-gateway controller (MGC), which establishes a pathway to the CSCF during the call or session. The interworking between SS7 (usually using ISUP) and SIP is provided by the signalling gateway (SG).

Application servers. A range of servers provides the control logic for the applications being provided by the IMS. Common examples of application servers (AS) are:

- Presence, which identifies the online status of users
- XML (extendible mark-up language) document management (XDM) provides the user with stored contact lists and other written information
- Telephony application server (TAS), which provides multimedia telephony control. This can be associated with other ASs, such as Presence and XDM as required.

Support systems. Support systems are required to provide a range of functions associated with the services being controlled by the IMS. For example, the charging and billing functions are provided by several systems, covering prepaid and post-paid subscribers, under the control of the policy and charging resource function (PCRF).

Media resource function. The media-resource function (MRF) provides a set of tones and announcements for use in multimedia calls, as well as conference bridges (i.e. mixing voice streams). It may be realised through a number of separate component systems all under a central control (MRFC).

More detailed descriptions of the IMS, its architecture, interfaces, and modes of working can be found in [16–18].

7.4   Summary

In this chapter we saw the close association of signalling with the control of call connections across the network. Firstly, the different forms of subscriber signalling were identified, ranging from the old fashioned use of a dial, the ubiquitous tone dialling system and the more advanced forms of common channel signalling from ISDN terminals and PBX. We then concentrated on the inter-nodal signalling, specifically the widely deployed ITU standard SS7. After considering the message flows involved in a simple telephony call across three exchanges, we examined the use of SS7 to convey data transactions, typically enquiries of a data base – the NCR form of signalling. We then considered the signalling system, SIP used within IP networks.

In Section 7.3 we examined the control aspects, initially looking at the exchange-control system in a telephone exchange. We then looked at how more complex calls, which require number translation or more call processing than normal telephony, can be handled by an IN. This subject included the use of interrogation messages to the SCP of the IN carried over NCR signalling through the SS No.7 network. We also noted the need for switching these NCR messages through STPs. The idea of multiple service providers gaining access to the control facilities of a network, initially defined by the Parlay Consortium, was then introduced. Finally, we examined IMS, the new control systems being deployed in IP networks, especially 4G mobile networks to provide a range of multimedia services, including telephony.

References

[1]   Manterfield, R.J. 'Common-channel signalling', *IEE Telecommunications Series No. 26*, 1991, Chapter 1.

[2]   Welsh, S. 'Signalling in telecommunications networks', *IEE Telecommunications Series No. 6*, 1981, Chapters 3–8.

[3]   Manterfield, R.J. 'Telecommunications signalling', *IEE Telecommunications Series No. 43*, 1999, Chapter 6.

[4]   Manterfield, R.J. 'Telecommunications signalling', *IEE Telecommunications Series No. 43*, 1999, Chapter 7.

[5] Tanenbaum, A.S. *Computer Networks*, 4th edn. Englewood Cliffs, NJ: Prentice Hall; 2003, Chapter 6.

[6] Wisely, D.R. and Swale, R.P. 'SIP – The session initiation protocol'. In Swale R.P. (ed.), *Voice Over IP: Systems and Solutions*, London, UK: IEE Publishers; *BT Exact Communications Technology Series 3*, 2001, Chapter 11.

[7] Valdar, A. and Morfett, I. 'Understanding telecommunications business', *IET Telecommunications Series No. 60*, 2015, Chapter 11.

[8] Valdar, A. and Morfett, I. 'Understanding telecommunications business', *IET Telecommunications Series No. 60*, 2015, Chapter 1.

[9] Redmill, F.J. and Valdar, A.R. 'SPC digital telephone exchanges'. Stevenage: Peter Peregrinus Limited; *IEE Telecommunications Series No. 21*, 1990, Chapter 15.

[10] Valdar, A. and Morfett, I. 'Understanding telecommunications business', *IET Telecommunications Series No. 60*, 2015, Chapter 2.

[11] Schiller, J. *Mobile Communications*. London: Pearson Education; 2000, Chapter 4.

[12] Turner, G.D. 'Service creation'. In Dufour I.G. (ed.). *Network Intelligence*, London: Chapman & Hall; *BT Telecommunications Series*, 1997, Chapter 8.

[13] Anderson, J. 'Intelligent networks: Principles and applications', *IEE Telecommunications series No. 46*, 2002, Chapter 2.

[14] Anderson, J. 'Intelligent networks: Principles and applications', *IEE Telecommunications Series No. 46*, 2002, Chapter 6.

[15] www.GSMA.com/identity/API-exchangewww.GSMA.com/identity/API-exchange

[16] Camarillo, G. and Garcia-Martin, M. *The 3G IP Multimedia Subsystem (IMS)*. Chichester: John Wiley & Sons; 2006, Chapter 3.

[17] Comarillo, G., Chakraborty, S., Peisa, J., and Synnergren, P. 'Network architecture and service realisation'. In Chakraborty, S., Frankkila, T.S., Peisa, J., and Synnergren, P. (eds.), *IMS Multimedia Telephony over Cellular Systems*, Chichester: John Wiley & Sons; 2007, Chapter 3.

[18] Handa, A. *System Engineering for IMS Networks*. Burlington, VT: Elsevier; 2009, Chapter 1.

Chapter 8

Data (packet) switching and routeing

8.1 Introduction

Previous chapters have concentrated on the communication of voice, primarily in the form of the telephony service as carried over the public switched telephone network (PSTN). We are now in a position to consider the various non-voice services, generally referred to as data, and the fixed and mobile networks that carry them. At the end of this chapter we will investigate the way that these data networks can also convey voice, and that leads us to consider the new generation of communication networks. We conclude by considering the trend in the technology used for data networking equipment.

8.2 The nature of data

Data is intrinsically different to voice. By definition, data is information which originates in the form of a digital representation – normally binary 1's and 0's and it therefore does not need to be converted to digital within the network, unlike voice which originates from the telephone microphone as an analogue signal. Data originates from terminals, such as temperature measuring devices, or from a laptop or computer, smartphone, tablet, and ATMs (i.e. 'automatic teller machines', not the data service described later in this chapter!) in the wall of a shop or bank. Applications generating data from these devices include the sending of electronic mails (e-mails), downloading from a website, transfer of information between files and other forms of enquiry, remote control of machinery and other forms of surveillance, etc. A data service is one that conveys data between the terminals and other devices. There are several key differences between the characteristics of voice and data and hence the different requirements for successful communications, as summarised below.

(i) Data applications may be tolerant or intolerant of delay on a communication network, known as 'latency' tolerance or intolerance. (Unlike voice, which is always latency intolerant.)

(ii) Data may be generated as a continuous stream or in bursts. (Digitised voice is continuous stream.)

(iii)   Data may flow at any speed, ranging from the extremely high billions of bits per second to the very low few bits per second. (Digitised voice is carried at 64 kbit/s.)

(iv)   Data is usually highly susceptible to error, every bit is precious. (Digitised voice is fairly robust to errors and even the loss or corruption of many bits is not perceived by the listener.)

(v)   There is a wide range of ways that the data can be organised, i.e., many protocols are used. (Voice encoding is highly standardised.)

(vi)   The data may be encrypted (as is digitised voice over mobile networks).

Circuit-switched telephone connections through a PSTN can be used to carry data services, indeed the earliest form of switched data service in the 1960s and 1970s was known as 'Datel' – data over telephone. However, the nature of circuit switching, with its continuous connection held for the duration of the call, does not provide the most economical communications medium for most types of data. Also, the relatively slow speed of call set-up can be a problem for short bursts of data. The specific needs of data services have, therefore, been addressed by communications networks based on packet rather than circuit switching. Not only does packet switching give better network utilisation and economics than circuit switching, but it also copes well with the speed variation and other special characteristics of data. The remainder of this chapter now considers packet switching for data. Later we will examine the treatment of voice as just another form of data service and look at the switching of voice over packets rather than circuits.

A brief note on terminology: The technical literature associated with data networks, describes the size of the chunk of data being considered either as a number of 'bits' (i.e. 1's and 0's) or in units of eight bits known as 'bytes'. Similarly, the speed of data transmission is described as either 'bits/s' or 'bytes/s' – obviously, the two differ by a factor of eight. (See also Appendix 1 for a summary of binary multipliers used in connection with computers.)

8.3   Packet switching

A data packet comprises of a string of bits with the front end formed into a header, giving the destination address and the remainder of the packet forming the payload carrying the actual data for transmission. The stream of data is thus deposited into one or more packets, each with an appropriate address, and the packets sent down a transmission bearer interleaved with packets from other sources. Figure 8.1 contrasts the packet approach with the continuous path of a circuit-switched connection. A useful analogy for packet switching is that of a postal service. Users put their letter into an envelope onto which they write the destination address. Similarly, the data is placed into the payload of the packet and the destination address inserted into the header. Like the postal service, other information may be included in the header, for example, originating (from) address, date of dispatch and a reference number. Different capacity envelopes are used in the postal case to handle the range of lengths and size of letters – and a similar arrangement exists for

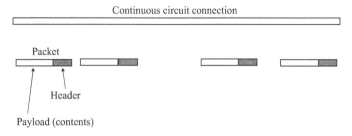

Figure 8.1   Packet and circuit switching

the data packet services, as we shall see later. Finally, the equivalent of packet switching is the role of the sorting office, that is, the reading of the envelopes and the depositing into the appropriate outgoing bags destined to be conveyed to the next stage of sorting. Different sizes of envelopes are accommodated by the postal bags and all the envelopes are intermingled or interleaved within the bags as they are transported by van to the next sorting office. Just like the postal service, data packets can get misrouted or lost – once this is detected the recipients then ask the sender for a replacement packet.

There are two fundamental ways that packets can be switched through a network, following either 'connection-orientated' or 'connectionless' modes of operation.

8.3.1   Connection-orientated packet mode

Figure 8.2 illustrates the connection-orientated mode. This works on the principle of establishing a so-called virtual path (VP) for the packets to follow. The three phases involved – VP set-up, data transfer, and then VP clear-down – are analogous to the setup, conversation and clear-down phases of a circuit-switched call. However, unlike the latter no actual path is established for the packets. Rather, a relationship is set up between all the packet switches involved in the virtual routeing across the network. In the example of Figure 8.2, during the set-up phase it is determined that all packets sent between terminal x and the server at y will have a virtual routeing following path label 43 from A to D, and identity 25 on route between D and E. Each of the designated packet exchanges, in this case A, D, and E, have the appropriate outgoing route identified in their routeing tables against VP 43 and VP 25, as appropriate. All packets flowing from between x or y with the label *43* in their header at A or D and '25' at D or E will follow this predetermined routeing for as long as the virtual routeing exists. (This is analogous to the sorter at the originating office scribbling '43' or '25' on the envelopes as appropriate and all subsequent sorters merely using this number, rather than reading the full address and looking up the appropriate way to forward the letter, and all subsequent letters would follow the same routeing across the country.)

The paths may be set up for permanent operation – known as 'permanent virtual circuit' (PVC) working; or they may be set-up whenever a data session is required – a service known as 'switched virtual circuit' (SVC). (In the simple explanation above, the term 'path' and 'circuit' are used interchangeably, although

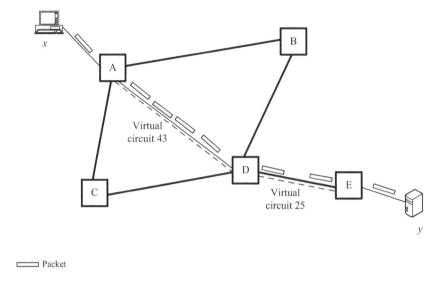

☐ Packet

Figure 8.2   Connection-orientated packet switching

as we see later for some systems, for example, ATM, 'paths', 'channels' and 'circuits' have specific meanings.) The earliest form of public packet data service, known as CCITT X 25 or just 'X25', used the PVC form of connection orientated operation (see Box 8.1). Other notable examples of this form of packet switching are Frame Relay (see Box 8.2) and ATM, which is described in more detail later.

8.3.2   *Connectionless packet mode*

In contrast, the connectionless mode of packet switching treats each packet in a stream completely independently within the network, operating what is known as a 'datagram' service. Figure 8.3 shows a network made up of packet switches, which are normally referred to as 'routers' in this mode of connectionless working, and a terminal x transferring a stream of packets to terminal y. The routers, A to E, handle packets from many sources, with the received packets queuing on their inlet routes. As the first packet from terminal x is received at router A the address contents of the header is examined and the outgoing route, say to router D chosen from a look-up schedule – i.e., the routeing table. In this way, the packet progresses through to Router E. The second and third packet may follow the same routeing, each router examining the full address in the header and making an independent decision on how to route the packet onwards. As the loading on the network changes with time due to the generation of packet flows from other users, the queues on the routers will fill and temporarily become unavailable causing different choices of routeing for subsequent packets from terminal x. Thus, the fourth packet might be routed $A–C–D–E$, and the fifth packet routed $A–B–D–E$. This independent handling of each packet means that the stream of packets between x and y could potentially all follow different routeings and consequently arrive at various times and out of sequence.

Box 8.1   X25 data service

The first connection-orientated public packet data network service was based on the CCITT (now International Telecommunications Union (ITU)) X25 international standard as implemented in many countries during the 1970s. It was introduced at a time when public networks were still based on analogue transmission, the noisiness of which tended to create errors on the X25 data stream, requiring high levels of correction. The X25 packet comprises a 3-byte (24 bits) header containing the virtual circuit number and packet acknowledgement and request sequence numbers, followed by a payload for the carried data. The maximum size of the packet is 128 bytes. Non-X25 terminals need to be connected to converters (packet assembler/disassembler (PAD)) before being linked to the serving X25 packet switch in the network, using a 48 kbit/s or 56 kbit/s modem driven digital stream over analogue and digital private circuits or dial-up over the PSTN. Although still in use in several parts of the world today, X25 has generally been superseded by Frame Relay or Ethernet service [1,2].

Box 8.2   Frame relay

Frame Relay packet networks are designed to exploit the low-noise capability of the digital transmission networks which are now prevalent around the world. The connection-orientated system, therefore, has very little of the error correction overhead of X25 and so can operate at much higher speeds, from 56 kbit/s up to some 45 Mbit/s. A wide range of data stream types can be carried in the variable payload of the frame relay packet, up to a maximum of 4 kbytes. The main application of frame relay network services is to provide interconnection between local area networks (LAN) on corporate sites, research stations and university campuses. The link from the customer sites to the Frame Relay packet switches is usually provided over leased lines, typically at 2 Mbit/s (or 1.5 Mbit/s in North America). The large variation permitted in the packet length and the consequent range in packet transport delays means that there is only a small amount of performance manageability possible with Frame Relay [1,3,4].

The most famous example of connectionless packet working is that of the Internet Protocol (IP), as used in the (public) Internet and corporate private intranets. The other notable example of connectionless working is switched multimegabit data service (SMDS), see Box 8.3.

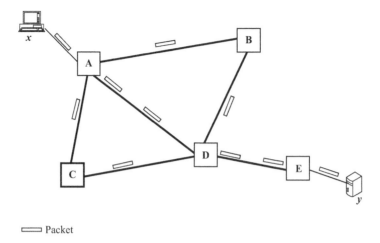

⟨══⟩ Packet

Figure 8.3   Connectionless packet switching

8.3.3   *Comparison of packet switching modes*

It is interesting to compare the merits of the two modes of packet switching. The connection-oriented approach provides a high degree of capacity and performance management of the services running over its switches. This is because the control mechanism will allow new VPs to be established only if there is sufficient capacity through the nominated switches. Connection-orientated packet switching may thus be characterised as incorporating an elaborate control mechanism – seen by some data network designers as a disadvantage – on the one hand, while on the other hand it has the distinct advantage of ensuring predictable managed performance. By comparison, the connectionless mode has relatively simple control, which is robust and able to cope with contention and failure of routers in the network, but it

---

Box 8.3   Switched multimegabit data service

This connectionless service is primarily directed at linking LANs, that is, working as MAN (metropolitan area network) and WAN (wide-area network), respectively. It is a relatively simple system with no provision for error correction and with no single standard version, so interworking between different manufacturers' equipment is difficult. The user data is encapsulated into a packet with some 40 bytes of header containing the destination and source addresses (at Layer Three of OSI (open system interconnect) Model). This packet is then chopped into 53-byte packets or frames for transmission at Layer Two. User Speeds range from 56 kbit/s up to 45 Mbit/s. Although deployment of SMDS is widespread, particularly within Europe, it has generally been overshadowed by the use of Frame Relay, particularly in North America [4].

does not provide any performance prediction or management. Not surprisingly, network operators try to get the best of both worlds by using connectionless packet systems running over connection-orientated networks! Another way of looking at this relationship between the two modes is to refer to the OSI (open system inter-connect) reference model (see Box 8.4) which places the connectionless packet switching at Layer Three and the connection-orientated mode at Layer Two.

Box 8.4   Open system interconnect reference model

The open systems interconnection model (OSI), which was devised by the International Standards Organization (ISO) in 1983, is the widely accepted way of delineating the functions that need to be performed by equipment involved in the communication of data [2,5]. Like many reference models it is based on the concept of layers, each of which performs functions or services for the layer above and is in turn supported by services provided by the layer below. By clearly defining the functions of the layers, many different manu-facturers and programmers can develop terminal equipment, communications equipment and software applications that successfully work together. Inter-actions between functions in the same layer at either end of a communications link are governed by appropriate sets of rules, the so-called protocols.

Figure 8.4 illustrates how the OSI model's seven layers apply to the communication between a data terminal (the client) and the server computer. Working from the bottom up, the functions of each of the layers are sum-marised below:

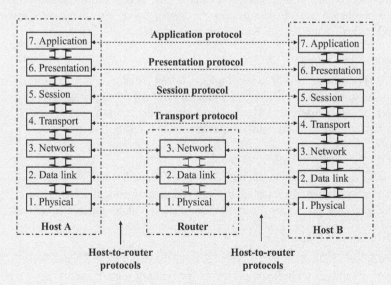

Figure 8.4   The open system interconnect reference model

Layer One: Physical

This layer is concerned with the actual transmission of the binary signal over the wires. Layer One specifications therefore cover the functions of each pin in the plug and sockets between terminals and communications equipment, and the shape and size of the electrical waveform in representing binary 1's and 0's. Actually, the physical transmission media, that is, the optical or metallic cables or radio links, are deemed to reside in a hypothetical Layer Zero outside of the OSI model.

Layer Two: Data Link

The main task of this layer is the safe and reliable transmission of the actual data messages or stream. It therefore concerns the formatting of the data into frames, packets or cells, which are manageable groups of bits which can be identified at the distant end and an acknowledgement returned. The Layer Two specification also includes error detection and flow control.

Layer Three: Network

It is at this level that the routeing over one or more links through switches or routers is managed. The specification includes the naming and addressing of network nodes.

Layer Four: Transport

This layer provides the management of the data communications on an end-to-end basis, masking any transmission actions at the network layer below. Thus, the Transport layer functions include the allocation of the source data into numbered packets, sequencing received packets into the correct order, packet-flow management, and re-transmission requests.

Layer Five: Session

This layer allows two distant computers to run a dialogue for the duration of the 'session'. The specification includes a protocol for ensuring proper hand-over between machines, avoidance of the same function being tackled on both machines simultaneously (known as synchronisation), etc.

Layer Six: Presentation

It is at this layer that the actual content of the information being communicated is formatted in a mutually understandable format for the two ends. The specification covers the syntax, semantics, and coding of the content.

Layer Seven: Application

The actual application being executed over the data link is defined at this top layer. Examples include the protocols for sending e-mails, accessing a web-page over the Internet, data file transfer, etc.

8.4   Asynchronous transfer mode

First, let us deal with the curious name of asynchronous transfer mode (ATM). It was coined in the 1980s by the engineering-standards fraternity, who wanted to differentiate the emerging system from the majority of existing data systems which were synchronous, that is, all actions initiated under the control of pulses from a digital clock, and the sending of idle bits in the absence of data. ATM was developed by the telecommunications industry, led by the network operators, as the universal switching system capable of handling all existing and expected future services, ranging from voice to video and all sorts of data: delay tolerant and intolerant, high and low speed. In fact, ATM was deemed to be the specification for broadband integrated services digital network (ISDN), the then-hoped-for future all-purpose service-switching mechanism for public networks, and the natural successor to the 'narrowband' ISDN which had recently been introduced to most PSTNs (as described in Chapter 6).

The principle of the ATM all-service capability is that an adaptation system allocates the various inputs to a set of fixed-sized 53-byte packets, known as 'ATM cells', as shown in Figure 8.5. As soon as each cell is filled it joins the queue for dispatch; slow speed inputs are carried in relatively few cells per second, while higher-speed inputs are carried at a proportionately higher cell rate. The user-network interface is an ATM multiplexor which performs the adaptation function, feeding the set of user data streams onto the ATM highway – this act of interleaving the set of ATM cells onto the highway is known as 'statistical multiplexing'. These ATM highways are carried through the Access and Core Transmission Networks to the serving ATM switch. Here, the cells are switched onto the VPs carried over the routes to other ATM switches, as described in Figure 8.2. Actually, the VPs are capable of being divided into virtual channels (VCs). Thus, for example a VP

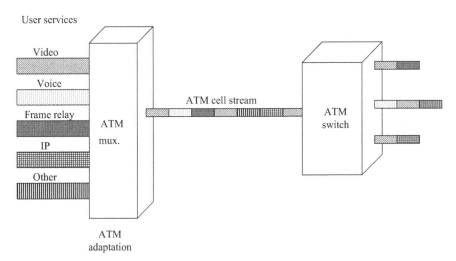

Figure 8.5   The multiservice capability of ATM

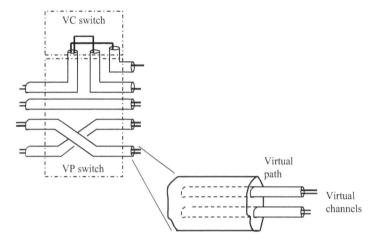

Figure 8.6   Virtual channels, virtual paths and switches

between two separated office buildings could contain several VCs providing connectivity between individual rooms in the two buildings, serving the two sets of accounts departments, financial management, human resources, and so on. An ATM switch therefore requires two switching stages, one for the VPs and a further stage for the VCs, as shown in Figure 8.6.

We can now look at the adaptation process more closely. The wide range of users' data services are divided into five categories for adaptation into payload of the ATM cells. These categories are defined as 'ATM adaptation Layers (AAL) 1–5', as shown in the table of Figure 8.7, which are used by the ATM system to create four classes of service (A to D) for the user. The great strength of ATM is that it the chosen class of service (CoS) is applied by the ATM switches to all cells passing through the network. This ensures that priority is given at the switches to cells carrying delay-intolerant services, for example. At the distant receiving end the ATM multiplexor extracts the data from the cells according to the choice of AAL and reconstitutes the service at the output tributary port to the user. Briefly, the service classes are:

Class A: Constant bit rate (CBR) service, in which the rate is guaranteed to be kept stable, for example, ordinary video or voice.

Class B: Variable bit rate (VBR) service with timing, meaning that although the rate may vary the ATM service will immediately dispatch the cell, for example, VBR video or voice.

Class C: VBR connection-orientated service, in which delay can be tolerated, for example, ordinary data transmission.

Class D: VBR connectionless service, this has the lowest priority within the ATM network, for example, data transmission of short messages (datagram). Network operators use ATM Class D to backhaul the ADSL (broadband) service traffic from the local exchange to the data network.

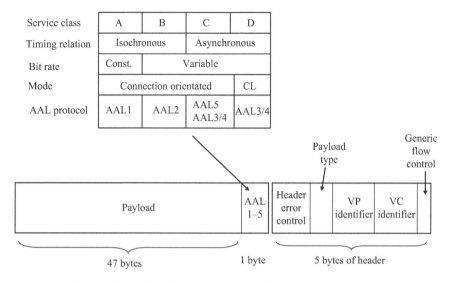

Figure 8.7   ATM service classes, AAL and cell structure

The CoS is indicated to the ATM switches during the VP set-up. The switch-control system will not allow a new VP to be established unless there is sufficient capacity available to guarantee that the requested class of service can be assured. Once the VP is established the switches respond only to the VP and VC 12-bit identifiers in the short header of the ATM cell, as shown in Figure 8.7. The ATM cell header also contains an 8-bit error detection pattern, an indication of whether the cell payload is user's data or messages relating to the VP set-up phase, and an indication whether cells can be jettisoned in the event of switch congestion. The number schemes for ATM networks, used during the set-up phase of the VP, are considered in Chapter 10.

Before we leave our consideration of ATM it is useful to consider how it is used in the operators' networks and private networks (see Figure 8.8). The key point to note is that an operator's ATM network is used to provide ATM service to its business customers, offering an ATM network-user interface (AAL-based) – so-called 'native ATM', as well as using the ATM infrastructure to support (at OSI Layer Two) the conveyance of other services provided by the network operator. Examples of the latter include IP service, video and TV links, broadband access, and voice. This use of the ATM infrastructure enables network operators to manage the transmission capacity within the Access and Core Transmission Networks through the setting up of VPs between nodes so that the various services carried receive the appropriate quality of service (QoS). It also represents an economical way of partitioning the transmission capacity of the networks. We will look at 'IP over ATM' later in this chapter. Business customers also use ATM within their corporate networks to provide managed data links between their office, research and factory sites.

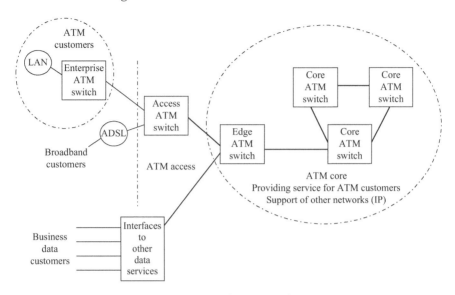

Figure 8.8   Applications of ATM

8.5   Internet protocol

The 'Internet protocol' (IP), as its name suggests, is the basic protocol used on the Internet (at Layer Three of the OSI model). However, the tremendously widespread adoption of this connectionless packet standard is due also to its applicability to so many situations. IP forms a simple means of transferring data between users on different types of computer and devices; it forms the common interface to many applications run on computers and, last but not least, it provides the connectivity infrastructure for corporate data networks (i.e. intranets). As we shall discuss later in this and subsequent chapters, IP-based technology can form the next generation of telecommunication networks. The operation of the IP service consists of the transfer of a users' data by IP packets, each of which is managed individually and passed progressively from router to router, using an IP addressing scheme, until the destination terminal or computer host is reached. There are currently two versions of IP being used in networks: Version 4 (IPv4) is the original and still most widely used, while Version 6 (IPv6) is being introduced as a replacement to take advantage of its greater addressing capacity and improved facilities [6]. The addressing aspects of both IP versions are considered in Chapter 10.

The IPv4 packet structure comprises a fixed header of 20 bytes and a variable payload of a maximum of 64 kbytes (although in practice this is normally limited to 1.5 kbytes bytes to avoid congestion within the network), as shown in Figure 8.9. The header contains the 32-bit source and destination addresses, as well as an indication of the length of the overall IP packet. Other fields in the header provide a checksum (for the header contents only), the version of IP being used, and a simple

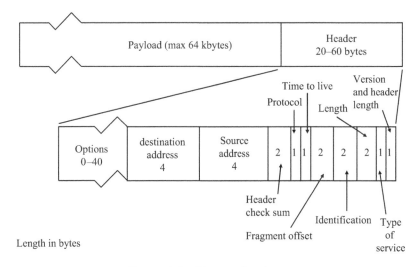

Figure 8.9   IPv4 packet format

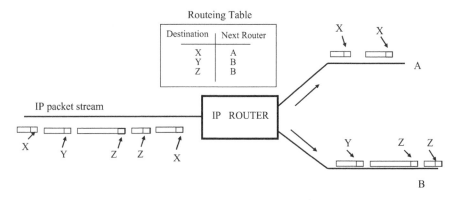

Figure 8.10   IP packet routeing

type of service indicator. An intriguing field called 'time to live' is used to set the maximum number of hops between routers before the packet is discarded; this is necessary to prevent the clogging of the routers by unsuccessfully delivered packets continuously circulating around the IP network.

As with any communication networks, there are two basic nodal functions in public and private IP networks: the edge routers supporting several access lines from the data users, and a set of core routers (usually of high capacity) providing the interconnectivity between edge routers and any other IP networks. The role of an IP router is shown in Figure 8.10. Packets arriving at the input ports of the router are queued, normally on a first-in first-out basis, and the header of each is examined in turn by the control system. A first-choice outgoing route to the next router is obtained from a simple look-up in the routeing table, using the appropriate part of the destination IP address. The packets then join appropriate queues associated with the output ports.

There are several ways in which the routeing table is generated and updated. Routers can discover the status of all the other routers in their network, that is, whether they are experiencing long queues, are out of service due to faults, and whether new routers have come on stream. Discovery protocols control the investigative stream of IP packets sent periodically by routers to each other (e.g. open shortest path first (OSPF)). In this way, a router-control system can build up a picture of the best way to route to particular destination routers by taking the least number of hops or travelling over on the least distance – the latter being determined by timing the round-trip of an exploratory packet. Routeing tables are automatically modified several times a day, reacting to changing circumstances in the IP network [7].

The terminals and hosts using an IP network may be part of a localised data network, that is, a subnetwork, within the workplace or even a residence (LANs are discussed later). There is therefore a need for an incoming IP packet addressed to the subnetwork to be associated with the appropriate terminal or host machine. This is done using one of several available protocols, for example, reverse address resolution protocol (RARP) and dynamic host configuration protocol (DHCP) [8,9].

A network of IP routers owned by an organisation is usually formed into a group known as an autonomous system (AS) for the purposes of management. So, for example an AS owned by a corporation may have a gateway to other IP networks, but it needs to limit the extent of users on other networks gaining access to the corporate network. Therefore, there are two algorithms required to determine routeings: one for routeing within the AS – known as interior gateway protocol (IGP) and for routeing between one AS and another – known as exterior gateway protocol (EGP). Figure 8.11 illustrates the relationships between IP autonomous domains.

In its raw state, IP does not actually provide a useful data conveyance service. Some additional functionality is needed at the transport layer (i.e. Layer 4 in the OSI 7-Layer Reference model) to manage the flow of IP packets. The two main contenders for this role are the Transmission Control Protocol (TCP) and the User Datagram Protocol (UDP). Figure 8.12 illustrates the role of TCP, which acts as a connection-orientated protocol between the sender and receiver endpoints (terminals or host machines) at Layer 4. None of the routers in the IP network are aware of the TCP activity. The main function of TCP is the numbering of IP packets, 'datagrams', at the sending end and the examination of these numbers at the receiving end – and requesting the retransmission of any packets not arriving within a specified limit. This number sequence is also used by the TCP facility at the receiving end to assemble received packets into the correct order. Finally, TCP provides a monitoring of the rate of flow of packets getting through the IP network and consequently adjusts the launch rate of IP packets. The TCP envelopes

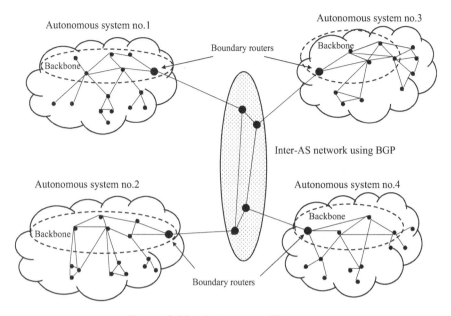

Figure 8.11   Autonomous IP systems

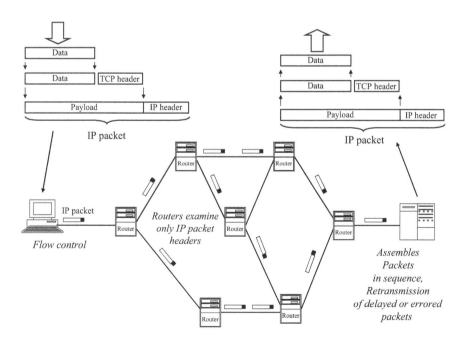

Figure 8.12   Transmission control protocol

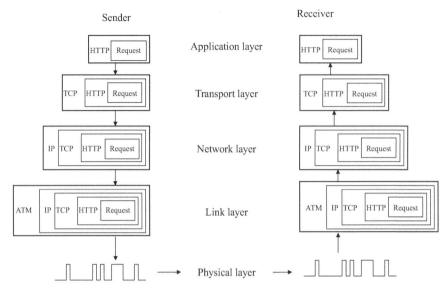

Figure 8.13   Message transmission using layers

the source data by a 20-bit header and the composite is then carried by the IP packet as a payload.

The UDP operates in a connectionless mode, as a datagram, providing some similar facilities to TCP but with the important difference that no re-transmissions of lost or errored IP packets is requested. UDP therefore does not incur the delays due to re-transmissions, making it suitable for real-time applications such as voice over IP.

It is instructive at this stage to try to picture all enveloping of packets within packets that occurs in an IP network. The example given in Figure 8.13 is for a request to see a web page using hypertext transfer protocol (HTTP) – this is the actual data packet coming from the application. This user data is then encapsulated in the TCP packet, which in turn is deposited into the payload of the IP packet. This may not be a one-to-one mapping since the TCP packet could be spread over several IP packets. In this example, we assume that the IP is carried over ATM, as described later in this chapter. Because of the small fixed size of the ATM cells, IP packets are typically chopped into several chunks of 53 bytes for transmission over an ATM VP. This stream of ATM cells may then be deposited into a virtual container within an SDH frame for transmission over the Core Transmission Network! At the distant end, in our example the web page host machine, the reverse process takes place with each of the packet/cells being extracted at each level, as shown in Figure 8.13.

One way of trying to make sense of all the many protocols involved in IP networks is the use of an architectural diagram, as shown in Figure 8.14. This shows all the appropriate protocols mapped against the seven layers of the OSI

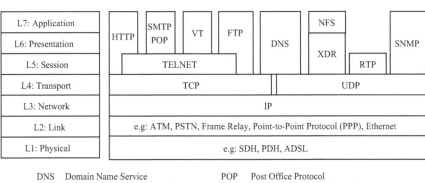

OSI   Internet

L7: Application							NFS		
L6: Presentation	HTTP	SMTP POP	VT	FTP		DNS		XDR	SNMP
L5: Session		TELNET						RTP	
L4: Transport			TCP				UDP		
L3: Network				IP					
L2: Link			e.g: ATM, PSTN, Frame Relay, Point-to-Point Protocol (PPP), Ethernet						
L1: Physical				e.g: SDH, PDH, ADSL					

DNS	Domain Name Service	POP	Post Office Protocol
HTTP	Hyper Text Transfer Protocol	RTP	Real-time Transport Protocol
FTP	File Transfer Protocol	SMTP	Simple Mail Transfer Protocol
IP	Internet Protocol	SNMP	Simple Network Management Protocol
TCP	Transmission Control Protocol	UDP	User Datagram Protocol
VT	Virtual Terminal	NFS	Network File System
XDR	eXternal Data Representation	OSI	Open Systems Interconnection

Figure 8.14   The Internet and OSI protocol reference models

Reference Model. At the physical layer are all the various transmission systems in the Access and Core Transmission Networks. At Layer 2, there is an array of data link systems, including ATM, Frame Relay and Ethernet LAN. The PSTN is also shown as a possible link mechanism because in the case of dial-up access to the Internet the switch path through the PSTN acts as a simple data link between the caller's computer and the ISP. (It should be noted that the PSTN is, of course, operating at Layer 3 while the call is being set-up.) IP resides at Layer 3. At Layer 4 the two transport protocols, TCP and UDP, support the latency tolerant and latency intolerant applications, respectively, as indicated in this architectural diagram of Figure 8.14. There is a vast array of protocols covering the many possible applications that may be run over an IP network – Internet or intranet – or an IP-based computer system, only a selection of which are shown.

8.6   The Internet

8.6.1   The Internet as a utility

Chapter 2 introduced the concept of the Internet, describing it as a constellation of (IP) networks. There is a vast range of these IP networks, some of which form an entire AS, others may be small parts of an AS. Unlike the public telecommunication networks, which are designed to international standards recommended by the formalised ITU, the design of the Internet is based on a consortium of users, academics and data equipment manufacturers known as the Internet Engineering Task Force (IETF). The latter have set recommendations for numbering, routeing, and all the various protocols at OSI Reference Model Layers 2–7, – although the IETF considers Layers 5–7 as a single entity for convenience.

It should be appreciated that the Internet has gone through rapid development: originally a US defence department project (ARPANET) conceived in the mid-1960s, going international with the first non-US node in University College London in 1973, with the addition of TCP in 1983, and becoming a widespread information-sharing network, mainly used by universities and research institutes around the World. However, it was the development of the HTTP, which allowed information in many forms to be easily shared, that led to the introduction of the now famous World Wide Web (WWW) in 1991 [10]. Undoubtedly, it was the utility of the Web in allowing access to information – text, pictures, sound and moving image (video) – resident on distant computer storage facilities that has made the Internet the phenomenon it is today. Without the Web, it seems improbable that the original academic and specialist Internet would have been adopted by the general populous as both a work and domestic facility. In fact, it was the sheer size of the problem of being aware of and gaining access to the enormous wealth of information stored on servers at libraries, business and research organisations, government agencies, all manner of commercial enterprises, etc., that led to the introduction of search engines. These have not only eased the problem of finding information, but have also provided commercial opportunities for companies providing an entry into the Internet/www, the so-called 'portals'. Of course, the biggest change to daily life has been caused by the wide-scale adoption of e-mail, beginning in the mid-1990s and now forming one of the major applications run over the Internet.

At the turn of the century use of the Internet took another leap forward as several mechanisms for ensuring security of credit card information were introduced. Customers were then able to make purchases over the Web: undertaking all the stages of perusing the catalogues, placing an order, paying, and tracking the progress of delivery of the goods. This, together with other developments, such as electronic signatures and public key encryption, has given rise to the derivative phenomenon of electronic commerce, better known as 'e-commerce'.

It must be emphasised that all the developments associated with IP, both technical and commercial, have two distinct separate, but related, areas of application: the public Internet and the private intranets. The same technology is used in both areas, but the control of who uses the facility and the commercial implications are significantly different. In the case of the Internet there is freedom of access to all the subscribers of ISPs and the loading on the routers and links can vary enormously, giving rise to variable and unpredictable performance (download times, etc.). However, the biggest problem with the open-to-all policy of the Internet is its vulnerability to malicious users sending e-mails with attachments that inject corrupting software into the recipient computers, i.e. a virus attack. When directed at major ISPs this can cause the complete locking-up of the routers causing a widespread failure in the Internet. On the other hand, the intranets are self-contained IP networks, using all the IETF standards for numbering, etc., in which all the users are members of a commercial or government organisation located at their offices, factories, laboratories or academic institutes. The performance of an intranet can be controlled by the organisation's communication department, since the network can be appropriately dimensioned for the well-defined population of users. Organisations create a protected environment for the intranet allowing entry or 'ingress' from the Internet only via a firewall – the latter being a server that critically examines every IP packet to apply an admission policy, rejecting those packets that do not conform.

8.6.2 Routeing through the Internet

As with all networks, the nodal and endpoints within the Internet (or intranet) need to be identified by an address or number in order to route the IP packets accordingly. The identification of either an e-mail subscriber or a web page is made using a human-friendly name (e.g. 'name@company.co.uk'). This name needs to be translated into a destination address that can be placed into the header of the IP packets. As a simplified example of this process we will consider the routeing of an e-mail across the Internet between two corporate intranets, as shown in Figure 8.15. The request to send an e-mail is detected by the server in the Local Network 1 (an intranet) and a brief exchange of IP packets takes place with the serving ISP, which request the address of the e-mail. The ISP router then needs to pass an enquiry IP packet to the nearest data base of names to seek the destination address. This data base is known as a domain name server (DNS) and there is a DNS protocol (shown in Figure 8.14) used for this interrogation. Once the

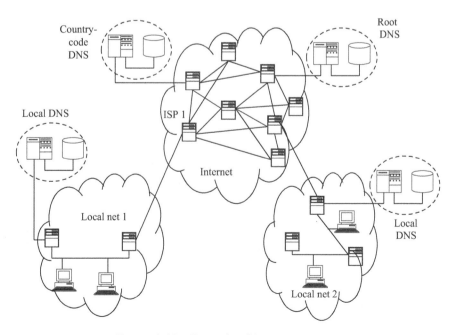

Figure 8.15   Example of Internet routeing

originating computer has received a packet containing the Internet address for the e-mail recipient it can then use that address in all the IP packets conveying the e-mail contents. Each of these packets is then routed through the Internet from ISP-1's router to the destination ISP router, and thence the recipient's computer on Local Network 2. (The role of DNS and IP numbering and addressing is considered more fully in Chapter 10.)

8.6.3   The Internet of things

The idea of device-to-device data communication has been well established for several years, especially in applications using radio-frequency identification (RFID) systems with bar codes [11]. However, the continual decrease in the cost of computing and sensors, as well as people's increasing acceptance and reliance on the Internet, gave rise around 2015 to the interesting concept of an 'Internet of Things' (IoT). There is a wide range of IoT applications exploiting the concept, covering smart living, manufacturing, transport, utilities and energy, and smart cities [11,12]. One smart living example is the wearing of IoT devices such as blood-pressure and pulse-rate monitors which automatically communicate with fitness-monitoring applications run on a smartphone or laptop/tablet, with status and alarms being sent to a local gym or doctors' surgery, as appropriate. Another often quoted example is the household's rubbish bin detecting each empty tin of baked beans, beer bottles, etc., being thrown away and the system automatically ordering additional supplies during the next online shopping session.

The significant enabler of IoT is the increasing population of so-called smart appliances, for example, web-enabled TVs, washing machines, cookers and domestic heating systems. Therefore, several low power wide area (LPWA) low data-rate wireless systems has been developed to enable connectivity between local devices, say within a home on a device-to-device basis, as well as ingress or egress to the Internet over the broadband hub of the fixed network or via a mobile handset [13]. (These wireless systems are briefly reviewed in Chapter 9.) Therefore the IoT is often operating just within premises, although there may be extension over the Internet to other sites or to give remote access to users.

Based on this description, we can deduce that an ecosystem for the IoT comprises of a set of sensors attached to the 'things', LPWA wireless communications, and some application-control software and local data storage. However, there is a further important aspect of the IoT, namely the analysis of all the data that accumulates from the devices and IoT applications. Given that on a domestic level there will be thousands of pieces of data generated by each household per month – which when gathered and analysed with the many other households generate 'Big Data' [14]. Companies or government agencies analysing this big data, even at the meta level (i.e. where the data is anonymous) can produce valuable marketing or security intelligence. In the business world, IoT provides the ability to track goods being transported globally, or monitor the flows on production lines in factories; and here, also, analysis of the big data gathered has value, for example, influencing forward planning, and real-time resource management.

8.7   Voice-over-IP

8.7.1   The concept of VoIP

The multi-service capability of packet networks has been exploited for the conveyance of voice for many years. Some frame-relay corporate networks were carrying internal voice calls (VoFR) as early as the 1980s [1]. Such deployments were small scale and specialised, with the poor quality being acceptable for corporate private networks because of the cost savings. More recently, voice traffic has been handled by ATM switches (VoATM), achieving much higher levels of quality, but still applied only to special situations. Then in the early 2000s British Telecommunication (BT) replaced many of their trunk exchanges with a hybrid system using circuit switching and ATM to create a 'next-generation switch' (NGS) [15]. However, we will now concentrate on voice-over-IP (VoIP) because this has now become the generally accepted successor to the circuit switching of voice.

Actually, there is a range of scenarios in which voice can be carried over an IP network, as shown in Figure 8.16. The earliest application was the use of a telephone handset, earphones and microphone plugged into a computer connected via the Internet to a colleague using a similar arrangement – 'Internet Telephony', shown as A-to-B in Figure 8.16. This arrangement requires both parties to be on line to the Internet simultaneously and for an e-mail pathway to be established for a conversation to occur. The quality is highly dependent on the instantaneous loading of the Internet and can vary from appalling to adequate. However, recent developments have led to a new generation of A-to-B VoIP applications (e.g. *Skype*) which have much improved quality because of the use of broadband access – and are a widely popular form of cheap (or even free) public telephony, as

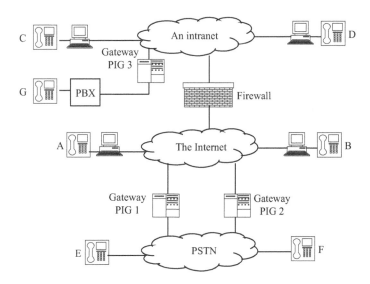

Figure 8.16   Voice-over-IP scenarios

described later. A corresponding arrangement applies to the corporate network situation (intranet telephony), in which the computer-handset terminals are connected via the company intranet – shown as C-to-D in Figure 8.16. This application has proved to be increasingly acceptable because the quality can be managed by appropriate dimensioning of the intranet, and there is opportunity for the voice and data infrastructure within office blocks to be combined, resulting in only one set of internal office wiring and the elimination of the (circuit-switched) PBX (private branch exchanges) [16]. Calls can also pass between the users on the intranet and the Internet, for example, C-to-B in Figure 8.16, with the limitation that the IP packets must be vetted by the corporate firewall.

However, the biggest opportunity for VoIP is the substitution of some or all of the PSTN by an IP alternative. This is achieved using a gateway, known as a PIG (PSTN-to-Internet gateway). Referring to Figure 8.16, calls from Subscriber E are switched in the PSTN as far as the PIG 1, where the call then progresses over the Internet to the exit PIG 2 where it terminates in the PSTN and subscriber F's line. There are several issues to bring out here.

(i)   The choice for use of the Internet for some of the call is made by Subscriber E dialling to an ISP and subscribing to a VoIP service.

(ii) The Internet routeing could bypass the long-distance part of a PSTN call routeing, or even one or more international routes around the world if subscriber F is in a foreign country. Since no charge is made for the use of the Internet, Subscriber E can enjoy a considerable saving in the total call charges.

(iii) The quality of the overall call connection will be dictated by the Internet portion, whose performance depends very much on its instantaneous loading.

(iv) The scenario could be extended to allow calls to be routed via the Internet to an intranet, enabling E-to-C communication, for example.

An alternative scenario for E-to-F communication is for one or more of the exchanges in the PSTN to be replaced by VoIP equipment, with an intranet rather than the Internet used to provide the IP networking between PIGs. In this situation, unlike the above examples, the subscribers may not be aware that their call is being handled fully or partially by VoIP equipment, with the tariff and quality appearing as a standard telephony offering.

Figure 8.17 illustrates how a call is set up between two PIGs following this alternative scenario. There are two links between the 'egress' (i.e. exit gateway) exchange and associated PIG: a signalling link carrying SS7/ISUP (see Chapter 7) which terminates on the SS7 signalling gateway (SG); and the 30 channels of PCM speech which terminate on the media gateway (MG). Now we can consider how a connectionless IP network is manipulated, or as the Internet fraternity say 'spoofed', into establishing a telephone call connection.

(i)   The call set-up is initiated by an initial address message (IAM) on the SS7 signalling link to the SG in PIG 1. This generates a request message from the SG to the media gateway controller (MGC).

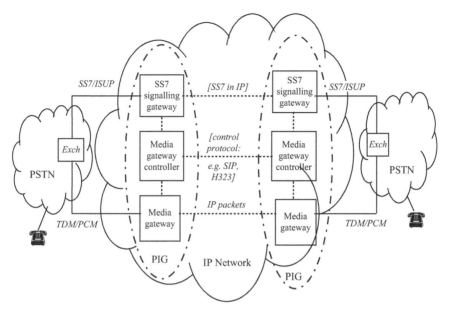

Figure 8.17   PSTN-to-IP Gateway (PIG)

(ii)   The MGC now has to examine the PSTN number of the called subscriber and determine the IP address of the appropriate egress PIG to serve this number. This would normally be done on the basis of the trunk or area code for a national call, and the country code for an international call. A collated data base would normally hold this information.

(iii)   The MGC now passes the IP address of the destination PIG to the MG, indicating the channel identity of the calling speech circuit.

(iv)   The MGC sends a message in an IP packet to the egress IP's MGC requesting the call set and indicating the destination PSTN number.

(v)   The egress MGC also instructs the SG to send an SS7 IAM to the associated PSTN exchange. A sequence of IAMs now progresses through the PSTN to set up the call. Once the 'address complete' SS7 message is received (indicating calling subscriber is receiving ringing tone) the SG advises its MGC of the call status.

(vi)   The egress MGC now sends an IP packet to the ingress MGC to advise that the speech reverse path should now be established over the intranet.

(vii) The egress MG now extracts each PCM byte from the called channel (the speech is already in digital format) and deposits them into the payload of successive IP packets, setting the ingress PIG address as the packets' destinations. A speech path now exists over the IP network (and the PSTNs) between called and calling subscriber and the latter can hear the ringing tone generated by the destination PSTN exchange.

(viii) Once the answer SS7 message is received by the egress SG, the speech path from the Ingress MG to the egress MG can be established – each PCM byte being deposited in the payload of subsequent IP packets addressed to the egress MG.

(ix)   Any SS7 messages that need to be passed between the PSTN without interrogation by the PIGs are passed in IP packets directly between the ingress and egress SGs.

Figure 8.17 also shows some of the protocols that are used between PIGs. The speech is carried in IP packets associated with UDP at the Layer 4 and real-time protocol (RTP) at Layer 5 to minimise transmission delay and provide as high a quality of perceived performance as possible. Control messages between the MGs are sent using SIP or H323 (see Chapter 7) or with the new MGC-to-MGC protocol. MGC-to-MG messages may use Megaco (MG control) or MG control protocol (MGCP) [17].

Consideration of the elements of Figure 8.17 will reveal that the switching of voice calls over an intranet or the Internet – using MGs, SGs, and controllers – is achieved without any switch-blocks! In effect, the IP network is performing the switch-block function. The essence of the call is really created by the software on the MGCs. For this reason, MGCs are often referred to as 'soft switches'.

8.7.2   VoIP over broadband

Probably the most profound development affecting the adoption of VoIP has been the introduction of broadband access over ADSL and VDSL for the residential and small business customers. This has given a cheap facility for carrying data from homes and offices, which not only provides wide bandwidth (typically up 2–10 Mbit/s downstream) but also an 'always on' capability. Since VoIP enables voice to be treated as data over an IP network, the broadband access facility can also be used for voice calls. This gives rise to the so-called 'VoIP over broadband'. There are several companies providing public telephony services using VoIP over broadband technology, each using their own proprietary designs – so full inter-working between the systems is difficult. Examples of VoIP companies include 'Vonage' and 'Skype'.

Overall, the VoIP over broadband services perform well, often with exceptionally good sound quality, as a result of using high-speed codecs which convert the voice to IP packets without needing the normal 4 kHz frequency filtering and the wider bandwidths of broadband compared to the 'narrow band' (i.e. 4 kHz and 64 kbit/s) PSTN. This high sound quality is, of course subject to the loading on the Internet, which if high can lead to loss of packets and hence degradation to the sound quality. There are also limitations to such services because they are essentially data, rather than switched-voice based. For example, calls are possible over most of the systems only when both subscribers have their computers on line (i.e. connected to the Internet), so calls must be on the basis of pre-booking a time or waiting until the other party comes on line. Nonetheless, since the broadband access to the ISP and hence the Internet usage is usually on a subscription and not usage basis, the voice 'calls' using VoIP over broadband are cheap or at no extra cost. For many people, such services are used in preference to the PSTN-based telephony.

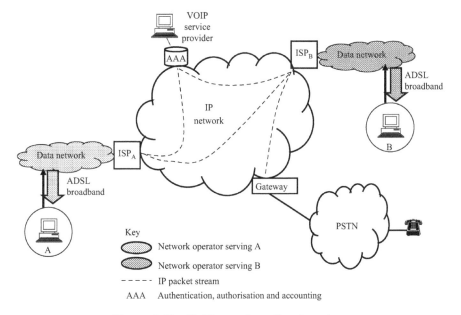

Figure 8.18   VoIP over broadband services

Figure 8.18 illustrates the principle of VoIP-over-broadband telephony. First, we consider the case where both subscribers, A and B, have their computers connected to a broadband Internet access services, using xDSL as described in Chapters 4 and 5. Thus, there is a data path established by the serving network operators from subscribers A and B to their respective ISPs. (This is described in the next section.) In the simplest form of service, the subscribers' terminals comprise computers with telephone headsets. Voice communication is then essentially provided by stream of VoIP packets between the computers of subscriber A and B, flowing over the path set up by the broadband connection at both ends and the connection to the Internet provided by the two ISPs, as shown in Figure 8.18. Two actions are required before this flow of IP packets can be initiated: firstly, both subscribers must have logged on to the VoIP service when they went 'online' (i.e. switched on their computer and logged on to the ISP) and secondly, the calling subscriber needs to be made aware that the called subscriber is currently on line. The latter may be achieved using the 'presence' indications available from instant messaging facility, that is, the fact that subscriber B is on line is indicated to all subscribers of the VoIP service, or at least those that have subscriber B in their personal VoIP-service address book. Setting up the call between A and B is then a matter of setting the appropriate IP addresses in the IP packet headers and a simple interchange of 'ring' and 'answer' message packets.

This direct data communication between A and B is often referred to as 'peer-to-peer' working, since there is no intermediate server involved in setting up the path. (*Note*: not all VoIP over broadband services work on a peer-to-peer basis.)

The initial registration of subscribers to the service is managed by an authentication, authorisation, and accounting (AAA) system managed by the VoIP service provider. This system monitors the status of each subscriber (i.e. whether they are currently online and logged on) and operates the presence indications, as well as providing accounting information for billing, if required. Figure 8.18 shows the IP packet paths for subscribers A and B logging on to the AAA system.

In the case where a call is to be made to a subscriber on another network, for example, the circuit-switched PSTNs or mobile networks, the IP packet stream needs to be routed to a gateway (i.e. a PIG), as shown in Figure 8.18. Such 'break out' calls are usually charged for by the VoIP service provider to compensate for the interconnect charges they incur from the PSTN, etc., operators. Similarly, calls from the PSTN or other networks are routed to the VoIP service provider's gateway so that an IP stream can be established to the called subscriber, using the principles described above for the PIG.

8.8   Network aspects: IP over ATM

Public networks need to provide connectivity for a wide range of services – switched voice and data, and un-switched leased lines. In practice, this means that capacity within the Access and Core Transmission Networks has to be partitioned so each service gets the bandwidth and throughput that it needs. Network operators therefore use a combination of ATM and IP, as well as synchronous digital hierarchy (SDH) (and perhaps plesiochronous digital hierarchy (PDH)), to carry the range of data services and, increasingly, voice services as well.

In the Access Network ADSL is carried over a single copper pair to provide a both-way broadband data circuit with the DSLAM located at the local exchange, as described in Chapter 5. In the case of VDSL and G.fast the DSLAM is located at a street or pole/DP, respectively with optical fibre back to the exchange building. The network configuration to support xDSL lines is shown in Figure 8.19. Network economies are achieved by introducing contention at the aggregation point, as provided by the DSLAM. Typical contention ratios are 20:1, 25:1 and 50:1 (compared with the 10:1 traffic concentration ratio at the subscriber concentrator switch-block in a digital local exchange). This contention is provided by an ATM switch within the DSLAM, in which the maximum cell arrival rate at the inputs from all the subscriber lines, is proportionately greater than the cell capacity on the output highway. Cells from the subscribers are queued, and if necessary discarded, to match the output cell rate. In practice, the subscriber lines are relatively lightly loaded with only a small proportion active at any one time so the contention at the DSLAM, causing a drop in throughput, is not normally perceived by subscribers. The ATM path is established from the ADSL termination on the subscriber's premises where ATM adaptation is performed through to the DSLAM input. The ATM cells are then switched on to VPs from the DSLAM and carried over SDH transmission links to the first ATM switch. Here the network operator is

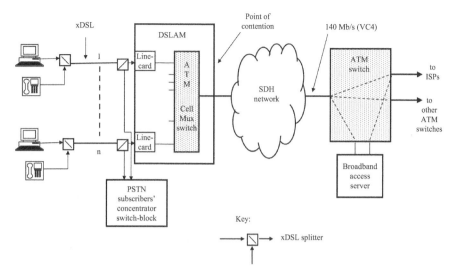

Figure 8.19   IP over ATM using xDSL access

able to intercept the cell flow to monitor and authorise the subscriber's use of the broadband service. Cells are then switched onto VPs directly to one of the many ISPs or on via other ATM switches to the serving ISP. The IP packet flow is, thus: between the computer on the subscriber's premises – over the ATM paths provided by the ADSL, through the DSLAM, the ATM switches – and the ISP. The ATM VPs marshal the flow of IP packets and establish manageable permanent paths from the many hundreds of xDSL aggregation points to the, typically, tens to a few hundreds of ISPs serving a country.

The IP-over-ATM is also used as the core infrastructure of a so-called multiservice platform (MSP), as implemented by many operators to provide their range of different data services economically. The concept is shown in Figure 8.20. The objective of an MSP is to use the multiservice capability of an IP-over-ATM network as the core supporting all the services, with equipment at the edges of the network to provide the individual services to the subscribers, presenting the appropriate protocols, interfaces and data rates. In this way, the aggregated traffic from the many access lines is carried over the MSP core. Any temporary or long-term variation in demand on an individual service (above or below forecast) is absorbed by the aggregated core capacity. A further advantage for a network operator is that the MSP should also be able to provide complete cross-country connectivity for any new low-volume data services at marginal cost. (The important concept of an MSP is covered in more detail in [18].)

The use of ATM in MSPs is progressively being replaced by Ethernet in the access portion and MPLS within the core, as described in the following sections.

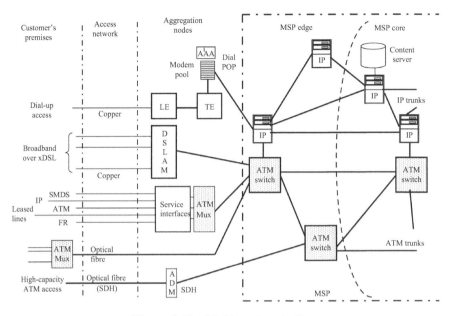

Figure 8.20   Multiservice platform

8.9   Multi-protocol label switching

Interestingly, the merits of ATM in providing a connection-orientated Layer 2 to support the connectionless IP packets at Layer 3 have been recognised by the Internet fraternity (IETF, etc.) in so far as they have invented a more IP-friendly substitute called multi-protocol label switching (MPLS). Essentially, MPLS uses labels to establish a VP for the IP packets to follow. Unlike ATM, however, MPLS does not chop up the IP packets into 48-byte chunks (hence the epithet of 'Data shredder from hell' used by IP advocates to describe ATM), rather it envelops the complete IP packet. The MPLS header, which is attached to the front of each IP packet, comprises the 20-bit label, a 3-bit class of service indicator, a 1-bit bottom-of-stack indicator and an 8-bit time-to-live indicator (a total of 4 bytes).

An example of the use of the MPLS forwarding of IP packets is given in Figure 8.21. The IP routers enhanced with the MPLS capability form domains within an IP network. The routers at the edge of the domains are known as label edge routers (LERs), while the routers within the domain are called label switch routers (LSRs). The required VPs, known as 'switched label paths' (SLP), are established using a set-up procedure similar to that used with ATM. Figure 8.21 shows the label switch routeing tables associated with three of the MPLS-enabled routers. There are two SLPs associated with the LER-1: packets destined for IP address *127.85.13.4* are switched to outgoing port 1 and assigned a label of *20*; IP packets destined for *205.76.23.1* are also switched to outgoing port 1 but assigned a label of *10*. All IP packets have the appropriate MPLS header (with label) added. At the second router, which is an LSR,

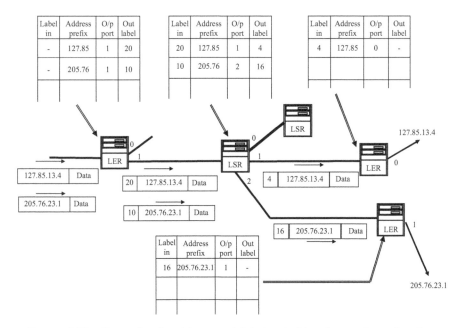

Figure 8.21 Example of multiprotocol label switching forwarding of packets

all MPLS encased packets with label *20* are given a new MPLS header with a label *4* and sent on output port 0. Similarly, packets with label *10* are re-labelled *16* and sent on output port 1. In this way, the various streams of IP packets are quickly routed through the network following the predetermined virtual path (SLP).

MPLS offers the network operator many advantages over simple IP, also called 'native IP', namely:

(i) Quality of service:
   - Segregation of QOS classes can be applied by the routers in the network by reference to the service classification in the MPLS headers.
   - The SLP set-up is made with reference to the required QOS.
   - Prioritisation is applied at each MPLS router based on class-of-service indicator.
   - Selective resilience can be applied in an IP network, through identifying the class of service of each packet.
(ii) Faster routeing:
   - The use of simple look-up tables based on the reading of labels in the MPLS header is a faster process than having to open up the IP packet and examine the full IP address. This is because label switching can be performed using hardware, whereas normal IP routeing involves the slower processing of software to deconstruct and decode the IP address. Indeed, it was through the development of higher-speed IP routers that the concept of MPLS evolved.

(iii) Traffic engineering:
- The use of explicit routes, that is, SLP, for the conveyance of IP packets means that the capacity of the network can be dimensioned and used predictably.
- Resources (i.e. capacity in the routers) can be reserved using the SLP set-up process.
- Rules, such as constraint-based routeing, can be applied to the establishment of SLPs.
- Overall, traffic engineering similar to that available with circuit-switched networks is possible.

(iv) Virtual Private Networks (VPN)
- Isolation of closed-user groups within a public IP network is provided by the SLPs, so enabling the establishment of a series of VPNs to be maintained. Figure 8.22 shows an MPLS enhanced IP network supporting VPNs for companies A, B, and C. These are usually referred to as 'IPVPNs'.
- Accountability is possible, since the IP packets relating to individual VPNs are segregated and easily monitored.

Network operators offer MPLS as a premium service to data customers, who in turn use it to create quality managed corporate networks (intranets, LANS, etc.). MPLS is also used by network operators within the public network to provide IPVPNs for corporate customers (as described above). Finally, the quality assurance of MPLS makes it an essential component for VoIP networks. As Chapter 11 discusses, IP with MPLS forms the basis of next-generation networks (NGNs), whose VoIP capabilities enable them to replace the existing PSTNs.

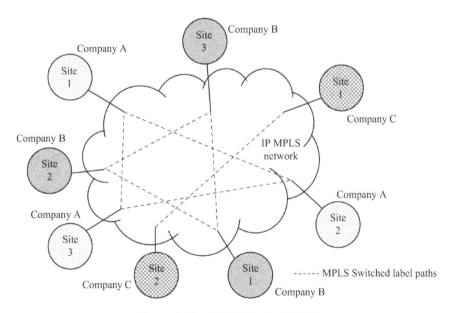

Figure 8.22   IP VPN using MPLS

8.10 Local area networks

8.10.1 The LAN concept

Rather than considering the external or public network aspects of data transport, we now address the way that data is carried between the terminals on people's desks and the computers, printers, servers, etc., in the workplace (e.g. office, campus, research centre and factory). The term 'local area network (LAN)' is used to describe the data networks which provide this facility. It is useful to consider briefly how LANs work because this helps an understanding of how customers generate the need for data services from a public network operator. Also, the Ethernet LAN specification is used as a standard interface between user's data terminals and a network operator's data service.

The principle of a LAN is that it uses a common highway or 'bus' (short for omnibus – meaning 'for all') to link all the devices. As with any network system, LANs need an addressing scheme to identify each termination on the bus, and a mechanism to allow the terminations to use the transmission capabilities of the bus one at a time. The LAN may be physically constructed using twisted pair cable, coaxial cable or optical fibre cable and configured as a single bus or as a set of buses in a tree, ring, or star topology. In practice, clusters of terminals are connected in star formation onto a hub device where they terminate on a common bus, as shown in Figure 8.23. In the case of many terminals, additional hubs may be added and interconnected via a LAN switch.

Several standard types of LAN have been developed and deployed since the first was introduced in 1976, namely: Ethernet, Token Ring and Token Bus [2].

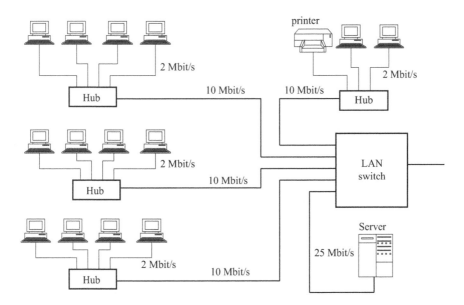

Figure 8.23   Local area network concept

Box 8.5   Institution of Electrical and Electronic Engineers 802 range of standards

The US-based Institution of Electrical and Electronic Engineers (IEEE) is the largest professional group in the world. Amongst its many activities is that of a standards group 802 addressing a range of LANs and associated systems. Many people are now aware of particular IEEE 802 standards, for example, '802.3' and '802.11', but there is a long list of outputs from the group, as shown below:

802.1 Overview and architecture of LANs
802.2 Logical link control
802.3 Ethernet LAN
802.4 Token Bus LAN
802.5 Token Ring LAN
802.6 DQDB (dual queue dual bus, an early form of MAN)
802.7 Technical advisory group on broadband technologies
802.8 Technical advisory group on optical fibre technologies
802.9 Isochronous LANs
802.10 Virtual LANs and security
802.11 Wireless LANs
802.12 Demand priority (proprietary LAN)
802.13 Spare
802.14 Cable modems (now defunct)
802.15 Personal Area networks, for example, Bluetooth
802.16 Broadband wireless (wireless MAN)
802.17 Resilient packet ring

Box 8.5 lists the range of LAN standards within the IEEE 802 series. By far the most popular and successful standard system is that of the Ethernet family (IEE802.3.x), as described in Section 8.10.2.

8.10.2   The Ethernet family

The distinguishing feature of an Ethernet system is that the mechanism for managing the joint use of the bus is based on collision control. This relies on each termination wishing to communicate sending a burst of data and its destination address within an Ethernet 'frame' onto the LAN. If any other terminals simultaneously send data the resulting 'noise' (mutilated signals) caused by the collision of two sets of data frames is detected by both terminations, which stop immediately. These terminations each wait for a short random period and then retry – normally, one of the terminals is first and the other waits for the bus to clear. This mechanism,

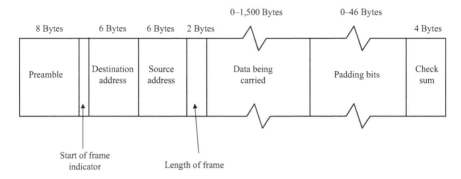

Figure 8.24   Ethernet LAN frame format

known as carrier-sense multiple-access (CSMA)/collision detection (CD), is analogous to a conference telephone call without a chairman, when two or more participants inadvertently speak at the same time, each stops and tries again later.
Actually, CSMA/CD is one of several protocols – known as media-access control
(MAC), part of OSI Layer 2 – used in data transport systems to control the access of
several terminals to a single resource.

Figure 8.24 shows the format of the Ethernet frame, mentioned above, which
is really a form of data packet. An 8-byte binary number acts as a so-called
'preample' to allow the recipient terminals to lock into the speed of transmission of
the received frame. The frame proper starts with the 6-byte destination and source
addresses and these are followed by the field for the actual data, which has a
maximum capacity of 1,500 bytes. There is also a small field to indicate the length
or type of the frame and a 46-byte padding field used to ensure that the minimum-
allowable 64-byte length of the frame is achieved. Errors in the received frame are
detected through examining the 4-byte checksum. More detail on Ethernet LANs is
available in the references [2,19–22].

There are commonly several speeds of Ethernet LAN: 10 Mbit/s, 100 Mbit/s
(Classical Ethernet) and more recently 'Gigabit Ethernet' has been introduced
working at 1, 10 or even 100 Gbit/s. The LAN provides a managed data path
between terminals and servers, allowing terminal-to-terminal, terminal-to-server
and server-to-server communication – effectively at Layer 2 of the OSI reference
model. IP packets then flow over the LAN path at Layer 3. The Ethernet LANs
usually have a gateway giving access to links to other LANs or to one of the public
data networks. The transmission capabilities of Gigabit Ethernet over optical fibre
cable make it a candidate for the Access Network, linking a LAN site, say at an
office, to the first network node, thus extending Ethernet beyond the customer's
premises. New generations of PON use the Ethernet interfaces [23]. Also, Gigabit
Ethernet is used within data centres distributing high-speed data via Ethernet
switches to the suites of servers, as described in Section 8.11.3.

The popularity of Ethernet as a LAN technology which is readily upgraded in
speed while remaining basically a simple design, has led to its rapid spread to ever-

increasing LANs and WANs applications. A further development of the Ethernet family was the inclusion of tags that enable virtual LANs (i.e. VLANs) to be supported in the IEEE 802.10 standard, similar in concept to the way that MPLS provides VPN capability. This ability to segment different traffic groups within a bearer also enables Ethernet to provide standard Layer 2 virtual circuits (known as EVC). This upgraded version of Ethernet is usually referred to as Carrier Ethernet (CE), reflecting its application to long distance transmission networks carrying IP traffic.

Although CE may be transported over MPLS, CE can take the Layer 2 role completely and map on to SDH/SONET, PDH or TDM transmission systems. However, in some applications the CE can be directly mapped to the optical-fibre transmission systems. An important application of CE is the provision of backhaul links between the cell sites and their parent Node B in long-term evolution (LTE) 4G mobile networks, as described in Chapter 9. This mobile system has an all-IP architecture, in which users' data and voice are carried and switched as IP traffic, the ports on all the equipment is Ethernet VLAN.

8.10.3   Corporate networks, MANs, and WANs

A corporate network comprising of several interconnected LANs at various sites within an area of some 3–30 miles is known as a MAN. The interconnection between LANs, often called the 'sub-network' may be realised using leased lines or a public data service, for example, ATM, Frame Relay or SMDS, which is provided by the network operator. A further concept is that of the WAN, which links widely separated LANs and MANs (typically over distances > 30 miles) [2,5,24].

8.10.4   Wireless LANs

In addition to the wired form of LANs, there is a family of wireless-based systems, known as Wireless LANs (WLANs), 'Wireless Ethernet', or colloquially 'Wi-Fi'. These LANs serve a group of terminals, typically smartphones, laptops or PCs, each with a network interface card (NIC) transmitting over a wireless link to a centrally located hub, the Access Point (AP), as shown in Figure 8.25. A special MAC protocol is used to cope with the requirements of wireless links, namely: carrier-sense multiple-access (CSMA)/collision avoidance (CA), which operates similarly to the LAN CSMA/CD except that the terminals carefully listen for exiting transmissions before transmitting to avoid collisions. Since all the terminals share the wireless capacity on an AP, larger capacities are achieved through having two or three APs, each operating on a different wireless carrier frequency. The APs are linked by cable to an Ethernet switch. The NICs can determine the appropriate AP to join by comparing the signal strengths of the various carrier frequencies within the WLAN. A collocated server is connected by cable to the APs. All communication is between the NICs and the AP, since the terminals are not able to link directly.

The WLANs were originally designed for use within the office environment, giving users a degree of freedom to connect to the office computing facilities without using cables. However, the progressive reduction in the costs of the

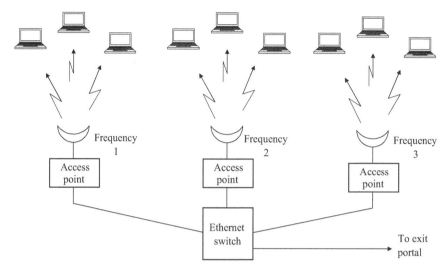

Figure 8.25   Wireless Ethernet LAN (Wi-Fi)

equipment and the fact that the system operates in uncontrolled wireless spectrum (i.e. no fees or licence) has opened up the WLAN applications far beyond the office environment. For example, a wide-spread application is the use of WLANs in the so-called 'hotspots', for example, hotels, railway stations, airport departure lounges, and coffee shops around town. Here, people can communicate to their corporate LAN or the Internet from their laptop computers, if equipped with an appropriate NIC, within the catchment area of an AP. This provides what is, in effect, a form of nomadic mobility, whereby a person on the move can communicate within the vicinity of any hotspot, but not when moving between them. In addition, WLANs are now installed in homes, linking PC s, laptops, printers, routers, etc., meeting personal and home-working needs. In Chapter 9 we consider how mobile sub-scribers may use a Wi-Fi hotspot for their data services when in the vicinity of a hotspot, the so-called 'Wi-Fi off-load' of the mobile networks.

There are several versions of WLAN, all in the IEEE 802 range (see Box 8.5), and the series is periodically expanded by new upgrades. Early systems meeting the 802.11b standard operated with wireless carriers at 2.4 GHz, offering user speeds up to 11 Mbit/s at a range up to 150 m. The more recent 802.11a, which uses the OFDM (see Chapter 4) and operates at higher carrier frequencies of 5 GHz, and consequently has a shorter range of 50 m, but offers user speeds up to 54 Mbit/s. The IEEE 802.11e introduces a QoS support for the LANs, acting as an enhance-ment to 802.11a and 802.11b. Other important versions include 802.11ac with user speeds up to 1.3 Gbit/s, and the 802.11ac Wave 2, which uses MU-MIMO (see Chapter 9) to achieve user rates up to 6.9 Gbit/s.

The data frame structure for the 802.11 range of WLANs is shown in Figure 8.26. As will be noticed, the 802.11 frame is similar to non-wireless LANs

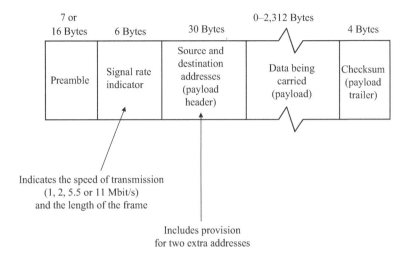

*Figure 8.26   A simplified illustration of the frame format of IEEE 802.11
wireless LAN*

except for the additional 'Address 3' and '4' fields which enable NICs to communicate with terminals or hosts on other wireless cells, and the use of duration (transmission time of the frame burst) rather than length of frame to indicate the variable amount of data being carried in the frame [19,25].

To complete the picture of wireless data transmission systems, we should note the emergence of the broadband wireless link defined in 802.16, also known as 'WiMax' [26]. This is really a high-capacity fixed link between a stationary user and a stationary server; many operators consider it an alternative to the provision of ADSL over copper cable to provide broadband to residential and small business users (see Chapter 4). We also note the short distance (up to 30 m) wireless system, known as 'Bluetooth', which is defined in the IEEE 802.15 standard. Bluetooth provides a convenient replacement for wires between equipment around the office or home, for example, the cordless headset. It is also widely used in vehicles and mobile handsets for short distance device-to-device communications.

8.11   Data services

8.11.1   Range of data services provided by Telco

Telecoms network operators (Telco) have been running data services in addition to their voice services since the 1970s. Initially, they were specialised niche services for the business, manufacturing and academic communities, and represented a small proportion of the total network traffic. Since the mid-1990s, however, non-voice traffic has dominated. Interestingly, Telco's are also referred to as multiple-service operators (MSOs) to reflect this fact. More significantly, data services are now serving consumers as well as the business market – with delivery over both mobile and fixed networks. As discussed in Section 8.6, it is the advent of the WWW and people's adoption of online shopping, social networking, downloading from the web, and streaming and downloading of audio and video content over their broadband access that has driven the rising demand for higher-volume and faster data service from the Telco.

However, probably the biggest changes to the world of telecommunications have resulted from the opportunity that the wide-spread nature of this data capability gives for non-Telco service providers to offer a variety of applications – a phenomenon known as over-the-top (OTT) services [27,28]. Examples of OTT services include:

- Web-based voice (VoIP), video and conferencing services (e.g. Skype, Vonage, GoToMeeting);
- Catch-up TV (e.g. BBC iPlayer);
- Streamed and downloaded video and films on demand (e.g. Netflix);
- Messaging (e.g. WhatsApp);
- Social network sites (e.g. Facebook);
- Recorded music streaming (e.g. Spotify);
- Video-sharing (e.g. YouTube);
- Photo-sharing (e.g. Instagram).

An important aspect for users is that because the OTT services are not managed by the Telco there is no guarantee of quality of network performance. The commercial model for the OTT services relies on a combination of advertising revenue and subscription or usage charging of the users. However, the Telco are restricted to just providing and charging for their broadband data platforms supporting the OTT services. Although, in some cases, Telco do lease content-distribution network (CDN) cache storage and distribution (see later in this section) to the OTT providers enabling them to offer their users some level of QoS.

To counteract the OTT threat some Telco's have introduced managed services to run over their networks – the prime example being IPTV. In addition, many Telco now also provide their own cloud services from their data centres over their core and access networks. We consider the network aspects of Telco's IPTV and cloud services in the following sections. Table 8.1 presents the main data services run by typical Telco, showing the variety of network platforms and service nodes involved.

8.11.2 IPTV

As the name suggests, IPTV relates to video content being provided and delivered to subscribers by Telco over their IP network with managed and guaranteed QoS. The video content is sourced either directly by the Telco or packaged with other content and then sent as an IP data stream over the core data network, Core Transmission network, and then the fixed broadband network to a TV set-top box (STB) in the subscriber's premises. There are two main categories of IPTV, namely: video on demand (VOD) and multistreaming of live (linear) TV. We will consider each of these.

Table 8.1   Data services provided over a Telco's platforms

Services	Access platforms	Core platforms
Broadband data access to the Internet (ISPs)	• Fixed: xDSL, G.fast, PON, FTTP, etc. or • Mobile: GPRS, 3G, 4G	Core Transmission network and Core data network
ISP role	–	ISP routers and DNS Core data network Core Transmission network
Text messaging	Mobile: GSM, GPRS, 3G, 4G	Text messaging unit on MSC Core Transmission network
IPTV	Fixed BB access: xDSL, PON, FTTP, etc.	CDN and VOD server; Linear TV headend; Core data network Core Transmission network
Cloud	Fixed BB access: xDSL, PON, FTTP, etc.	Data centre Core data Network Core Transmission network
Leased lines	Fixed NB and BB access: xDSL, PON, FTTP, SDH, Ethernet	DXC network Core Transmission network
VPN	Fixed BB access: xDSL, PON, FTTP, Ethernet	Core Transmission network Core data network with MPLS

VOD. A VOD service enables subscribers to view their choice of video content, such as feature films, music videos, documentaries, etc., whenever they wish. Subscribers may directly view the content on a TV set associated with the Telco-provided STB, on a PC, or other devices. Therefore, a VOD service requires centrally located video storage to provide a library of content, a management system to enable authorised subscribers to select their viewing on demand, together with an appropriate network delivery system. A simplified schematic diagram of the VOD system is shown in Figure 8.27.

The library of video content is contained in several video storage centres (computer high-capacity storage with server) and managed by a VOD server as shown in Figure 8.27. Prior to storage the video content needs to be encoded into suitable digital format, as described later in this section. Once registered with the Telco for VOD service, subscribers can select video content from a menu of programmes and item listings displayed on their TV screen, using a remote control. The content requests are formatted by the STB and sent as an IP-based message to the website of the VOD-subscriber management system, which after authentication actions the sending of the selected programme by setting up a VOD session. The subscriber management system also holds the usage records for billing and customer service enquiries. This system is based on the standard AAA server used in IP networks (see Section 8.18). An important feature of running a VOD service is that of digital-rights management (DRM), namely the ownership of the digitalised content and its distribution.

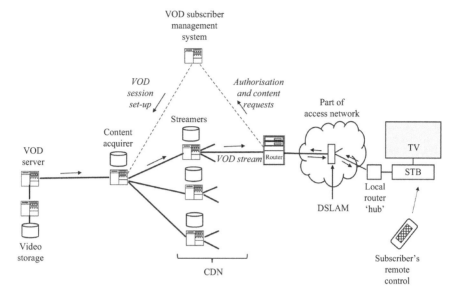

Figure 8.27   Video on demand delivery using content-distribution network

We now need to consider the challenge of delivering many individual video programmes (films, etc.), each running for 100–200 minutes or more, separately to a huge population of VOD subscribers on demand. Clearly, if all IP video content was streamed from a single point through the network to all the active IPTV subscribers around the country there would be severe loading on the Data and Core Transmission backbone networks, with consequent difficulty in maintaining QoS. The solution the Telco adopts is the use of a content-distribution (or delivery) network (CDN). In general, CDNs are deployed in large-scale data networks where individual selections from fixed content are to be distributed to a widely dispersed set of subscribers at different times. VOD networks rely on a set of localised video storage, known as streamers where video content is cached, as shown in Figure 8.27. Each streamer serves its own catchment area of subscribers. By ensuring that the streamers contain the most requested video content, most of the IPTV delivery can be over just the relevant access network and so does not unnecessarily load the backhaul and core parts of the data network. Therefore, only the seldom-requested non-cached content needs to be supplied from the central main VOD server, so minimising the capacity required in the Core of the network and the chances of degradation to QoS through packet loss within the network.

Linear TV. In providing linear TV service, the Telco is essentially acting as a TV broadcaster, by delivering over the IP broadband network one of a few TV channels in real time simultaneously to a large population of subscribers. The challenge of delivering a potentially huge continuous flow of IP packets carrying the same contents across the data network is addressed by using IP multicast streaming. This is a general network technique for dealing with delivering one

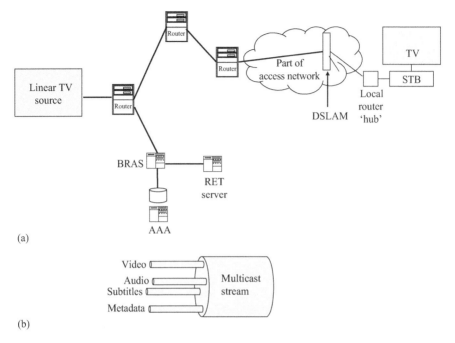

Figure 8.28   Linear TV multicast delivery. (a) Linear TV distribution networking. (b) Multicast stream assembly

message to many end points. The source IP packets which are to be broadcast to users within a group (e.g. subscribers who currently wish to watch a particular TV channel) are replicated at routers close to the users, so enabling the minimum number of packets required from the source. Thus, the packet flows follow a tree-like structure fanning out through the network. This technique requires a multicast protocol to instruct the appropriate IP routers to replicate the content in each IP packet in the stream and forward the copies on to the destination addresses of participating subscribers.

An alternative technique is used to deliver linear TV to mobile devices, known as adaptive bit rate (ABR), in which the video source is encoded at multiple bit rates and the recipient device chooses the best fit to the available received down-stream bandwidth.

A simplified illustration of linear TV using multicast IP is shown in Figure 8.28(a). This diagram shows the same access configuration as for the delivery of VOD, but the backhaul and core of the network uses the multicast enabled IP routers rather than the CDN. The AAA, which provides the subscriber management for the linear TV service, receives programme selection messages from the subscriber's STB, as described for VOD, and initiates the programme streaming from the linear TV source to the STB. The linear TV source assembles the live video (e.g. football match) into the appropriate format for carrying over the IP network with other data traffic.

The broadband capability of the core data (IP) network is managed by a BRAS (broadband remote access server, also known as BAS). At the start of the linear TV session the BRAS allocates an IP address for the delivery of the packets containing the linear TV content which are then sent using RTP and the multicast technique. Finally, a copy of the live TV content is cached in a retransmit (RET) server ready for transmission of any packets lost (due to corruption or dropping) as requested by the relevant STB. This approach requires some buffering at the STB so that the retransmitted packets can be correctly inserted into the output stream. (Alternatively, some IPTV networks use forward error correction (FEC) rather than using RET.)

Each IP multicast stream emanating from the Linear TV source carries one TV channel, which contains four components known as 'elementary streams'. These streams, as shown in Figure 8.28(b) are as follows:

- Video content
- Audio content
- Subtitles
- Metadata, that is, programme identification, etc.

The elementary streams are split into 'transport stream (TS) packets', which are packed in groups of seven into the IP packets forming the multicast stream [29].

Formatting the IP packet stream. For both the VOD and linear TV services, there is the huge challenge of converting the video and audio content into a manageable stream of IP packets for dissemination to the appropriate subscribers. Firstly, in the case of direct analogue-to-digital encoding within a TV camera, the video content is in the form of a high-speed digital IP stream, depending on the resolution of the TV scanning rate, pixel density and frame rates (e.g. 8 Mbit/s for high-definition (HD) TV). The digital content for transmission through the Telco's network is minimised by using predictive techniques which exploit the redundancy between TV frames, with full restoration of the video stream at the STB. In addition, the digitalised content needs to be assembled with the audio, subtitles and meta-data content, then added to the other TV channels before being encrypted so that only authorised recipients can watch the video content. The full process is summarised in Box 8.6 and Figure 8.29.

8.11.3   Cloud service and data centres

The generic term 'cloud' refers to the storage or processing of users' data (e.g. documents, photos, videos) within the network – its actual location being opaque to the users. This facility offers not only a convenient form of off-site back-up of computer contents, but also the ability to access a person's suite of stored material and application programs from any location using mobile or fixed communication. A range of Cloud services are taken by both business customers and consumers [27]. Since users rely heavily on the cloud services for their computing and back-up storage, it is essential for them to be run over a dependably reliable data network, ideally one with widespread coverage. Also, importantly the users need to have total trust in the integrity and longevity of their cloud service supplier.

Box 8.6   Carrying TV and video content over IP

The stages of the process as set out in Figure 8.29 and the protocols used are as follows:

- Image sampling: The TV camera produces an analogue video signal through scanning and image sampling adhering to TV standards, that is, SD (standard definition), HD, 4k, Super Hi-Vision (UHDTV).
- Encoding: The raw video digital signal is digitally encoded using bit-rate reduction techniques, for example, MPEG-2, MPEG-4.
- Encapsulation: The various elementary streams are encapsulated in video stream groupings using, for example, MPEG2 transport stream, MPEG4.14 (MP4).
- Metadata: The additional information to be associated with each video stream is encapsulated using one of several proprietary systems, for example, CableLabs, TV-Anytime, DVB(P)SI, MXF, XML.
- Encryption: The content is secured from unauthorised viewing using an encryption scheme, for example, Wide vine, CENC (Common Encryption), Verimatrix, MS PlayReady, Marlin, MagraVision.

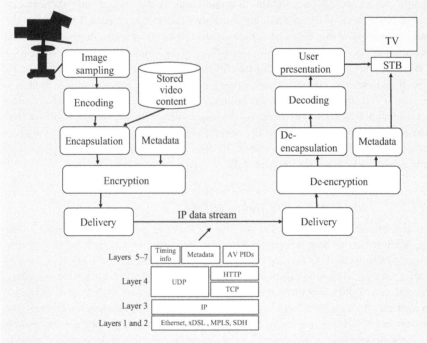

Figure 8.29   The process of carrying TV programmes over an IP network

- Delivery: The packed content forms an audio-visual packet ID (PID) carried through the IP network using multicast or CDN and L4 protocols, for example, RTP and UDP, RTMP, DASH, etc.
- User presentation: The received video content is then formatted by the STB to suit the form of viewing device, for example, HTML5, CEA-2014, HTML/CSS, Adobe Flash.

Network operators are therefore well placed to provide such services. However, there are many OTT service providers also offering their version of cloud using the Telco's network.

Cloud services are provided by data centres (DCs) containing many high-capacity computer servers and mass storage [30]. Access between the users and the DCs is provided through the fixed and mobile networks and the core data IP networks, as shown in Figure 8.30. Normally, the Telco own and operate the DCs and all the data- and access-network infrastructure. In some cases, the Telco uses the DCs for their own data applications, such as billing and payroll, as well as for cloud services. The OTT Cloud service providers own and operate the DCs, but depend on the Internet and leased lines and fixed and mobile access networks from the Telco.

Figure 8.30 also shows the typical configuration within DCs, in which a mesh of Ethernet switches linked by 100 Gbit/s provides flexible access to the suites of

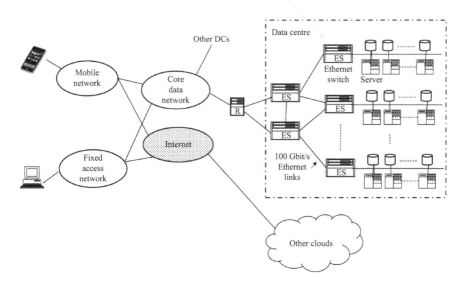

Figure 8.30   Cloud and DC, as provided by a Telco

servers and storage. This high-speed Ethernet network operates at OSI Layer 2, providing semipermanent VPs over which the users' IP packets flow at Layer 3 to the servers.

The DCs are usually contained within warehouses and large office buildings (e.g. 20,000–40,000 m^2), with the many servers held in long suites of equipment racks interconnected by optical fibres supported in overhead racks. The significant amounts of electrical power (e.g. 10–25 MVA) are distributed to the server and storage-suites through under-floor cabling. Such high levels of power consumption create a cooling challenge within the DCs, usually addressed by widespread deployment of chillers along the equipment suites. Of course, the fact that so much of the customers' data is being processed or stored by the cloud service sets the requirements for security and reliability at the highest levels. These are typically met as follows [31]:

- The optical fibre links into the DCs, which terminate usually at the 10–40 Gbit/s rate on the server suites, are carried in cables entering the building through at least three separate entry points. This gives a high level of transmission capacity diversity, which when associated with 'separate routeing', that is, physical separation of the cables in the Access and Core Transmission Networks – gives the highest achievable transmission availability (see Chapter 5).
- The mains power distribution around the DC is usually carried on a fully-duplicated basis.
- Use of rapid response standby power generators, which in the event of mains failure will automatically start up, together with battery back up.
- High standard of physical security protecting entry to the DC building.

8.12   NFV and SDN

8.12.1   *Developments to data network technologies*

From around 2012 there was an important shift in the technologies used within data network equipment. The drivers for the change were the need to cope with ever-increasing volumes of data processing and storage, rapid reconfiguration to accommodate changes of traffic levels, and cost reduction by using commercial off-the-shelf (COTS) software where possible. Two distinct, but closely coupled, technologies have been developed, namely: network function virtualisation (NFV) and software-defined networks (SDN). Both technologies have been deployed in several areas of data networking, but mainly in DCs [30].

8.12.2   *NFV*

The idea of virtual-machine working is a well-established technique in large-scale computing centres to make efficient use of the server equipment. Several separate application programmes can be run over the same machine (i.e. computer) – each programme being isolated from the others so that to it appears to be run on a

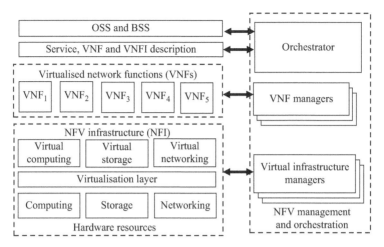

Figure 8.31   ETSI's NFV general architectural model

dedicated machine – a process known as virtualisation. (This is comparable with a VPN, in which the users appear to have independent private networks and are not aware that they are sharing a common IP network resource, as discussed in Section 8.9.) NFV exploits the capability of creating data network functionality by using standard computers, storage, and routeing hardware devices. A bank of these devices can be used to provide data network functions, like packet replication or routeing, for as long as required and then the same devices can be reused for other functions. The virtualisation is achieved by extracting all but the basic functionality to a central control level, so that the basic resources can be used in a variety of ways as required. This approach also allows rapid re-allocation of capacity between applications, so spreading the data networking load across the bank of basic resources [32,33].

The European Telecommunication Standard Institute (ETSI) has defined a general architectural model for NFV, which is shown in Figure 8.31. The bank of resources is shown at the bottom left of the model, split into computing, storage and networking. The various functionalities of these resources are captured in virtualised form, so that the set of resources become the NFV infrastructure (NFI). A portfolio of virtualised network functions is held in readiness for use as determined by the service needs of the DC or network node. Each of the levels – infrastructure, functional, and service control – is managed by NFVI managers, VNF managers, and Orchestrator, respectively.

8.12.3   SDN

The idea behind SDN is that much of the control activity of data equipment can be held in software form, leaving the control of more basic functions to be held in 'firmware' (i.e. semi-permanent very fast memory). Thus, in an MPLS enabled router, the simple forwarding based on a label-routeing table can be done under firmware control at a rapid rate. This leaves the more-complex routeing of non-MPLS IP packets, which may require some processing, to be controlled from central software controllers (with consequent longer execution times). In effect, this splits the router function into control and data planes. Several data-plane resources could be controlled by a single remote software system. This gives the ability to also shift the loading of traffic across hardware/firmware resources in the data plane. As mentioned earlier, there is a natural fit between NFV and SDN, and so they are frequently deployed together – the SDN providing the ability to flexibly control the functionality of hardware/firmware resources in the (NFV) virtualisation process [32,33].

8.13   Summary

In this chapter we looked at the variety of packet technologies providing the wide range of data services used by both business and residential customers. After considering the differences between voice and data services, we then noted that several packet systems are capable of providing voice switching. We looked at ATM in detail, noting that the VPs established enabled the QoS offered to customers to be predicted and assured. However, the main focus of the rest of the chapter addressed the various aspects of IP and why it has proved to be ubiquitous and popular as a network service as well as the main interface to user computer applications. Consideration of the Internet provided an insight into the applicability of IP and, with the advent of the WWW, the massive impact on everyday life at work and home.

An important next step was the examination of how IP can be used to switch voice and hence become the contender for replacing the circuit-switched PSTN. Importantly, we considered the joint working of ATM and other Layer 2 technologies with IP. In particular, we looked at the new service offerings and network traffic engineering made possible by MPLS. The latter, with VoIP, forms the basis of the so-called NGNs, as discussed in Chapter 11.

We also considered the role of LANs, particularly Ethernets, in serving the communication between terminals and servers within an office environment and generating the need for public data services as provided by network operators. We noted that a recent extension of the LAN concept is the wireless LAN systems in public places, for example, coffee bars and airports, and most importantly in domestic premises. Ethernet is also used as a high-speed data interface for business (or enterprise) data networking.

The description of the technology and techniques used in data networks was then applied to two examples of Telco-provided services: IPTV and Cloud. These descriptions introduced the concepts of CDNs and DCs in data networking. Finally, the impact NFV and SDN have on cost reduction and flexible use of data network resources was considered.

References

[1] Goleniewski, L. *Telecommunications Essentials*. Boston, MA: Addison-Wesley; 2003, Chapter 7.

[2] Tatenbaum, A.S. *Computer Networks*, 4th edn. Amsterdam, The Netherlands: Prentice Hall; 2003, Chapter 1.

[3] Shepard, S. *Telecom Crash Course*. New York, NY: McGraw-Hill; 2002, Chapter 6.

[4] Dennis, A. *Networking in the Internet Age*. New York, NY: John Wiley & Sons; 2002, Chapter 6.

[5] Dennis, A. *Networking in the Internet Age*. New York, NY: John Wiley & Sons; 2002, Chapter 1.

[6] Davidson, R. 'IPv6: An opinion on status, impact and strategic response', *Journal of the Communications Network*, 2002;**1**(1):64–68.

[7] Chao, H.J., Lam, C.H., and Oki, E. *Broadband Packet Switching Technologies: A Practical Guide to ATM Switches and IP Routers*. New York, NY: John Wiley & Sons; 2001.

[8] Tatenbaum, A.S. *Computer Networks*, 4th edn. Amsterdam, The Netherlands: Prentice Hall; 2003, Chapter 5.

[9] Loshin, P. *TCP/IP Clearly Explained*, 2nd edn. San Diego, CA: Academic Press; 1997, Chapter 20.

[10] Tatenbaum, A.S. *Computer Networks*, 4th edn. Amsterdam, The Netherlands: Prentice Hall; 2003, Chapter 7.

[11] Boyle, D., Kolcun, R., and Yeatman E. 'Devices in the Internet of things', *Journal of the Institute of Telecommunication Professionals*, 2015;**9**(4): 26–31.

[12] Beart, P. 'Evolving to an ecosystem of IoT applications', *Journal of the Institute of Telecommunication Professionals*, 2015;**9**(4):15–20.

[13] Crane P., Putland C., and Harrold, C. 'Enabling wireless IoT networks', *Journal of the Institute of Telecommunication Professionals*, 2015;**9**(4): 32–37.

[14] Fisher, M. and Davies, J. 'Creating Internet of things ecosystems', *Journal of the Institute of Telecommunication Professionals*, 2015;**9**(4):9–14.

[15] Newman, R., Robinson, G., and Turner, N. 'Evolution of the BT PSTN trunk network using next-generation switches', *Journal of the Communications Network*, 2002;**1**(1):69–74.

[16] Catpole, A.B. and Middleton, C.J. 'IP telephony solutions for the customer premises'. In Swale, R. (ed.), *Voice over IP: Systems and Solutions*. Stevenage: Institution of Electrical Engineers; *BT Exact Telecommunications Technology Series No. 3*, 2001, Chapter 4.

[17] Rosen, B. 'VOIP Gateways and the MegaCo architecture'. In Swale, R. (ed.), *Voice over IP: Systems and Solutions*. Stevenage: Institution of Electrical Engineers; *BT Exact Telecommunications Technology Series No. 3*, 2001, Chapter 7.

[18]   Valdar, A. and Morfett, I. 'Understanding telecommunication business'. Stevenage: Institution of Electrical Engineers; *IET Telecommunications Series No. 60*, 2015, Chapter 5.

[19]   Tatenbaum, A.S. *Computer Networks*, 4th edn. Amsterdam, The Netherlands: Prentice Hall; 2003, Chapter 4.

[20]   Ross, J. *Telecommunication Technologies: Voice, Data and Fiber-Optic Applications*. Indianapolis, IN: Prompt Publications; 2001, Chapter 4.

[21]   Goleniewski, L. *Telecommunications Essentials*. Boston, MA: Addison-Wesley; 2003, Chapter 8.

[22]   Dennis, A. *Networking in the Internet Age*. New York, NY: John Wiley & Sons; 2002, Chapter 4.

[23]   James, K. and Fisher, S. 'Developments in optical access networks'. In France, P. (ed.), *Local Access Network Technologies*. Stevenage: Institution of Electrical Engineers; *IET Telecommunications Series No. 47*, 2004.

[24]   Ross, J. *Telecommunication Technologies: Voice, Data and Fiber-Optic Applications*. Indianapolis, IN: Prompt Publications; 2001, Chapter 1.

[25]   Dennis, A. *Networking in the Internet Age*. New York, NY: John Wiley & Sons; 2002, Chapter 8.

[26]   Patachia-Sultanoiu, C. 'Deploying WiMAX certified broadband wireless access systems', *Journal of the Communications Network*, 2004;**3**(3): 105–116.

[27]   Valdar, A. and Morfett, I. 'Understanding telecommunication business'. Stevenage: Institution of Electrical Engineers; *IET Telecommunications Series No. 60*, 2015, Chapter 1.

[28]   McCarthy-Ward, P. 'The growth of OTT services and applications', *Journal of the Institute of Telecommunication Professionals*, 2016;**10**(4):8–14.

[29]   International Organization for Standardization (ISO 13818-1). (2007). 'General coding of moving pictures and associated audio information'. [online] Available from http://www.iso.org.

[30]   Rivera, S. 'NFV takes telecom by storm', *Journal of the Institute of Telecommunication Professionals*, 2014;**8**(2):18–22.

[31]   Valdar, A. and Morfett, I. 'Understanding telecommunication business'. Stevenage: Institution of Electrical Engineers; *IET Telecommunications Series No. 60*, 2015, Chapter 9.

[32]   Stallings, W. *Foundations of Modern Networking – SDN, NFV, QoE, IoT and Cloud*. Upper Saddle River, NJ: Pearson Education; 2016, Chapter 1.

[33]   Chayapathi, R., Hassan, S.F., and Shah, P. *Network Functions Virtualization (NFV) with a Touch of SDN*. Boston, MA: Addison-Wesley Professional; 2017, Chapter 1.

Chapter 9
Mobile systems and networks

9.1 Introduction

Mobile phones are now such an everyday feature of life throughout the world that in total they far outnumber the population of fixed telephone lines. For many, the mobile phone is the preferred means of communication. Perhaps more significantly, the mobile phone instrument is usually a smart device which enables voice, video and data services and applications to be presented together with touchscreen human interaction. The mobile networks therefore need to convey both voice and data content. This chapter explains how a mobile phone system works and the relationship between mobile and fixed networks. We begin by considering the nature of a mobile system. After a simple review of how a telephone call can be carried over a radio link, we examine the various cellular mobile network systems, and conclude by looking at how the mobile and fixed networks in the future may begin to merge.

9.2 Characteristics of mobile networks

Most people would characterise a mobile phone as something that uniquely relies on the use of radio. Indeed, mobile phones do use radio – but, not all systems using radio are actually mobile. An example of non-mobile usage of radio is the point-to-point radio links between exchanges and users' premises, as described in Chapter 5. There are, in fact, several characteristics in addition to the use of radio that together distinguish mobile from fixed telecommunications systems.

9.2.1 What is special about mobile networks?

The key characteristics of mobile networks are described below.

Tetherless link. A so-called tetherless link is used between the mobile handset and the exchange to enable the required freedom of movement. Whilst in theory there are several technologies that could provide a tetherless link, including infrared, laser light, etc., invariably a radio link is used. In effect, this radio link performs the role of the local loop in a fixed network.

Need for handset identification. Taking a fanciful view, one could imagine that if a technician standing at the termination of the local loop cable at the entry to the fixed exchange (i.e. at the main distribution frame (MDF)) took hold of a copper pair and gave it a great jerk, at some distance away the telephone on the subscriber's desk would move! Whilst clearly not true, this story does illustrate the point that the identity of any telephone on a fixed-line network is determinable from its point of entry to the exchange. By contrast, mobile networks do not have this fixed physical relationship to enable identification of the (telephone) handset. All mobile networks therefore need a mechanism to identify their subscribers' handsets.

Need to track the location of users. Users of mobile phones may move extensively across the domain of the serving network, or indeed move into the catchment area of a foreign mobile network – the latter action known as 'roaming'. A key attribute of a mobile network is therefore a mechanism to determine the instantaneous location of its users' handsets, and to track this location during a call. This is known as 'mobility management'.

Need for a complex handset. The handset for a mobile network not only acts as a telephone terminal, but it also has to provide the unique mobile functions, including those that are undertaken in the line card of a fixed telephone exchange, as well as the transmission and reception of radio, power management and the provision of handset identification. Consequently, the mobile handset is significantly more complex and expensive than a fixed-line telephone. It is interesting to note that the ubiquitous GSM handset contains processing power equivalent to a Pentium microprocessor! Yet this complex device is now viewed by many users as a fashion accessory – and the success of mobile service offerings are highly dependent on the attractiveness, that is, the look and feel, of the handset. Indeed, many handsets now contain a variety of other functions, such as cameras, video recorder, MP3 players, and diary, contact list, calculator, etc. (For convenience, we will use the term 'handset' to mean a mobile terminal whether it is hand-held or incorporated in some other device, such as a hands-free unit in a car or a plug-in card to a computer lap top. There are several terms used for a handset within the industry, including 'mobile station' (MS) and user equipment (UE).)

Use of complex commercial model. Due both to the nature of their services and the more competitive commercial and regulatory environments pertaining at the time when they were introduced, mobile networks tend to operate within a more complex commercial model than the longer-established fixed networks. This means that there are usually several different companies – or 'players' – involved in the provision of mobile services, each undertaking one of the following roles:

- Mobile service providers (MSP), who sell to the user, have the billing relationship and provide the overall service management;
- Mobile network operators (MNO), who build and operate the mobile network;
- Mobile virtual network operators (MVNO), who appear to the users to be a full mobile network operator and service provider with a distinctive brand – although they actually do not own any network, rather leasing capacity from an MNO instead.

The resulting commercial relationships (who serves whom and how the money flows) can become complex, particularly when there may be several MNVOs and MSPs, all dependent on a single MNO. Of course, when considering data and web access services via a mobile handset this complexity increases greatly with the addition of Internet service providers (ISP) and information providers [1].

Need for Specialised service support. The more complex nature of the services available from a mobile network, for example, roaming abroad, data calls, and the separation of the service provider role from the network operator role – requires specialised service support to handle customers' enquiries, ordering, fault handling, and billing.

9.2.2   A simplified generic model of a mobile system

We can now consider a simple model of a generic mobile network based on these characteristics as shown in Figure 9.1. The heart of the mobile system, indeed some would say the crown jewels, is the set of radio frequencies allocated exclusively to the operator: the radio system. A mechanism is required to control the access of the many handsets to the constrained set of frequencies in the radio system, a process known as multiple access (MA), as described later. (In fact, this control of access to the radio resource is an example of the generic function of 'medium-access control' (MAC), used in networks whenever a transmission medium is shared across many users.) Figure 9.1 shows the functions of mobility management, and identification (and authentication) at the exchange, together with the switching and routeing to other mobile or fixed networks. Finally, as described above, our generic picture of a mobile system needs to show the service support function (including network operations and maintenance). We will look at all these functions later in this chapter. But first, a simple consideration of how radio works is appropriate.

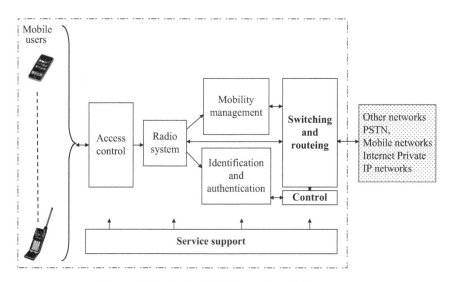

Figure 9.1   A simplified generic mobile system

9.3   How does radio work?

The concept of the use of radio to provide transmission paths for telecommunications is introduced simply in Chapter 4 (see Figures 4.6 and 4.8). Radio paths of vibrating electromagnetic energy (waves) are set up between transmitting and receiving antennas. Different forms of radio propagation are possible depending on the frequency of the electromagnetic vibrations, which can range from about 3,000 Hz (pronounced Hertz, the unit for vibrations per second) to around 300,000,000,000 Hz (300 GHz, pronounced 'gigahertz') – see Chapter 4 and Box 4.1. For the purposes of considering mobile systems we will concentrate on radio frequencies in the microwave range, that is, 300 MHz (300 million Hz, pronounced 'megahertz') to some 3 GHz, in which the radio waves follow a line-of-sight propagation method.

Radio systems use an electromagnetic signal in the shape of a sine wave, as shown in Figure 4.1 of Chapter 4. The most efficient form of antenna of such sine waves is a metallic pole, known as a 'dipole', whose length is half the wavelength of the sine wave. In the case of a mobile network using frequencies at, say, 2 GHz, the wavelength (λ) is given by dividing the speed of the radio wave (which is the same as that of light), 300,000,000 m/s, by the frequency (2,000,000,000 Hz) equalling 15 cm. So the efficient antenna size is $\lambda/2$, i.e. 7.5 cm long. However, for convenience and compactness the less-efficient but smaller antenna size of $\lambda/4$ (i.e. 3.75 cm), or the even less-efficient $\lambda/8$ (i.e. 1.88 cm) is used in mobile phone handsets and car mounted antennas, as appropriate.

Figure 9.2 illustrates the principle of radio propagation [2]. The electrical signal is passed through the dipole antenna causing an electromagnetic field to exist

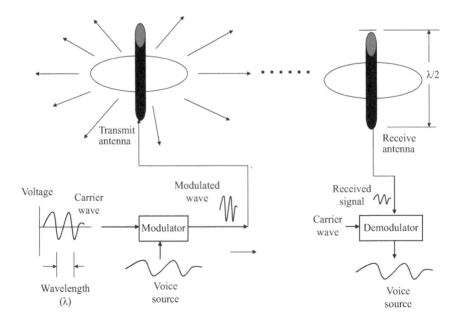

Figure 9.2   Radio propagation

around the length of the dipole. It radiates the radio wave equally in all directions at right angles to its length, but there is no radiation at the two ends of the dipole. The radio wave travels through the air just like a light wave – in fact, of course, light is also an electromagnetic wave, but vibrating several thousand times per second faster than the radio wave. A similar dipole placed some distance away will be 'excited' by the propagated radio wave causing a corresponding but much attenuated electrical signal to be generated across the dipole and passed to the receiving equipment, where it is amplified or regenerated.

In the ideal conditions for radio propagation, that is, free space, without any air or physical matter in the way, the radio signal will experience attenuation (i.e. a reduction in magnitude) at the rate of the square of the distance travelled – known as 'free space loss'. This loss is due to the transmitted energy having to be spread across an ever-increasing size of wave front as it radiates out from the antenna (similar to the surface of an expanding spherical balloon).

However, in practice mobile networks operate in cities, towns and countryside where there are a range of buildings, metal bridges, railways, vehicles, hills, trees, rivers, etc., all of which cause additional loss to radio transmissions. The various causes of the additional losses to free space loss are as follows:

- **Blocking** by large buildings, walls, etc. The loss becomes progressively worse at higher radio frequencies.
- **Reflection** by obstacles such as buildings or large metallic building cranes, the strength of the reflected signal depends on the degree of absorption by the obstacle.
- **Refraction and scattering** by roofs, trees and mountain tops, etc.
- **Multi-path** effects caused by the radio wave reaching the destination via many paths, each experiencing different degrees of reflection and absorption and different path lengths, as shown in Figure 9.3. The net effect is that multiple versions of the transmitted signal are received staggered in time and at differing strengths. If severe this can cause the received radio transmission to be too mutilated to be reliably recovered.

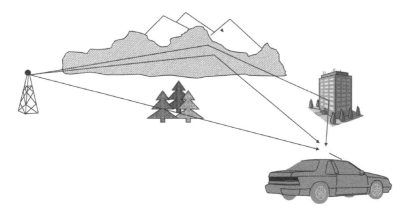

Figure 9.3   Multi-path propagation

- **Rain absorption and scattering** causing attenuation that varies with the weather – the heavy density of water droplets in a fog can have a pronounced effect on the radio signal strength. The effect worsens with increasing frequency of the radio wave.

Other propagation effects experienced by radio waves in a mobile network include:

- Doppler shift in the frequency of the radio signal due to the receiver moving relative to the transmitter, the effect worsening as the speed increases. (A commonly experienced example with sound waves is when the pitch of a train's whistle changes as it rushes towards then away from someone standing at a station.) The distortion to the received frequency caused by the Doppler shift is clearly an issue for mobile users in a moving car or train; consequently, the faster a mobile user travels the poorer the quality of the received signal.
- Electrical noise caused by interference from other radio sources, such as other mobile phones, public radio communication systems, discharges from faulty or unsuppressed electrical systems, etc.

Designers of communication systems using radio therefore have to cope with the loss in signal strength (attenuation) due to free space loss, absorption, scattering, etc., as well as the spreading and distortion to the signal shape due to multi-paths and Doppler shifting. This diminished and distorted signal then has to be extracted from a background of radio noise also picked up by the receiving antenna. The development of mobile telecommunication systems has been characterised by a succession of improvements to the way that the radio paths give ever greater usable capacities by ingenious ways of injecting and then recovering voice and data waveforms, despite the attenuation and distortion to the radio carrier.

So far, we have just considered the propagation of a single sine wave along the radio path. However, the communication channel needs to be superimposed onto this basic sine wave, the latter then becomes the 'carrier' wave. That is to say, the communications channel waveform 'modulates' the carrier wave, also shown in Figure 9.2 and as described in Chapter 4. Of the many modulation techniques available, digital mobile phone systems such as GSM mainly use a form of frequency shift keying (FSK), in which the digital signal to be transmitted switches the carrier between two frequencies [2]. The more recent third generation mobile systems use the alternative technique of phase modulation (PM), in which the encoded (spread spectrum) digital signal causes the carrier's waveform to jump ahead of its sinusoidal sequence in discrete steps. (The now-obsolescent analogue mobile systems used frequency modulation, in which the frequency of the carrier is varied in sympathy with the analogue speech waveform.)

As described above, the radio-carrier frequencies available to mobile operators are a valuable and strictly limited resource, which are allocated by national governments either on a quota basis or through auctioning. The range of frequencies made available to the mobile operators must comply with the overall management of the radio spectrum in that country, which in turn conforms to internationally agreed allocations administered by Wireless Regulation Congress (WRC) (part of the International Telecommunications Union (ITU)). Mobile networks are allocated bands of radio frequencies variously at around 450 MHz, 900 MHz, 1,800 MHz, 1,900 MHz, 2,000 MHz (i.e. 2 GHz) and 2.1 GHz. The exact frequencies allocated vary between countries according to local circumstances and depending on the needs of other radio users, such as cordless telephone systems (e.g. digital enhanced cordless telecommunication (DECT) system). In addition, radio users such as wireless local area networks (LANs) use unregulated parts of the spectrum, in which there are no specific allocations by governments.

A large part of the operational management of a mobile network is the allocation of the set of frequencies owned by an operator to the various parts of its network – a function known as 'radio planning'. This needs to take account of the natural terrain of the country or area served by the network, sources of interference, etc., as well as the expected distribution of demand for calls and data services. There is also a need for all the mobile operators in a country to coordinate their radio planning activities to ensure that there is minimal radio interference between their respective networks. (This is necessary, even though all operators are allocated separate ranges of frequencies, because all radio systems are capable of generating diminished but detectable radiation in adjacent frequency bands as a result of faulty equipment, abnormal weather conditions, etc.) We will now look at how a mobile operator makes maximum use of its allotted frequency bands.

9.4   Cellular networks

Although small capacity radio-telephone car systems were in use as early as the 1950s, large-scale mobile networks really started during the 1970s with the introduction of cellular wireless technology, based on concepts developed in Bell Laboratories [2,3]. The initial cellular mobile systems were based on analogue transmission: AMPS in the United States, NMT in the Nordic countries, and TACS in the United Kingdom. During the 1990s the so-called second generation of cellular systems based on digital transmission were introduced: a wide range of systems in the United States (e.g. D-AMPS, TDMA, CDPD, CDMA one), PDC in Japan, and GSM initially in Europe and subsequently most of the World. In most countries the analogue mobile systems have now been replaced by second-generation (2G) digital systems. This move was driven by the improved user experience – better quality transmission, privacy from eavesdropping through the inherent encryption used in digital systems – and for the operators an improvement in the use of spectrum. An interim generation of mobile systems, the so-called 'Generation 2.5', providing small capacity packet switching for data services, was installed alongside existing 2G systems during the late 1990s.

Subsequent generations of mobile networks, that is, 3G and 4G, have progressively increased the available data speeds (mainly in the downlink) in the radio access system, as well as introducing high-capacity packet switching for data services [4]. In parallel with the increasing data capability of the mobile network systems there have been huge advances in the handset technology, with the inclusion of touchscreens and increasing levels of processing power and data storage capability – leading to the advent of 'smartphones'. As we will discuss later, the developments of the mobile network systems through the generations also encompassed changes in the architecture and the way that voice calls and data sessions are handled. We will also see that in order to give contiguous coverage and to provide good quality voice as well as data connectivity when rolling out 3G then 4G systems, network operators rely on the widespread presence of the older second-generation equipment (e.g. GSM) for the switching of voice calls. Thus, even though 3G was introduced in the early 2000s and 4G from 2012, most national networks comprise a mixture of mobile generations. We will consider each of these in more detail, beginning with GSM since it forms the basis of many of the cellular network concepts.

Figure 9.4(a) illustrates the basic concept of a cellular mobile radio network. The radio network is configured as a honeycomb of cells (shown in diagrams as hexagons for convenience, although in practice they are a variety of shapes), each served by a radio mast. Separate small bands of radiofrequencies are allocated to each cell. However, by careful management of the strength of the transmitted power from each radio mast to a level that is just sufficient to serve the area of its cell, the power received in the next-but-one cell will be so low as to appear to mobile handsets as low-level indecipherable noise. This means that the same set of frequencies can be used in non-adjacent cells right across the network. It is through this frequency reuse that mobile operators have been able to construct the widespread high-capacity networks throughout the World using relatively small amounts of the radio spectrum.

Several cellular frequency patterns are possible, giving 4, 7, 11 or 21 repeat patterns. Figure 9.4(a) shows a seven-cell repeat pattern, the most commonly

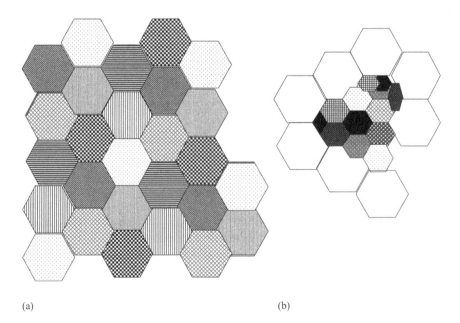

(a)   (b)

Figure 9.4   Cellular structure of a mobile radio network. (a) Full frequency re-use in a 7-cell structure and (b) micro- and pico-cells

adopted by mobile network operators. The required size of the cells depends on the size of the frequency band available and the forecast number of simultaneous calls from within that cell during the busy period. Therefore, cell sizes will vary: typically, from around 1.5 km radius in a town centre to some 15 km radius in a rural area. However, in areas of heavy concentrations of demand for calls (i.e. hotspots), for example, in a shopping mall – small (350 m radius) or very small (50 m radius) cells, known as 'micro-cells' or 'pico-cells', respectively, are used, as shown in Figure 9.4(b) [5].

For a cellular system to work there are several essential functions required, namely:

(i)   A power-management system controlling the strength of the transmitted signal from the antenna serving the cell, and the transmitted power of the handsets in the cell.

(ii) An automated hand-over system to enable handsets to move between cells during a call.

(iii) The ability to locate and deliver calls to and from any handset in any cell.

(iv) A common control-channel available continuously in all cells to all handsets.

We can now consider the basic architecture of a cellular network, as shown in Figure 9.5. Each of the cells contains a radio mast with transmitting and receiving radio equipment and a local control system operating at the allocated send and receive set of frequencies for that cell; this assembly is known as a 'base station' (BS). These BSs are connected by a fixed terrestrial transmission link to a central base-station controller (BSC) using SDH or SONET over optical fibre cable or a microwave radio system (as described in Chapter 4). The BSC manages the temporary allocation of radio channels to handsets on demand in each of its dependent cells through the use of a control channel with each of the base stations. (The various

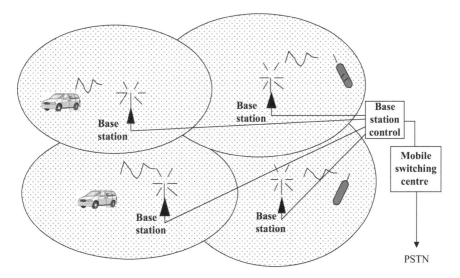

Figure 9.5   Architecture of a cellular network

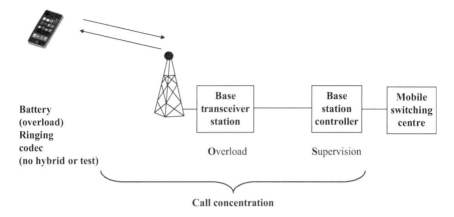

Figure 9.6   BORSCHT functions in a mobile system

ways that the radio capacity in each cell is made available to the mobile handsets is described in the section on access mechanisms later.)

The send and receive speech paths from the handsets are then extended from the BSC over a terrestrial transmission system (cable or radio) to a centrally located mobile switching centre (MSC). Typically, an MSC would have several BSCs in its catchment areas, each of which has a few up to a 100 or so dependent cells (and hence Base stations). A full national mobile network might comprise some 20–100 MSCs, depending on the size of the country.

Comparing the simplified generic view of a mobile system of Figure 9.1 with the architecture of a cellular network in Figure 9.5 we can see that the access control and radio system are provided by the bases stations and base station controllers. The mobility management, identification and authentication of the handsets are provided by systems within the MSC, as is the control, switching and routeing of calls. The MSC acts as a combined local and trunk exchange within the mobile network: switching calls between handsets in its catchment area, and routeing calls to other MSC in the operator's mobile network or to the public switched telephone network (PSTN) and other mobile networks, as appropriate.

From a traffic switching point of view, looking at Figure 9.6 we can see that the equivalent of the concentration function provided by the subscriber concentrator switch-block in a fixed network is undertaken by the base station serving the cell. Here, radio channels are allocated to handsets only on demand when a call is in progress – this constitutes traffic concentration between the many handsets and the small number of radio channels serving the cell. Clearly, as in any traffic-loss system (see Chapter 6), in the event of all radio channels being occupied in a cell no further handsets will be able to initiate or receive a call until an existing call is cleared.

We can take this consideration of the architecture of a cellular network a stage further by identifying the location of the functions necessary for supporting telephones. In Chapter 6 (see Figure 6.5) these functions are captured by the acronym of 'BORSCHT'. In a fixed network, with the normal analogue local loop (copper pair), the BORSCHT functions are all provided by the subscriber line card in the concentrator switch block of the local exchange. However, as Figure 9.6 shows, in a cellular network these functions are spread between the mobile handset and the MSC. In the handset there is a battery (B), a ringing tone device (R) powered by the battery, and part of the supervision function (S), and also the codec (C) providing the analogue-to-digital (and vice versa) conversion of the voice. No hybrid (H) is required since the communication from the handset is on a completely 'four wire' basis, with separate send and receive channels throughout the mobile system. At the BS, BSC or MSC there is no provision of any per-user equipment, so unlike a fixed-network local exchange there is no equivalent of the test (T) or electrical overload protection (O) functions.

From the above, we can conclude, therefore, that the functions undertaken by equipment that represents some 70% of the cost of a fixed-network local exchange (see Chapter 6) are provided by the user's handset in a mobile network. This means that the mobile handset is not only much more expensive than a fixed-line telephone instrument, but also the technical specification is more complex and exacting since the handset must perform many of the critical network functions. As we shall see later, the mobile handset also contains extremely complex and powerful digital-signal processor microchips, which manage the interaction with the BS and the control of its radio transmission power. It is worth noting that the handsets are sold in a competitive consumer market and they are produced by a variety of companies, many of whom are not telecommunications network equipment manufacturers. Often, the relatively high cost of the mobile handset is hidden from the users by the service providers devising tariffs that subsidise the handset cost by increased call or usage charges spread over a period of 6, 12 or even 18 months.

9.5 Access mechanisms in cellular networks

As mentioned in the generic description of a mobile system, a mechanism is required to enable the handsets to gain access to the radiofrequencies for the duration of a call. The mechanism must allocate this capacity in both send and receive directions – known as 'duplex' working. There are several ways in which the band of frequencies allocated to a cell can be made available on demand to the handsets currently in the cell. These so-called multiple-access schemes are based on the three types of multiplexing regimes introduced in Chapter 3: frequency-division multiplex (FDM), time-division multiplex (TDM) and code-division multiplex (CDM). The choice of which multiple-access scheme to deploy depends on their technical characteristics and different efficiencies in the use of the spectrum, the maturity of the appropriate technical standards at the decision time and, to some extent, commercial and political considerations.

(i) **Frequency-division multiple access (FDMA)**: In which the allocated frequency band is divided into a set of frequency blocks, as in an FDM system. When required, handsets are then allocated to a free send-frequency block from the FDM stack on a predetermined or random basis. It is convenient to have the receive frequency block determined by a simple separation, that is, the send and receive frequency blocks are paired; this is known as

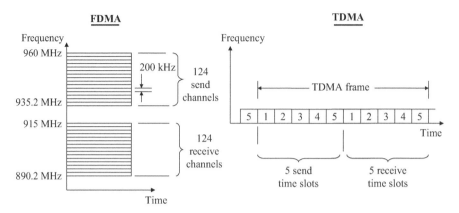

Figure 9.7   Multiple access and duplex arrangements for mobile networks

frequency-division duplex (FDD). An example is shown in Figure 9.7, in which 124 frequency blocks of 200 kHz are stacked as a send group (occupying 925.2–960 MHz) and 124 frequency blocks are stacked as a receive group (occupying 890.2–915 MHz) [6].

(ii) **Time-division multiple access (TDMA):** In which one-half of the allowance of frequencies for that cell are shared across a set of time slots, as in a TDM system (Figure 9.7). When required, a time slot is allocated to a handset for the send direction. The other half of the allocated frequencies are used in a TDMA system for the receive direction, that is, using frequency division to give duplex working (FDD).

(iii) **Code-division multiple access (CDMA):** In which one-half of the allowance of frequencies for that cell are continuously made available to a set of send channels, each separated by the application of a different digital code, as in a CDM system. Again, the second half of the frequencies may be used for a CDMA system in the receive direction (FDD working). When required a send and a receive channel are allocated to a handset.

(iv) **Orthogonal frequency division multiple access (OFDMA):** In which the traffic is carried over a set of subcarriers, rather than just one as in the above-mentioned systems – the OFDM system (as described in Chapter 4) and the multi-access is achieved by continuously allocating the various user channels to the available frequency resources. Thus, the OFDMA technique is essentially an FDM-type scheme which varies with time so that the maximum use is made of the available frequency spectrum by the various users as their demand and the radio transmission conditions vary. The application of OFDMA within the long-term evolution (LTE) system is described in more detail later in this chapter.

In addition to the frequency division method of duplex working, mobile systems also use the alternative time-division duplex working (TDD). Here the send direction digits are sent in a burst followed by a burst in the receive direction – and so on alternately, giving rise to its popular name of 'ping pong' working.

9.6 The GSM system

9.6.1 Background to GSM

The term 'GSM, which for many people is now a household word, was originally a designation for a standards definition group within CEPT (see Appendix 1) – the 'Groupe Spéciale de Mobile' – set up in the 1980s to specify a common second-generation mobile system for Europe. The key attribute sought was the ability for mobile users to roam freely between European countries, enjoying a constant level of service and with charges recoverable by the host network operator from the parent operator. In fact, since its launch in the early 1990s the GSM standard has been a spectacular success, not only across Europe but also across all of the World's continents. Even in the United States, where the widespread deployment of several incompatible proprietary second-generation-mobile systems resulted in an inability to roam nationwide or even regionally, GSM has been adopted belatedly as the only universal system. Consequently, the term 'GSM' has been redesignated as 'Global System for Mobile'. In view of its importance as a second-generation system we will consider its architecture and means of working; it also gives a lead into the descriptions of 2.5G, 3G and 4G systems later in this chapter.

The original GSM networks launched in Europe operated at 900 MHz, then a second wave of GSM networks using the 1.8 GHz (1,800 MHz) band, called 'PCN' (personal communication networks) or 'GSM1800' were introduced. The handsets contain the radio transmitters/receivers (known as 'transceivers') capable of operating at 900 MHz and 1.8 GHz – 'dual band' – so that users can move freely between both types of networks. Generally, in order to be universal, handsets also have the capability of operating at the 1.9 GHz band, as used by GSM (or personal communication services (PCS)) operators in the United States and elsewhere; these are referred to as 'tri-band'.

The main features of GSM are as follows:

- Automatic handover between cells. This enables a user to continue a call uninterrupted as they pass through a succession of cells, for example, when the user is travelling by car. The call will, however, drop out at the boundary of two cells if there are no spare channels for the call to be handed over to in the new cell.
- Automatic roaming between operators' GSM networks in different countries. This requires a commercial agreement to exist between the parent and host operators so that charges can be recovered between them. When a mobile user enters a country where there are several operators, all of whom have commercial agreements with the parent operator, the choice of host allocated to the user is either on a random or sequential basis.
- In order to minimize the amount of radio spectrum used per channel in the GSM cells, normally speech is encoded at 13 kbit/s, that is, much lower than the standard 64 kbit/s used in the PSTN (see Chapter 3). Whilst this gives a noticeably degraded but acceptable sound quality for the users under normal conditions, such low-bit rate coding does incur perceptible delay – due to the time taken by the codec to process the speech signal – and a vulnerability to adverse conditions, when the performance deteriorates sharply.

- Data calls can be established giving 9.6 kbit/s data rate (e.g. as used in the earlier dial-up to the Internet arrangements).
- Short message service (SMS) enables users to 'text', using standard telephone keys, handsets on any GSM network with whom a commercial agreement has been set up by the parent operator. As described later, this low-speed message service, originally specified for engineering supervisory use, is carried in the signalling links between GSM exchanges. SMS was initially an unexpected commercial success with users, but now forms a major source of extra revenue for mobile operators.

9.6.2   GSM system description

We will now work through a simple description of a GSM system by considering the block-schematic diagram of Figure 9.8 which presents the standard GSM architecture. Beginning from the left-hand side, the user's handsets, known as 'Mobile Stations' in the GSM standard literature, are linked via the radio channels to the cell base-transceiver (i.e. transmitter-receiver) station – the BTS. A standard interface is defined between handsets and the BTS, so that any type of handset adhering to this so-called 'Air Interface' may be used on the GSM network.

As described earlier, the BTSs are linked by a fixed transmission system to the parent base station controller (BSC). However, at the entrance to the BSC the speech channels are converted (i.e. transcoded) from the low-bit rate of 13 kbit/s, used over the radio links, to the high-quality standard 64 kbit/s rate which is used throughout the rest of the non-radio parts of the GSM network and, of course, any fixed networks that might be included in the call.

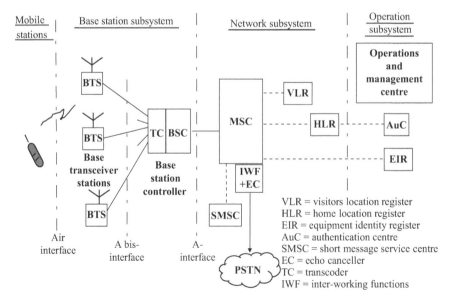

Figure 9.8   Architecture of the GSM network system

Each of the base station controllers are then linked by a fixed terrestrial transmission system (optical fibre or microwave radio, as appropriate) to its parent MSC. (It should be noted that the terrestrial fixed links between the BTS, BSC and the MSCs, may be provided by the mobile operator or carried over leased lines provided by a fixed network operator.) As Figure 9.8 shows, the 'A interface' is defined between BSCs and their MSC – again, to enable a mobile operator to use compliant BSC and MSC equipment from different manufacturers.

The MSC itself is essentially a digital switching unit, as described in Chapter 6. Thus, the MSC comprises typically a TST digital switch-block, through which a series of circuit-switched 64 kbit/s connections are established for the mobile calls. Control is provided by a multiprocessor system with some degree of dispersed regional controllers using micro-processors on the individual modules, as described in Chapter 7. As with the fixed-network exchanges, signalling in the MSC is provided by SSNo.7 systems, but using the mobile message set of the mobile application part (MAP), rather than the fixed-network message set of ISUP or TUP (see Chapter 7). The mobility management functions are provided by adjunct systems associated with the MSC, as follows.

Home location register. The home location register (HLR) is the heart of the mobility management of the GSM network. This is the central database containing the definitive information of all the subscribers registered on the GSM network. Each MSC has a direct link to the single, centrally located HLR. The HLR is supported by an authentication centre (AuC) which holds the details of the registered SIM (subscriber information module) cards allocated to users. The SIM is a card measuring some 1.0 cm × 2.5 cm which, when stimulated, generates a unique code which can be compared with the details held in the AuC. The SIM card is fitted into the handset, which then becomes part of the GSM network; users may change to another handset by swapping the SIM card over. Thus, the SIM in the handset provides the MSC with an authoritative indication of the identity of the user, just as the point of entry of the local loop does in a fixed network – as described at the beginning of this chapter. (Modified versions of the SIM are used in 3G and 4G networks, as described later.)

Visitor location register. Each MSC has a visitor location register (VLR), which is used to record the information relating to any handsets currently in the MSC catchment area. These handsets are either registered on the HLR of the GSM network operator, or they are visitors from other GSM networks and so their registration is with their parent operator's HLR.

Interworking functions and Echo cancellers. One or more of the MSCs in a GSM network also acts as a gateway for calls to and from the PSTN. The main interworking function (IWF) that needs to be provided here is the conversion between the mobile (MAP) and the fixed set (ISUP) of SSNo.7 signalling messages, as described in Chapter 7. There is also a need for echo-cancellation (EC) equipment at this gateway to compensate for the relatively long delays introduced by the 13 kbit/s low-bit rate encoding of the mobile handsets. The EC equipment normally copes with all but very long delays – the latter causing difficulties for telephone users on that call.

Equipment identity register. The equipment identity register (EIR) is a database held by a GSM operator containing the status of all the mobile handsets registered with that network. The serial numbers of the handsets are placed in one of three categories: white for authorised, grey for needed to be kept under surveillance, and black for unauthorised. In this way, stolen or faulty handsets are prevented from using the GSM network.

Short message services centre. The short message services centre (SMSC) co-located with the MSC provides a message-switching function for SMS messages. This is essentially a store-and-forward system associated with the SSNo.7 signalling of the GSM network. The SMS messages from a handset, although addressed using the recipient GSM mobile telephone number, are not in fact switched as telephone calls within the MSC. Instead, the SMS messages are passed by the BSC to the SMSC directly or via a semipermanent path through the MSC switch-block. At the SMSC, the SMS messages are stored and examined one at a time; they are then sent to the destination handset on the GSM network, or another operator's GSM network, as appropriate, carried as messages in the SSNo.7 network between MSCs. Clearly, before the text messages can be dispatched, the location of the destination handset needs to be determined – this is provided by the location-management system as used for telephone calls, and described in Section 9.6.2.

Operation and network management centre. The operations and maintenance subsystem contains the computer-based support systems used to manage the GSM service. These systems provide the following functions for the MSC:

- Data build for the call routeing table in the exchange-control system
- Management of software loads onto the exchange-control system
- Alarm management
- Usage traffic statistics
- Management of billing data.

In addition, the subsystem includes the AuC and the EIR.

9.6.3   Location management in a GSM system

The key attribute of a mobile system is, of course, the ability to determine the location of all active handsets on the network. GSM, like all second-generation systems, relies on the use of a common-control radio channel permanently available for all active handsets to access. The location-management process described below is initiated as soon as a handset is switched on (i.e. 'made active') and continues until it is switched off. (Although the radio control channel is permanently available, 'send' and 'receive' radio channel pairs are allocated only when a call is to be made, as described earlier in this chapter.)

Handset initiated within the same MSC area. Whenever a handset is switched on it sends an initial signal message over the always-available control channel. The serving BSC then forwards the message together with an indication of the identity of the cell-transceiver currently serving that the handset – in effect giving the location of the user, to within the cell or sub-cell area – to the VLR at the serving MSC. This message is simply the subscriber identity number, as given by

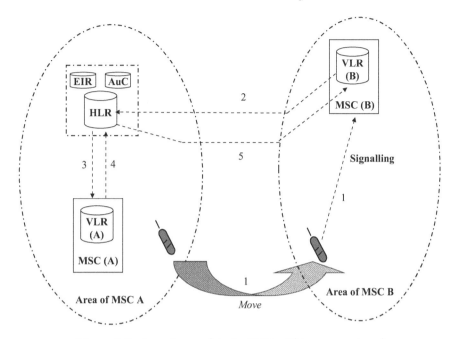

Figure 9.9   Location update in GSM within same network

the SIM card. The record in the VLR indicates that this terminal is already registered as being in this cell, BSC, and MSC areas. Thus, all that is required is for the status of the handset to be recorded as 'active'. The VLR then sends a signal over the control channel to the handset indicating it is now an active participant on the GSM network, capable of making and receiving calls and SMS messages. This indication is in the form of the name of the GSM service provider (which, of course, may be different to the network operator in the case of a MVNO) appearing on the screen of the handset.

Location update with handset in a new MSC area. If the handset is switched on in a different MSC area to when last active, the initiating signal will be passed over the control channel to a new VLR. On receipt of this signal the VLR determines that this is a 'visiting' handset and commences the location-update procedure, which is illustrated in Figure 9.9. This begins with an enquiry message comprising of the SIM identity sent by the new VLR, that is, VLR(B) – to the HLR for the GSM network – carried over SSNo.7 signalling links between the new MSC, i.e. MSC(B), and the HLR (shown as No. 2 in Figure 9.9).

The HLR first determines the status of the subscriber's registration, for example, whether they are authorised to make calls and whether any special service features apply. The HLR then sends a message (No. 3) to VLR(A), that is, the one on which the terminal was last registered, seeking any update status information and advising the VLR that the terminal has moved to VLR(B). The VLR(A) sends a signal message (No. 4) back to the HLR advising of updated information, and that the temporary storage of this terminal's details have been cleared.

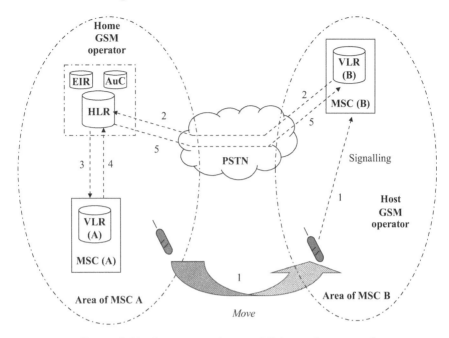

Figure 9.10   Location update in GSM on other network

The HLR now sends a final message (No. 5) to VLR(B) detailing the relevant status and service features information relating to the terminal. An active participant signal is then sent by VLR(B) to the terminal.

Location update with handset in the serving area of another GSM network operator. An important variant on the location-update procedure described above is where the handset is moved into the serving area of another GSM network operator, often this is in another country. In principle, the procedure is the same as above, except that VLR(B) will not have links from its MSC to the parent HLR since the latter is part of another GSM network. The VLR(B) to HLR signalling (messages No. 2 and No. 5) must therefore be carried between the two GSM networks ('host' and 'home') over the SSNo.7 signalling network as shown in Figure 9.10. Normally this capability is provided by the PSTN, or two PSTNs with a section of international network in the case of host and home networks being in different countries. Exceptionally, certain GSM networks may have direct traffic routes between them which carry these SSNo.7 messages.

9.6.4   Mobile call in a GSM network

Once the location-update procedure is completed the serving VLR has all the necessary location and status information and the handset is ready to make or receive calls. After keying the digits the caller (A) presses the send button on the terminal, which initiates a call-request message containing the destination number to the serving MSC. The MSC then handles the call in a similar way to a fixed-network

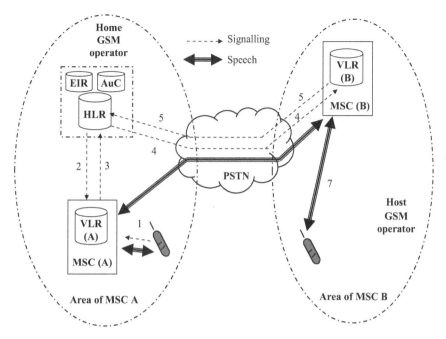

Figure 9.11   Mobile call in GSM system

telephone exchange, with the dialled digits examined by the exchange-control system. If the called number is part of the numbering range of A's GSM network, then the MSC interrogates the HLR to determine the current location of the called handset. The HLR indicates the identity of the destination VLR currently supervising the called handset so that the originating MSC can switch the call directly to the destination MSC/VLR. The called handset is then set to ring by a signal message from the destination MSC; but unlike a fixed network telephone, the ringing is generated by the handset itself. (This gives the user the ability to personalise their ringing sound by either selecting from a pre-set range built into the handset or downloading prepared tones from an information provider.)

Figure 9.11 illustrates the slightly more complex example of a call between two registered users of a GSM mobile network (home), but handset B is temporarily located in the area of another GSM network (the host). As a result of the location-update procedure described above, the home HLR contains the identity of the host VLR currently supervising handset B. Thus, on receiving the dialled digits the home MSC has the message interchange with the HLR (No. 2 and No. 3) described before, but now the HLR needs to communicate with the host VLR on the distant network in order to complete the call. This is normally achieved by signalling over SSNo.7 PSTN network between the two GSM networks (No. 4 and No. 5). Once the host VLR has indicated that the terminal B is active and free (message No. 5) the call can be completed by establishing a standard circuit-switched connection over the PSTN, including an international section if appropriate. This PSTN call connection is routed using a telephone number temporarily applied to that call from a pool allocated to the host MSC for this purpose. On receipt of this call from the PSTN, the host MSC is able to initiate ringing of mobile B (No. 7).

9.6.5   Cell hand-over and power management

As described at the beginning of this chapter, it is necessary for the power of the radio transmitter (at the BS) in a cell to be adjusted to a level that enables good reception by the handset up to the boundary of the cell; then the level should rapidly deteriorate at greater distances. Similarly, the transmitted power from the handset needs to be set at just sufficient level to be adequately received at the base station's transceiver. A GSM network's cellular structure critically depends on efficient power management.

Since the handset is likely to be on the move for much of the time, the GSM power-management system needs to check and set continually the radio transmission power level of active handsets. The variation in required power depends not only on distance from the cell's BS, but also on the local reflective and absorptive environment – which can change in only a few metres of handset travel. Typically, power management control signals are sent to active terminals every 5 minutes or so.

The GSM handset not only monitors the signal strength from its BTS, but also from the BTSs of the adjacent cells, and sends back an indication of the signal strengths to the respective base stations (BTSs). Thus, in the case of a moving handset approaching a cell boundary, once the radio channel received by the adjacent cell becomes the stronger a handover is initiated. Where the two cells BTSs are controlled by the same controller (BSC), it can manage the handover. This involves the send and receive speech paths from the MSC to be switched at the BSC from the link to Base Station 1 (BTS1) to BTS2. The uninterrupted call, then continues via the radio send and receive channel combinations allocated to BTS2. Clearly, in the case where all of the send/receive channels are currently occupied in cell 2, the call cannot be sustained and it has to be abandoned – 'call drop out'.

Figure 9.12 illustrates the GSM handover arrangements in the situation where the adjacent cells are dependent on separate BSCs, in which case the common-parent MSC controls handover; and where the BSCs are on separate MSCs, in which case the two MSCs work together to manage the handover.

9.6.6   GSM frame structure

Finally, in considering the GSM system we will briefly look at the way that the radio capacity is allocated to users making calls – namely, the GSM frame structure. Actually, GSM uses a combination of the FDMA and TDMA access mechanisms described earlier in this chapter. The arrangement is illustrated in Figure 9.13. Firstly, the spectrum owned by the network operator is split into a send and receive set of frequencies, known as 'uplink' and 'downlink', respectively, each of which is frequency-division multiplexed into a group of GSM frame channels of 200 kHz width (In the example of Figure 9.13, the 124 upframe channels occupy 935 MHz to 960 MHz and the 124 downframe channels occupy

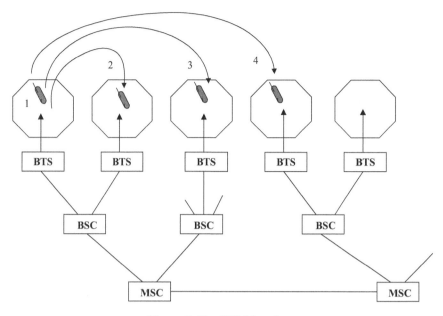

Figure 9.12   GSM handover

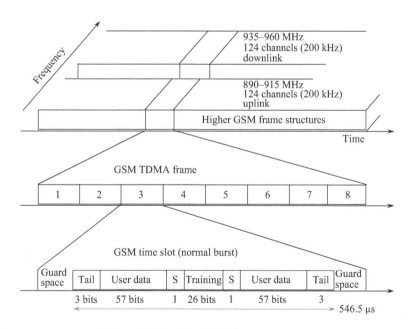

Figure 9.13   The GSM frame structure

890–915 MHz). Each of the GSM channel frames is then divided into eight time slots using time-division multiplexing. Each call is allocated an up and down time slot from the up and down channel frames. We may, therefore, describe the multiple-access arrangement for a GSM system as firstly a TDMA of a 200 kHz channel, which itself is in FDMA to the available spectrum. Finally, the uplink and downlink directions are separated by frequency bands, creating FDD. The scheme may therefore be described as 'TDMA/FDMA/FDD'. The time slots derived in this way carry not only the speech, or 'traffic channels', but also various types of signalling and control channels.

Each GSM time slot, whether carrying traffic or control channels, occupies 546.5 μs (~0.5 ms) and carries 114 bits of encoded speech or control information. In addition, there are tail bits which indicate the start and finish of the time slot, together with 26 so-called 'training' bits, which are used to compensate for radio transmission impairment. Since, as we discussed in Chapter 3, speech with 4 kHz bandwidth must be sampled at 8,000 times per second, that is, once every 125 μs, the TDMA time slot actually contains a 'burst' of the equivalent of four consecutive speech samples. For system-management purposes, the TDMA 8-time-slot frames are grouped into 26 or 51 to form 'multiframes', which in turn are grouped into 50s to form 'superframes', 2,048 of which fit into a 'hyperframe'. These frames are similar in concept to the TDM frames used in synchronous digital hierarchy (SDH) and plesiochronous digital hierarchy (PDH) transmission systems described in Chapter 4. More details of the GSM system are available in Mouly and Pautet's authoritative book [7].

So, to summarise, the voice of the caller on a GSM network is carried over the radio portion as a series of burst of bits in a time slot, which is part of a TDMA frame, which itself occupies part of the frequency range used in the cell. This combined frequency and TDMA arrangement may seem unnecessarily complicated, but it does provide an efficient use of the precious radio spectrum. In addition, the TDMA system is flexible so that the time slots can be used in other ways to create new higher-speed services, for example, general packet radio services (GPRS), as described in Section 9.7.

9.7   General packet radio service

In the beginning of this chapter, the move to providing packet-based data within mobile systems was described. The GPRS is the interim packet-based system – the so-called 2.5 generation (2.5G) – which was added as an overlay to many of the existing GSM networks. This gave mobile operators and users the opportunity to develop and use packet-based services at moderately increased speeds in advance of the widespread deployment of the more-complex full-facility and higher speeds of 3G and 4G networks.

Generally, the GPRS mobile terminals (e.g. handsets) enable the users to make normal GSM-like voice calls as well as packet-based data sessions to the Internet or private corporate networks (intranets). However, in addition a dedicated range of data-only GPRS terminals are available which plug directly into a PC or laptop computer, thus making the computers act as the mobile terminal. The design philosophy of GPRS is that it should be added to an existing GSM network, exploiting as much of its installed capability as possible. Thus, GPRS handset/terminals use the GSM (TDMA/FDMA/FDD) radio-access system.

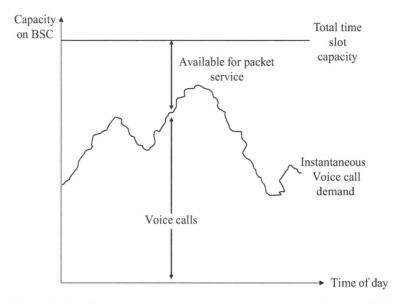

Figure 9.14   Allocation of spare capacity for packet switching in GPRS

At the base station controller, voice calls from GPRS handsets are passed to the GSM MSC for circuit switching and they are carried as standard calls within the GSM network, as described above.

When a packet session is initiated by the GPRS terminal the (GPRS-enhanced) BSC passes the 'call' to the serving GPRS support node (SGSN). There is another important difference with the packet 'call' (i.e. 'session'), and that is the ability for it to use more than one pair of up and down time slots in the radio access link. In fact up to eight timeslots can be concatenated in order to give the user higher data rates. The allocation of the additional time slots to a data session is made on a real-time basis, tracking the instantaneous spare capacity on the radio multiple-access system managed by the BSC. Priority is given to the demand for time slots by voice calls. Thus, the speed of packets throughput during a GPRS packet session will vary continuously between one and eight (maximum), depending on the number of spare time slots available at the BSC. This concept is shown in Figure 9.14.

We can now consider the specific GPRS network which handles the data packets, shown in Figure 9.15. The main component in the network is the serving GPRS support node (SGSN), which provides the packet switching capability. A GPRS network comprises a number of these SGSNs, typically co-located with the GSM MSCs, each of which serves a catchment area of several BSCs. The link between each BSC and the SGSN is provided by a frame-relay packet system, accessed via a packet communication unit (PCU) (see Chapter 8). Thus, virtual paths are established (at Layer 2 of the OSI 7-layer model) for each of the packet sessions from the GPRS terminals to the parent SGSN; IP packets to and from the terminals then flow (at Layer 3 of the OSI model) over the Frame Relay virtual

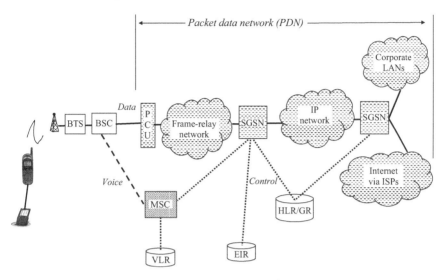

Figure 9.15   GPRS outline

paths to an IP router in the SGSN [8]. Figure 9.15 shows how the voice and data traffic from the BSCs is separately routed via the MSCs and SGSNs, respectively.

The GPRS network of SGSNs is linked by an IP network, usually by core transmission using ATM over SDH transmission on optical fibre or microwave point-to-point radio, as described in Chapter 8. One or more of the SGSNs also acts as a gateway from the GPRS network to all the external networks – these are designated gateway GPRS support nodes (GGSN). In practice, a large number of ports are established at the GGSN with virtual path links provided by network operators to a range of ISPs, corporate LANs, and the public data networks, as shown in Figure 9.15. The GSM network's VLR-MSC and the HLR provide the mobility management for the GPRS overlay network, as described earlier in this chapter.

The introduction of a packet-based mobile system, such as GPRS, raises the need for a different form of charging for data service compared to the circuit-switched voice calls on GSM (mobile) and the PSTN (fixed). The main difference is in the fact that a packet system enables an always-on status for the GPRS terminal whilst it is active. Bearing in mind that mobile terminals are typically kept active almost continuously, packets can flow from the terminal via GPRS to an ISP for Internet access at any time. The charging is therefore based on a flat rate, irrespective of packet activity, or alternatively the charging is based on packet or data throughput. An example of the latter is the charging for Mbytes of data carried per month.

There are several other digital second generation cellular systems which have been deployed – a selection of which are briefly reviewed in Box 9.1

Box 9.1   Other cellular systems

HSCSD. The basic GSM network can be upgraded to give higher-speed data transmission using the high-speed circuit-switched data (HSCD) system. This uses addition software in the handset/mobile terminal and MSC that enables up to four of the TDMA traffic channels to be concatenated to give data throughput speeds up to 43.2 kbit/s, compared to the 9.6 kbit/s data call available over standard GSM [8,9].

    EDGE. The enhanced data rates for global evolution (EDGE) system, which is essentially an enhanced version of GPRS, is viewed as a step towards full 3G by some operators and an early 3G system by others – as shown in Figure 9.16. It offers data rates up to 384 kbit/s through the aggregation of TDMA time slots [9].

    DECT. The DECT system is a European-wide standard cellular system used for cordless telephones (in the office and home), as well as campuses, factories, railway stations, etc. DECT is designed primarily for indoor use within cells of up to 300 m, but typically around 30 m radius, and with high density of users [8].

    TETRA. The European standard for terrestrial trunk radio (TETRA) system offers business and service communities (e.g. taxi services, vehicle fleets, police service) shared access to a set of radio frequencies for voice and data services. The system has a similar architecture to GSM [8].

9.8   Third generation mobile systems

9.8.1   Background to 3G

Whilst the 2G and 2.5G mobile systems offer good quality voice plus some degree of data switching, the Third Generation (3G) mobile system was designed specifically to support a full range of multimedia services. This encompasses good quality voice, packet access to the Internet and corporate LANs, video, interactive games, etc., with higher data rates. As described earlier in this chapter, the deployment of mobile networks throughout the World has followed various progressions between second-generation systems to 2.5G systems then 3G systems. The progression of standards following the individual paths within the United States, Japan, and Europe is shown in Figure 9.16. Thus, the generally used term 3G covers three different standards, all offering similar facilities, but with little compatibility between them. The European mobile operators favour the universal mobile telecommunication system (UMTS) which, as its name suggest, was seen as the natural successor to GSM and with similar expectations about World-wide adoption. The other two major standards are CDMA 200 mainly for the United States, and TDS-CDMA in China.

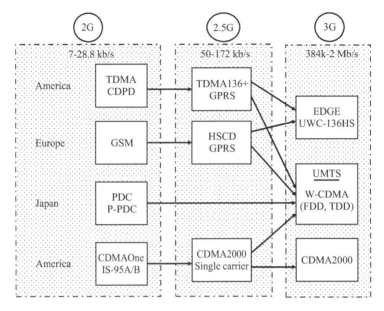

Figure 9.16   Mobile standards evolution from 2G to 3G

9.8.2   Universal mobile telecommunications system

The UMTS is a unified system which supports multimedia services, global roaming, service portability and Internet-style addressing (in addition to telephone numbering). It supports a range of new multimedia terminals with high-speed user rates depending on the extent of the user's movements: up to 2 Mbit/s for stationary terminals; up to 384 kbit/s for pedestrian movement; and up to 144 kbit/s for vehicular movement. The principal feature of UMTS is that it relies on a new radio access system, known as the UMTS terrestrial radio access network (UTRAN), which provides the higher-speed multimedia capability. The modulation and access mechanism is based on a wide-band form of code-division multiple access (W-CDMA) using either frequency or time-division duplex working (i.e. FDD or TDD, respectively) [10].

The UTRAN uses the set of frequencies allocated to 3G services (in the United Kingdom: 900 MHz and 2.1 GHz), which in many countries were auctioned by national governments to the mobile operators. Wideband-CDMA gives an efficient exploitation of available spectrum, supporting a variety of stationary and moving terminals, requiring differing data rates. It has the unusual characteristic of the so-called 'cell breathing': whereby, the transmission power from the base station is varied to cope with the instantaneous density of users' traffic, so causing the extent of the cell to increase or decrease. This is done to ensure that all terminals within the cell receive an adequate transmission quality (e.g. as measured by the signal-to-interference noise ratio) at all times. (As described in Chapter 3, all CDMA handsets within a cell involved in a call/session use the same set of frequencies continuously, with individual codes enabling each handset-to-base station link to be separated.)

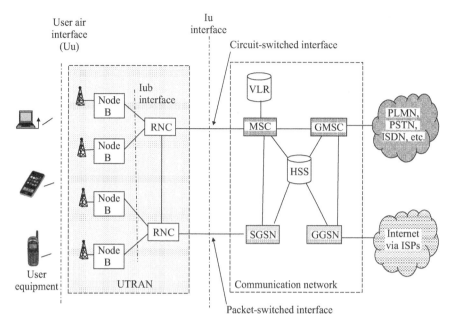

Figure 9.17 UMTS system architecture

The architecture for UMTS is shown in Figure 9.17. The UTRAN system comprises Node Bs, which are like the BTSs of a GSM network, and radio network controllers (RNC), which provide similar functions to the BSCs of GSM networks. Like GPRS, the two types of call are separated at the interface of the radio access (UTRAN) to the communication network (CN) where two paths of switching are provided. Voice traffic is extended by the RNC to the (usually existing GSM) circuit-switched MSC. Voice calls interconnect with the PSTN or other circuit-switched mobile networks via one of the MSCs, acting as the gateway (GMSC).

Packet-based traffic is sent by the RNC to the IP router in the serving GPRS support node (SGSN). IP packets from the users' terminals are therefore carried over the radio access system (UTRAN), via the ATM-based virtual paths from the RNC and through the IP routers in the SGSN. Egress to the ISPs and Internet, corporate LANs, etc., is provided via the gateway GPRS support node (GGSN), as described earlier for GPRS.

So, to summarise: the initial architecture for the European 3G system, UMTS, is based on a new radio access network (UTRAN), but with the mobile management and voice switching handled by the GSM standard MSC and the HSS/VLR system (note the HSS incorporates the HLR and AuC functions of the GSM network); whilst the data traffic is handled by the IP packet-based GPRS serving node. Voice traffic interconnects with other mobile and fixed networks via standard circuit-switched exchange interfaces, using SS No.7. Data traffic interconnects with the external data networks and ISPs via an IP interface [11].

9.9 Fourth generation mobile systems: LTE

9.9.1 Background to 4G

The widespread rapid adoption of smart phones from about 2010 onwards across the World greatly contributed to the major behavioural change in people's use of mobile phones. Rather than just providing voice calls with some small amounts of data use, they became the single essential personal device for all communications – especially social networking, messaging, texting, voice calls, and use of apps (together with other non-cellular network applications such as satellite navigation, Wi-Fi, music and video playing, and games). The consequent demand for efficient and cost-effective mobile service for increasingly fast rates of data therefore set the market-place drivers for the next generation of mobile cellular networks – i.e. '4G'. This generation is centred on establishing an IP data platform with high-speed mobile access, in which speech is treated as a form of data service. In fact, the 4G system does not have any voice circuit-switching capability. As we see later, in terms of its technical specification the 4G system architecture is essentially a mobile version of the Next Generation Network (NGN) defined by the ITU, and described in Chapter 11.

The 4G system specification has been defined in a series of phases spread over several years, as is normal with technical standards for telecommunications networks. The objective is to keep pace with both usage demands and the needs of network operators in line with developments in the various technologies, for example, batteries, screen display, wireless and antenna, IP transmission, high-speed packet access, and service-control systems. As well as providing the higher data speeds and lower latency (i.e. delay) demanded by users, the move to 4G is addressing the operators' needs for improved efficiency in the use of the radio spectrum (a valuable resource) in terms of bits carried per Hz, and hence lower unit costs. There have been two streams of global developments moving towards the internationally agreed ITU-R 4G objective: the LTE system specified by the 3GPP consortium and described in Section 9.9.2, and the WiMAX system specified by the IEEE, which is briefly described in Box 9.4 at the end of Section 9.9.

9.9.2 Long-term evolution system overview

As the name suggests, the LTE system is based on evolutionary developments of the UMTS 3G system, each phase of specification being closer to the objectives set by the ITU for the 4G mobile system [5]. Despite the manufacturers' and network operators' desire to use the prestigious '4G' label, the early versions of LTE were deemed by the standards community as transitional and given titles such as '3.75G' and '3.9G'. However, the mature version of LTE (i.e. Release 8) is generally acknowledged to be sufficiently enhanced for it to be considered truly a fourth-generation system since it represents a significant advance on 3G. (Although there is still a view that the more recent LTE-Advanced system (Release 10) is more justifying of this term.) The main attributes of LTE are:

(a) Multiple bandwidths specified for the radio links ranging from 1.4 to 20 MHz, allocated using OFDM multiple access (OFDMA). (OFDMA is described in Section 9.5.)

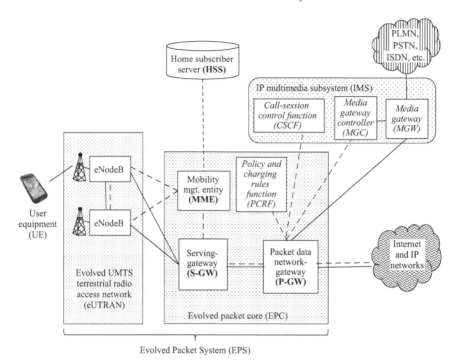

Figure 9.18   LTE system architecture

(b)   User data rates reaching some 100 Mbit/s in the down links and 50 Mbit/s in the Up links.

(c)   Low latency (in the region of 20 ms – that is, some 2 to 3 times better than in 3G networks).

(d)   A wide range of multimedia services are provided, with quality of service (QoS) management of the data flows.

As described above, the LTE architecture is based on an all-IP platform using systems evolved from UMTS (3G). This is illustrated in the simplified representation of Figure 9.18. An important concept within LTE is the separation of the packet flows into those carrying user data (including speech), shown as solid lines – and those carrying control information – shown as dotted lines. Often these two different sets of functions are referred to as the 'user plane' and 'control plane', respectively. The LTE architecture includes all the necessary components of a cellular mobile system, as described earlier in this chapter, namely: cellular radio access system, mobility management, subscriber information management, service control, traffic routeing, and inter-connection with other networks. We will briefly consider the components shown in Figure 9.18.

9.9.3   LTE user equipment

The LTE UE may be a handheld smartphone or an appropriate plug-in to a computer, laptop or tablet. In all cases, the UE comprises of the mobile termination

(MT) which provides the air interface to the evolved UTRAN (eUTRAN), and the terminal equipment (TE) which provides the user data protocol management – including the analogue-digital conversion (i.e. codec) and IP packet assembly for voice calls. Since the LTE uses an all-IP network, all transport-level interfaces are Carrier Ethernet (CE) (see Chapter 8). The radio air interface between the UE and the eUTRAN is based on OFDMA, using sub-carriers 15 kHz apart. This multiple-access method is based on OFDM, as described for xDSL digital transmission in Chapter 4, but it has the important distinction that the groups of users are allocated only some of the subcarriers for some of the time. We look at the OFDMA allocation process later in this section.

In addition to providing access for many users to the radio channels, OFDMA within the LTE system also improves the spectral efficiency (in terms of bits carried per Hz) [12]. Unfortunately, OFDM does have the disadvantage of generating a widely varying power level due to the vector addition of the powers from the various subcarriers – characterised by a high peak-to-average power ration (PAPR). This sets the need for expensive highly linear power amplifiers/ transmitters to prevent QoS degradation. These devices are relatively easily provided by the network operator at the eNodeB. However, due to the need to keep the costs low, this is not a practical solution for the subscriber's UE. Therefore, a modified version of OFDM is used instead in the UE for the up-link, in which a further stage of signal processing (fast Fourier transform) is applied to the OFDM signal. This produces a single carrier FDMA (SC-FDMA) signal with a more constant waveform power for transmission, allowing cheaper amplifiers/ transmitters to be used in the UE [12].

Finally, the transmitted signal is put on to the radio carriers in the eUTRAN using the quadrature-phase-shift-keying (QPSK) modulation technique (see Chapter 4) in the form of 16 QAM or 64 QAM (where the number of bits taken as a group in the phase-shift keying modulation process is 16 or 64, respectively).

9.9.4   eUTRAN and eNodeB

The eUTRAN is composed of radio antennas and their co-sited radio transceivers forming the evolved NodeB's (eNodeB), which not only perform the UMTS NodeB functions but also the radio power-control and carrier-allocation functions. So, compared to the 3G architecture the eNodeB is an intelligent node providing much of the functionality of the network controller. This configuration reduces the latency of the LTE system compared to the 3G arrangement. The LTE architecture also offers an additional saving in latency by enabling the eNodeBs to send user-data directly over transmission links to neighbouring eNodeBs, so providing speedy hand-over and subsequent control plane transmission between neighbouring cells.

The eNodeB works closely with the mobility-management entity (MME) to perform the functions of managing the cellular network, in particular, the physical layer roles of modulation and demodulation, channel coding and decoding, radio resource control (RRC), management of handover between cells, as well as the standard Data link (Layer 2) functions, such as error correction.

9.9.5 The LTE air interface

Antenna. The data capacity achieved over the radio links of the eUTRAN is increased by using multi-input and multi-output (MIMO) antenna systems. These exploit the multi-path characteristic of radio transmission (described earlier in this chapter) through the use of clusters of two or more antennas at the eNodeBs and, where possible, at the UEs also. This arrangement uses spatial diversity and techniques known as pre-coding and space-time coding to increase the data rate on the channels, particularly those operating with poor signal-to-noise ratio (SNR). The MIMO concept, which is also used in WiMAX as well as some Wi-Fi applications, is briefly described in Box 9.2.

Radio spectrum. A set of frequency bands have been identified by the 3 GPP consortium for use by LTE networks. These are around 1.4, 1.5, 1.7, 1.8, 1.9, 2.0, 2.1, 2.3, 2.4, 2.6 GHz and there is another group around 700 and 800 MHz (i.e. 0.8 GHz). Box 9.3 lists the full 3 GPP set of bands and the applicable regions in the world for LTE Advanced (the development of LTE, as discussed later). The allocation, of course, must serve both uplinks and downlinks and the two frequency ranges are either coupled – known as 'paired' or independent – known as 'unpaired'. The higher range of carrier frequencies allows greater bandwidth or capacity to be allocated per cell than for the 800 MHz range. However, the

---

Box 9.2 MIMO

A useful starting point in explaining MIMO as used in LTE and WiMAX is to consider the standard spatial multiplexing arrangement, in which a base station with two transmitting antennas sends two wireless beams to a receiver UE with two antennas (see Figure 9.19(a)). If both sets of antennas are spaced $\lambda/2$ (carrier frequency) apart the fading on one path will be compensated by a stronger signal on the other path. This technique of quality control is often called space diversity and can be associated with space-division multiple access (SDMA) systems.

If the two receiving UE are far apart (in terms of $\lambda/2$) then the two sets of wireless paths could use the same set of carrier frequencies and both UEs benefitting from two parallel paths, in effect doubling the traffic-carrying capacity (see Figure 9.19(b)). However, for this arrangement to work – that is, the two wireless paths being additive rather than cancelling – some pre-processing of the send signals is required at the eNB. This technique is usually known as MIMO (multi-input and multi-output) and in the case shown where two or more UEs are involved is referred to as 'multi-user' (MU) MIMO.

Finally, the spatial separation can be extended into the so-called 'beam forming' where several parallel wireless paths from a set of antennas at the eNB are able to direct all the wireless energy into narrow beams. The technique relies on all other paths (using the same set of carriers) to the other UEs cancelling each other out so that only the selected beam is effective, which is achieved by using a calculated set of phase shifts applied to the non-operative paths (see Figure 9.19(c)). Another set of phase shifts are applied during the next time slot to activate the path to UE2 – hence steering from UE1 to UE2. More information is available in [14–16].

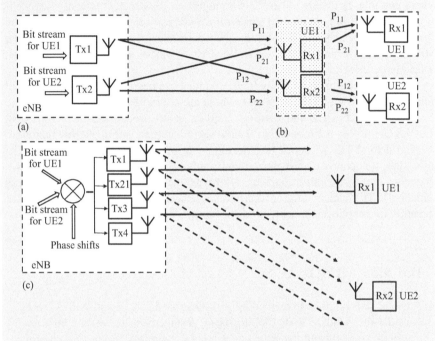

Figure 9.19   MIMO and beam steering. (a) Spatial multiplexing, (b) downlink MU-MIMO, and (c) beam steering

absorption loss also increases with frequency so the cell sizes for the higher frequencies are correspondingly lower. Operators, therefore, use the higher frequencies for deployment in the smaller more traffic-dense cells (usually serving the urban areas), whilst the lower traffic-capacity but larger radius cells of the lower 800 MHz range is best suited to the rural areas. In the United Kingdom, Ofcom awarded five 4G licences for the 800 MHz range (capacities: 2 pairs of 5 MHz, 2 pairs of 10 MHz); and in the 2.6 GHz range (capacities: 1 pair of 15 MHz, 2 unpaired of 25 MHz, 1 pair of 20 MHz, 1 pair of 35 MHz) [13].

OFDMA and resource allocation. The Access control system (see Figure 9.1) for LTE is based on OFDMA to distribute the available radio channel resources to the active users. The available spectrum for the mobile cell is therefore split into a set of sub-carriers, which in the LTE system are spaced 15 kHz apart. Each sub-carrier suffers different levels of fade during radio transmission, so the data traffic from each active subscriber is spread across only those sub-carriers currently experiencing acceptably low levels of fade. The allocation is managed by breaking the system's spectrum and time into a pool of resource elements, each covering one

Box 9.3   Operating bands for LTE-advanced [15,17]

Operating band	Uplink (UL) frequencies (BS receive/UE transmit) MHz	Downlink (DL) frequencies (BS transmit/UE receive) MHz	Duplex mode	Applicable regions
Paired bands				
1	1,920–1,980	2,110–2,170	FDD	Europe, Asia
2	1,850–1,910	1,930–1,990	FDD	North and South America, Asia
3	1,710–1,785	1,805–1,880	FDD	Europe, Asia, North and South America
4	1,710–1,755	2,110–2,155	FDD	North and South America
5	824–849	869–894	FDD	North and South America
6	830–840	865–875	FDD	Japan
7	2,500–2,570	2,620–2,690	FDD	Europe, Asia
8	880–915	925–960	FDD	Europe, Asia
9	1,749.9–1,784.9	1,844.9–1,879.9	FDD	Japan
10	1,710–1,770	2,110–2,170	FDD	North and South America
11	1,427.9–1,447.9	1,475.9–1,495.9	FDD	Japan
12	698–716	728–746	FDD	United States
13	777–787	746–756	FDD	United States
14	788–798	758–768	FDD	United States
15 and 16	Reserved	Reserved	–	–
17	704–716	734–746	FDD	United States
18	815–830	860–875	FDD	Japan
19	830–845	875–890	FDD	Japan
20	832–862	791–821	FDD	Europe
21	1,447.9–1,462.9	1,495.9–1,510.9	FDD	Japan
22	3,410–3,500	3,510–3,600	FDD	–
Unpaired bands				
33	1,900–1,920		TDD	Europe, Asia
34	2,010–2,025		TDD	Europe, Asia
35	1,850–1,910		TDD	North and South America
36	1,930–1,990		TDD	North and South America
37	1,910–1,930		TDD	–
38	2,570–2,620		TDD	Europe
39	1,880–1,920		TDD	China
40	2,300–2,400		TDD	Europe, Asia
41	3,400–3,600		TDD	United States

OFDMA sub-carrier's bandwidth of 15 kHz and one OFDMA symbol interval of 71.4 μs. Each symbol represents 2, 4, or 6 bits of the encoded traffic channel depending on whether QAM, 16-QAM, or 64-QAM modulation is used, respectively. The resource elements are grouped into resource blocks of 12 sub-carriers

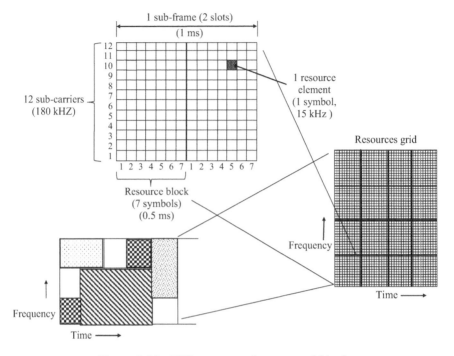

Figure 9.20   LTE resource elements and blocks

Table 9.1   Resource blocks available in LTE radio channels

Channel Bandwidth	1.4 MHz	3.0 MHz	5 MHZ	10 MHZ	15 MHz	20 MHz
No. of resource blocks	6	15	25	50	75	100
Occupied channel bandwidth	1.08 MHz	2.7 MHz	4.5 MHz	9.0 MHz	13.5 MHz	18.0 MHz
No. of sub-carriers	72	180	300	600	900	1,200
Bandwidth efficiency	77.1%	90%	90%	90%	90%	90%

(180 kHz) occupying seven symbols (i.e. 0.5 ms), as shown in Figure 9.20. The base station within the eNodeB schedules the radio resource capacity to active subscribers on the basis of a resource grid of blocks [18]. The minimum allocated bandwidth to a UE is just one block – that is, 180 kHz for 0.5 ms duration.

Figure 9.20 also illustrates how the resource blocks within the resources grid can be allocated to individual active users over time (for convenience in the diagram each user is shown as a different pattern). The resource algorithm is designed to continually adjust the allocation to make best use of the available resource blocks as the traffic load (in terms of the number of active users and their bandwidth demand) and radio transmission conditions vary.

Table 9.1 shows the six possible radio channel bandwidths for LTE systems and the consequent number of resource blocks that can be allocated. The actual bandwidth occupied by the resource blocks is less than the radio spectrum due to the overhead required for guard bands, synchronisation and signalling within the OFDMA system [18].

LTE channel logical architecture: At this stage of the LTE description, it is useful to consider a logical architectural view of the structure of the various channels which map the active users' data traffic/control signals to the radio channels within the eUTRAN cell. It is convenient to envisage three horizontal levels, which correspond to Layers 1, 2, and 3 of the OSI 7-layer model (see Chapter 8), which together map the user and control flows to (and from) the radio channels. The various channels flow in a vertical direction carrying the users' data traffic/control signals through the layers where encoding, mapping and modulation, etc., takes place, as appropriate. The set of logical channels fall into two groups: those carrying control information (from the control plane) and those carrying users' data (from the users' plane). Some of the channels are used only for uplink (UL), some for downlink (DL) and some are both way. As is normal with mobile systems, there are channels that all UEs may use (e.g. to request that a call or session be set up); and other channels that are dedicated to active users during the session/call once this has been established. Briefly, the roles of each layer are as follows:

- Layer 3 is known as the 'logical layer' and it provides 'radio link control' (RLC) mapping of traffic and control signals from UE or the NodeB (UL and DL) to and from the MAC layer (see Channel 8).
- Layer 2 is known as the 'Transport Layer' where the functions associated with the MAC layer are carried out – namely error control of the message and cell formatting and flow control.
- Layer 1 is known as the physical layer. This is where the standard layer 1 activities take place – for example, the modulating of the user data and control information onto the radio carriers, ready for transmission via the antennas (and correspondingly for the reverse direction of transmission).

The positioning of the logical, transport and physical layers in the context of the full LTE system protocol stack is shown for the radio link over eUTRAN (i.e. between UE and eNodeB) and the (usually) terrestrial link over land line between eNodeB and the MME in Figure 9.23.

Duplex modes. As described earlier the LTE system has been specified as a progressive evolution from the various 3G systems deployed by operators around the world. Therefore, the LTE specifications contain a range of options which aim to ease the deployment of this new generation mobile system within existing networks. One example of this is the range of duplex modes that the eUTRAN can accommodate. Some operators would wish to use the frequency-division (FDD) mode with paired spectrum, whilst operators having only unpaired spectrum would use the time-division (TDD) mode. (The appropriate duplex modes for the LTE bands are also shown in Box 9.3.) There is also a half-duplex option (i.e. only one direction of transmission at a time) for use with paired spectrum, which is a combined FDD and TDD mode that separates uplink and downlink between the UE and eNodeB in both time and frequency. This hybrid mode enables less complex processing in the UE, which is an advantage for multi-mode mobile handsets. Figure 9.21 illustrates these LTE duplex working modes in a simple time-frequency graphical form [18–20].

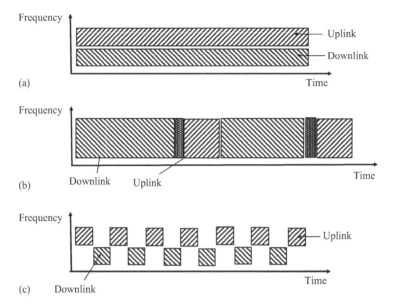

*Figure 9.21   Categories of duplex working in the eUTRAN of the LTE
system. (a) Pure FDD, (b) pure TDD, and (c) mixed FDD
and TDD (half duplex)*

Figure 9.21 also illustrates the other advantage of using pure TDD, namely the asymmetrical allocation of time capacity between uplink and downlink. There are several configurations which can be selected to suit the predominant type of user-traffic being carried within the cell – for example, if the major use is the downloading from websites this would favour the down capacity being the greater. The config-uration being deployed by each cell is 'advertised' in its control information.

Mapping encoded content to radio channels. The physical wireless infra-structure within the eUTRAN provides a flexible way of carrying the variable amounts of (data) traffic on the uplink and downlink within the cells. Two basic frame structures are defined, both occupying 10 ms, as shown in Figure 9.22. Type 1 frames, which are for use with FDD systems, comprise 10 sub-frames of 1 ms each divided into two slots of 0.5 ms. The required traffic capacity is mapped on to the slots for transmission in multiples of individual OFDM subcarriers (grouped into resource elements, as discussed earlier). Type 2 frames are for use with TDD systems. The flexible allocation of capacity between the uplink and downlink is accommodated using a 0.275 ms special sub-frame, comprising pilot time slots (PTS) associated with the downlink and uplink and a 0.05 ms guard period. This special sub-frame acts as a spacer between the downlink and uplink carried in the 7 traffic sub-frames (separated by small guard intervals) as shown in Figure 9.22 [17,18,21,22].

9.9.6   The EPC

As Figure 9.18 shows, the evolved packet core (EPC) provides the main traffic switching and routeing control functions for both user data and control data within

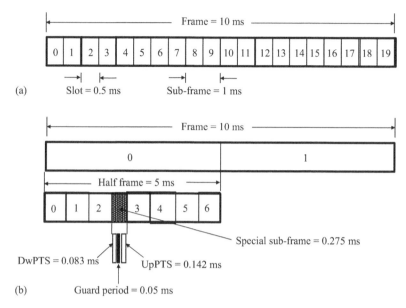

Figure 9.22   Frame structures for eUTRAN in LTE system. (a) Type 1 frame and (b) type 2 frame

the LTE system. The EPC comprises a network of IP routers and servers providing the serving gateway (S-GW) functions linked to service-control elements such as the MME, together with the gateways to outside packet data networks (P-GW). Briefly the roles of these EPC nodes are as follows [23].

The serving gateway functions. The S-GW provides the IP packet-routeing connectivity between the eUTRAN and the rest of the EPC. Importantly, it acts as an anchor point in the control plane for a call session when the UE moves between eNodeB's, providing connectivity between the different eUTRANs. It also provides access from eNodeBs to the relevant packet gateway (P-GW) to enable connectivity with other packet networks.

The packet gateway functions. As well as acting as an anchor point for data sessions with an external data network, the P-GW provides flow control and other IP-packet management (e.g. deep-packet inspection for virus blocking), as well as any charging functions if required.

The mobility management entity functions. The prime role of the MME is the mobility-management control of the dependent eNodeBs. It therefore controls the subscriber's interaction and data sessions using messaging within the control-plane, covering such actions as user authentication and establishing data sessions. As Figure 9.18 shows, the MME is directly linked over the control plane to the Home subscriber server (HSS) for the temporary and permanent subscriber information together with authentication (as in the HLR and AuC of the GSM system, respectively).

The policy and charging record functions (PCRF). The PCRF within the EPC provides the important information needed by the network operator to charge for the call or data sessions. As its name implies, the rules and tariff structure are defined in the PCRF, which together with the details of each call/data session (held in a charging data record, CDR) can be used by the operator's accounting and charging systems to generate bills. Although resident in the EPC, the PCRF system interacts not only with the MME but also the IMS (see below).

Home subscriber server. The HSS, which provides the functions of the HLR, VLR and AuC, as described earlier for GSM in this chapter, holds permanent data on each subscriber registered with the network, as well as acting as a temporary store for visiting subscribers' real-time status information held for the duration of the call or session. The HSS, which is outside of the EPC and may in fact be part of an existing 3G network with the same service area and subscribers, is accessed via the control plane from the MME.

9.9.7   IP multimedia subsystem (IMS)

The LTE system has been defined as an evolution from the preceding generations of 3GPP-defined mobile systems. This evolution allows network operators who already will have invested significant amounts of money in building widespread mobile networks across a country or region to deploy LTE alongside the existing 2G and 3G networks. Such an approach requires the continued ability of operators to offer existing services over the LTE and for good interworking between all their networks. So far, we have discussed only the control by the MME of data sessions, with use of the HSS and the PCRF. More complex services, including voice calls, require an additional control system, (which is strictly speaking external to the LTE architecture), namely IMS.

The IMS is a subsystem that provides control of multimedia services, which although specified originally for IP-based mobile networks is now considered suitable also for fixed IP networks. It comprises a set of dedicated servers linked by SIP signalling (see Chapter 7), and connected via the user and control planes to the components within the LTE system. An outline of the IMS as shown in Figure 9.18 is briefly described below. We look at IMS as a centralised control system in more detail in Chapter 7.

Call-session control function (CSCF). The CSCF holds the control software providing voice calls within the LTE system and handles calls to or from other types of networks (e.g. the PSTN). It therefore has to create within the IP-based LTE network a telephone call whose characteristics closely match those that subscribers experience normally from circuit-switched PSTN or 2G/3G mobile exchanges. The CSCF is therefore controlling a set of VoIP sessions, as descri-bed in Chapter 8. The term VoLTE (voice over LTE) is used to describe the telephony service provided by the LTE system using IMS (or similar). Chapter 7 gives more detail on how the IMS controls voice calls over LTE (i.e. VoLTE) using the CSCF.

Media gateway (MGW) and Media Gateway Controller (MGC). An important role of IMS is the gateway it provides for traffic between the LTE system and external networks. The MGW provides interworking for the user plane between the IP packets of the LTE network with the TDM/PCM circuits of the external digital circuit-switched networks. It also provides interworking for the control plane between the SIP signalling of the LTE network and the SS7 signalling of the external circuit-switched networks, as shown in Figure 9.18 (see also Figure 8.17 in Chapter 8).

The MGC provides the control functions in the MGW (as described in Chapter 8) in association with CSCF of the LTE network.

9.9.8   Circuit-switch fall back

As mentioned at the beginning of this section, the LTE system has been defined as an all-IP architecture and therefore essentially acts as a broadband data platform. Without the additional external control and interworking capabilities of the IMS (or something similar) the 'native' LTE system is unable to handle telephony-type voice calls. (Of course, voice can be carried if a user establishes a VoIP session (e.g. with Skype) where the LTE system just carries the voice as user-plane data and does not provide call control.) However, the full specification of the IMS was developed after the basic LTE networks. Consequently, the delay in developing economical IMS systems capable of handling the heavy loads of call control in a large public mobile LTE network meant that network operators began their deployment of LTE without IMS. Therefore, an interim means of handling telephony for the LTE users was required – lasting several years before VoLTE could be retrospectively introduced.

The most common interim method of dealing with voice calls over LTE was to drop back from the LTE network and hand the call to a co-sited 3G network with its established circuit-switched MSC. This is known as 'circuit-switch fall back' (CSFB) [24]. The UE are normally multimode, capable of connecting to LTE, 3G, and 2G cell sites. The LTE users would thus be able to make and receive telephone calls over their mobile handset (UE) in the normal way. Observant users will notice that the '4G' shown on the status bar of their smart phone changes to '3G' (or 'H', etc.) whenever they make or receive a voice call. Once the call has cleared down the UE is re-established with the LTE network (and '4G' is displayed on its status bar).

9.9.9   The LTE protocol stack

It is instructive to consider the protocol stacks for an LTE network, which for clarity are shown separately for the control and user data planes in Figures 9.23 and 9.24, respectively. These protocol stacks demonstrate the complex assembly of encoding and enveloping that takes place on the user's data and the control data as it is carried over the radio access (eUTRAN) through the eNodeB and then through the terrestrial network to the MME or through the EPC [25]. Figures 9.23 and 9.24 are essentially showing the functional architecture for LTE mapped against the OSI 7-layer model (see Box 8.4 in Chapter 8). By contrast, Figure 9.18 is more like a physical architectural view.

Starting with the UE at the top left-hand corner of Figure 9.23 we can see that the IP packets of control information are enveloped into RRC messages before being fed into the packet-data convergence protocol (PDCP), where encryption takes place. As we see later, the user data is handled differently within the PDCP layer before being combined with the control data packet stream. The combined flow of packets

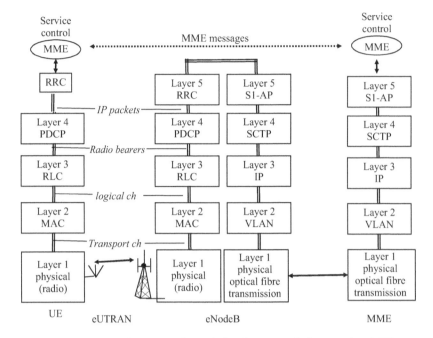

Figure 9.23   The protocol stack for the control plane within LTE

from the UE is then associated with a radio bearer and mapped onto a logical channel using the radio-link control (RLC) protocol. The packet traffic is usually segmented at this point for flow management over the radio link with automatic repeat requests (ARQ). The flow is then enveloped with a medium access control (MAC) identity for standard Layer 2 data management (Chapter 8) before being coded for the radio transmission and mapped onto the antenna (using MIMO).

At the other end of the radio link within eNodeB a reverse sequence of extracting data from the enveloped packets occurs. The control data then follows a similar protocol stack before being sent to the MME as a Carrier Ethernet stream over a physical landline, or over a microwave point-to-point radio link in remote areas, as shown in Figure 9.23. Layer 2 on these links would typically be VLAN (Chapter 8). At Layer 4 the LTE terrestrial system uses stream-control transmission protocol (SCTP) as an alternative to TCP. S1-AP is a defined interface for LTE.

Figure 9.24 shows the full set of protocol stacks for user data from the handset (UE) through the LTE network to the user's application server. The eUTRAN portion is like that shown for the control plane earlier, except that the user's data is carried in TCP or UDP to provide the Layer 4 transport linkage (see Chapter 8) between UE and the server (i.e. end-to-end). The IP/TCP (or IP/UDP) stream is then applied to PDCP, where the typically longer user data packets are compressed to save radio bandwidth between the UE and eNodeB. Now we can see the major purpose of the EPC: the user's data is carried transparently across the LTE's core IP network using a tunnelling technique. The protocol used is the GPRS tunnel protocol (GTP), in the case of the user plane this is referred to as GTPu. The tunnelling

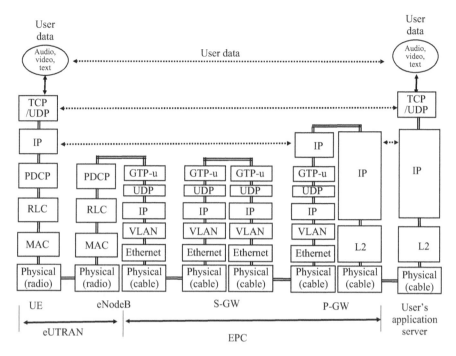

Figure 9.24   The protocol stack for the data plane within LTE

is achieved by adding the IP address of the egress node from the EPC, so that the user's data (in the form of IP/TCP/PDCP) can be carried through the routers of the S-GW to the appropriate egress P-GW, as shown Figure 9.24.

The tunnelled user-data is then carried over UDP on IP streams between the EPC routers, using the GTP-added IP address. The transmission between the eNodeB and S-GW and S-GW to P-GW is usually based on VLAN and Carrier Ethernet over optical fibre cable across the EPC. At the egress node (i.e. the P-GW) there is a standard interface to IP networks outside of the EPC network, and the IP stream of user-data is carried over appropriate Layer 1 and 2 technologies (e.g. SDH over optical fibre and MPLS, respectively) to the user's application server using its IP address. As Figure 9.24 shows, the user's data therefore flows end to end between the UE and the server. The reverse sequence of events applies to the IP data flow from the user's application server to the UE.

9.9.10   LTE advanced

As part of the progressive evolution of the LTE system, 3GPP have specified a set of enhancements which are detailed in Release 10 and subsequent releases. The primary aim is to achieve peak user-data rates of at least 1 Gbit/s downlink (500 Mbit/s uplink). One way this capacity can be achieved is through grouping up to five 20 MHz radio carriers to achieve a 100 MHz radio bandwidth – a process known as 'carrier aggregation'. Other techniques include enhanced MIMO and the use of radio relays [26].

Box 9.4   WiMAX

Worldwide interoperability for microwave access (WiMAX) has been defined by the IEEE as a set of standards in the 802.16 range for various forms of wireless delivery of broadband service within metropolitan area networks (MANS). Initial versions of the standard addressed the needs of fixed line-of-sight (LOS) broadband wireless access, operating in the 10–60 GHz carrier range – providing an alternative to wireline (i.e. CATV, xDSL and optical fibre) systems. This was extended by the IEEE802.16c version which included non-line-of-sight (NLOS) operation (2–11 GHz carrier range). A mobile version (IEE802.16e) was finalised in 2007, and recognised by the ITU as a 3G system. In 2010 a new version – IEE802.16m was accepted by the ITU along with LTE-advanced as meeting the criteria for 4G mobile systems.

At the physical wireless interface level there is similarity between WiMAX and LTE, in terms of support of both FDD and TDD (with asymmetry and flexible capacity allocation) and the use of OFDMA. However, WiMAX has a wide range of specification options allowing different modulation techniques to cope with mobile, fixed LOS and NLOS. The same base station switches between modulation techniques as required. The wide range of carrier frequency operation and other features mean that, in practice, definitive sets of options are listed by the WiMAX Forum. Although little used in its mobile application outside of North America, it is used as a competitive alternative to wireline broadband access in several parts of the world [27,28].

See also Section 4.4.12 in Chapter 4.

9.10   Multi-generation mobile networks

In this chapter we have described 2G, 3G and 4G mobile systems in isolation. However, in practice the deployment of each generation of mobile networks takes several years as network operators tackle the difficulty of finding acceptable sites for antenna masts, plus the considerable cost and time taken to provide coverage through the target service area, together with the backhaul and core network. Usually, the national regulator imposes coverage targets when spectrum is allocated to the operators. Also, given the rapid adoption of mobile as a medium for telephony and data, web access, etc., the growth in user demand for bandwidth capacity and the need for ubiquity in service areas, operators have added each generation of mobile to the existing infrastructure. This creates a spread of 2G, 3G and 4G mobile cell sites across the country – in many places the different technology cells fully or partially overlap, whilst elsewhere there are pockets of just 2G or 3G coverage. Of course, it is the overlapping of 4G cell sites with 2G or 3G that enables CSFB to be used, as described above. This mix of mobile systems also

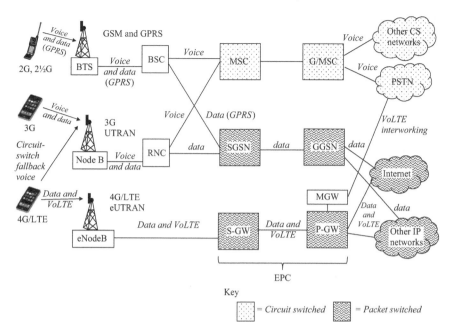

Figure 9.25   The routeing of voice calls in multigeneration set of mobile networks

means that mobile handsets (e.g. smartphones) need to be multimode and multiband, capable of providing the radio air interface of all three systems and coping with the range of carrier frequencies deployed.

9.10.1   The synergies and interworking between various mobile systems in multi-generation mobile networks

The mixture of mobile systems co-existing across the country is illustrated in simplified form in Figure 9.25, showing the different circuit- and packet-switched mobile nodes and external networks. This shows how the voice is sent via the MSC network within GSM, with the voice and data from GPRS and 3G being sent over the MSC or GSN networks, respectively; and the routing of data and VoLTE over the all-packet EPC of the LTE network. It also shows how CSFB telephone calls from a 4G/LTE UE are passed to the co-sited 3G cell, and hence through the circuit switched MSC. Figure 9.25 also shows how all mobile networks need access through their gateway nodes to the PSTN, other fixed CS networks, the Internet, and other CS and IP mobile networks.

9.10.2   Wi-Fi offload

An important part of today's broadband data scene is that of Wi-Fi (IEEE 802.11, see Chapter 8), which is used within most households and offices for the distribution of broadband access service from the termination of the fixed-network operator's external transmission system to the various user terminals. Increasingly, as mentioned at the beginning of Section 9.9, mobile UEs – especially smart phones – are also enabled to receive Wi-Fi transmission. The choice of whether the data is carried over the 3G or 4G mobile network or over the Wi-Fi can be made by the user via a system preference setting on the UE. However, the mobile operator may also use a control signal to set the UE to send or receive broadband data via the home/office Wi-Fi zone, as a way of reducing the data traffic loading over the mobile network. This so-called 'Wi-Fi offload' facility can be part of the mobile operator's strategy for managing the quality of service in areas where there are consistently high levels of broadband data which would otherwise swamp heavily loaded cellular capacity. It is a commercial decision for the mobile operator to use such offloading as a temporary measure whilst awaiting network capacity upgrades or as a permanent economic solution.

Of course, during offload it is the fixed network operator serving the Wi-Fi zone rather than the mobile operator that gains the user revenue for the data carried. On the other hand, where possible the users may prefer to send their data traffic from their 4G smart phones over Wi-Fi because of the lower tariffs on the fixed network. This is especially true for public 'Wi-Fi hot spots', such as cafes, airport lounges, hotels, etc., where the use of the Wi-Fi for high-speed data download to their smartphones or laptops may even be free.

9.10.3   Managing capacity and coverage in mobile networks

The mobile network operators' planning teams are continually reviewing the coverage of the service areas and the capacity available for the generated and received traffic. The techniques involved in cellular-network planning are covered in several of the reference texts [29–32], as well as Chapter 6 of the companion book, [33] where the user traffic permanently increases beyond the design limit of an establish cell the operator may achieve extra capacity by adding to the number of radio carriers available to the cell, or by splitting the cell, or introducing micro- or pico-cells, as described in Section 9.4.

Increases in the coverage of the mobile serving area are normally achieved by installing new cell sites, which involves adding new antennas as well as the backhaul transmission capacity. A further technique in the case of LTE is the use of repeaters and relays to extend the coverage of a cell beyond the existing periphery. This is achieved by installing a repeater antenna which terminates Uplink and Downlink to UEs in the extended coverage area. The signals are amplified at the RF (i.e. radio carrier) level and re-transmitted to the eNodeB, appearing as if it had come directly from the UE, and vice-versa for the down-link. The two sets of signals from the repeater appear as the strongest of several multipath signals at the UE and eNodeB. Unfortunately, the efficacy of this approach is limited by the fact that noise and interference as well as the original signal are amplified at the repeater. Now more usually in LTE networks the alternative relay method is to be used, where the received signals are stripped of the noise by being decoded down to baseband level and then re-encoded for transmission. LTE cell extension using relays can accommodate either FDD or TDD working [26,34].

9.10.4 Fixed-mobile convergence

Because of their differences in nature, mobile networks tend to be managed separately from fixed networks. This separation is due to several factors. Firstly, many national regulators apply different licence rules to the mobile network operators than to the fixed network operators. The technical standards for many aspects of mobile networks are different to those applied in fixed networks. Also, the quality of performance – and, indeed users' expectations – also differs for the two types of networks, as does the commercial and service management. However, for most users it is the terminals that most obviously differentiate their perception of fixed and mobile networks. Although, there are also differences in tariffs, telephone numbers, and service features applying to the two types of network.

However, this separation is not static. The progressive improvements in mobile handsets and the declining price of voice calls have driven many users to prefer the use of mobile networks even for calls inside buildings, where fixed telephones are readily available. This so-called 'mobile substitution of fixed calls' has created declines in traffic and revenues for the fixed networks. On the other hand, the advent of wireless LAN hot spots (see Chapters 4 and 8), which are part of the fixed data network, give freedom of movement to broadband users within the coverage areas. Users moving between hot spots therefore experience semi-mobile or 'nomadic' services. Thus, the demarcation between fixed and mobile domains has many overlaps.

Fixed-mobile convergence (FMC) is a generic term that describes services which span both fixed and mobile networks. They use a single dual purpose handset/terminal, which is linked to either the fixed network – say, at home and at the office – or a mobile network when out on the move away from home and office. An obvious example of FMC is the case described in Section 9.10.2 where the UE selects 3G or LTE when off premises and a pre-registered Wi-Fi when in the home or office. Full FMC is achieved when the UE seamlessly shifts between fixed Wi-Fi and the mobile network without disturbing the voice call or data session [35]. There are a variety tariff arrangements that can apply. It is important to note that with FMC it is the service to the user that is 'converged' – the fixed and mobile networks supporting the service remain separate.

Another related movement is the integration of fixed and mobile networks (FMI) through the use of common technical standards, and the deployment of completely integrated equipment, where appropriate. An example is the use of IMS for both LTE mobile and fixed networks, so providing an important feature of the 'next generation networks' (NGN) [36], see Chapter 11. This FMI is being pursued to give technical and operational advantages, as well as to provide some FMC services.

9.10.5 The fifth mobile generation (5G)

Many of the features described for LT – namely, the rising user demand for increasing bandwidth for data applications and services together with the progressive improvements in computing and wireless technology – are driving the specification of the next generation of mobile networks: 5G. The ITU have outlined a set of objectives for 5G, known as International Mobile Telecommunication system (IMT) 2020. The characteristics of 5G fall into five main areas [37]:

  (i) Increased user-broadband rates of some 100 Mbit/s with peak rates up to 10 Gbit/s
 (ii) Very low latency
(iii) Efficient exploitation of the wireless spectrum and the use of mm waves (i.e. carrier frequencies around 100 GHz, see Box 4.2 in Chapter 4)
 (iv) Spectrally efficient signals [38,39]
  (v) Seamless handover between various (fixed) wireless access systems and the 5G mobile network (i.e. true FMC).

9.11 The wireless scene

It is useful to consider the services provided by mobile networks and their evolution in the context of all the uses of wireless in communications. In addition to the cellular mobile systems described in this Chapter, the wireless scene includes the following:

*   Bluetooth, offering speeds up to some 0.5 Mbit/s, is designed for data or voice transmission over a short range (maximum 10 m) within buildings. A typical application is the provision of hands-free wireless telephone headsets, or direct connection between two mobile terminals over a short distance.
*   DECT (digital electronic cordless telephone), the European standard for cordless telephones providing a digital wireless link within buildings, with a range up to about 100 m.
*   Wireless LAN (IEE 802.11) and HyperLAN, whilst primarily offering local or hot spot data communication, do also provide freedom of movement within some 100 m radius (see also Chapter 8).
*   Low power wide-area (LPWA) radio is a new range of wireless systems addressing the cheap low speed wide coverage needs for device direct inter-connection within the home or office (Internet of things (IoT)).
*   Low-power low data wireless systems are designed for very short distance (typically below 1 m) communications. Several systems have been developed usually for specific applications, such as near-field communications (NFC) which are used by smart phones to provide e-money applications like 'touch' payments. (Actually, NFC is based on induction rather than wireless commu-nications.) Other systems have been developed for device-to-device commu-nications (i.e. IoT). The IEEE have specified Zigbee (IEE802.15) for what is known as 'personal area network' (PAN) applications.

Figure 9.26 illustrates the areas of application of the various wireless systems plotted against user-rate (in Mbit/s) and the degree of mobility provided. There are many areas of overlap between the systems which leads to potential competition between competing suppliers. However, there are important differences in the range, power consumption and costs between the various systems which is not shown in this Figure. Also, Figure 9.26 is indicative only, since this is a rapidly developing area of communication technology.

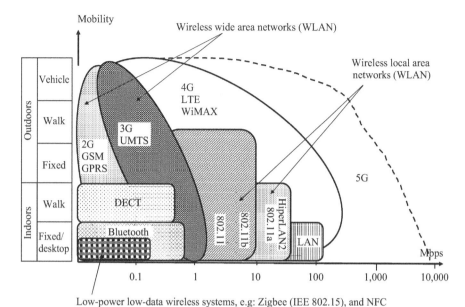

Figure 9.26   The wireless scene

It is interesting to note that the modern smartphone contains not only the range of wireless transceivers covering the various mobile standards, but also Wi-Fi (as described early), as well as Bluetooth (for headset links and links to media centres within in the user's car, etc.), and GPS microwave satellite links for 'satnav' applications.

9.12   Summary

In this chapter, we have looked at the basic structure of cellular mobile networks and noted their evolution through first, second, 2.5, third and fourth generations. As part of this review the basic principles of radio transmission, as used in the access portion of cellular networks, were described. The architecture and method of working of the ubiquitous GSM system was examined in some detail, covering the location-management function, handover between cells, and the use of standard circuit-switching within the MSC. We then considered the introduction of packet data service, initially through 2.5G systems such as GPRS grafted onto the 2G networks, and more recently through the rollout of the 3G and 4G networks. Importantly, we saw that the 4G system is essentially a mobile version of the NGN concept, being an all-IP network. Finally, the use of mobile systems was put in the context of the total use of radio systems in providing different degrees of mobility, reach and user speed – leaving an intriguing question around the nature of the 5G mobile systems, as well as the future of fixed-mobile convergence.

In view of the vast number of abbreviations introduced into this review of mobile systems, Box 9.5 provides a ready reference. These abbreviations are included in the main listing at the end of the book.

Box 9.5　Summary of abbreviations used in mobile networks

AuC = Authentication centre
BSC = Base station controller
BTS = Base transceiver station
CSCF = Call session control function
CSFB = Circuit switch fall back
DECT= Digital enhanced cordless telecommunications
BSC = Base station controller
EDGE = Enhanced data rate for global evolution
EIR = Equipment identity register
eNodeB = Evolved NodeB
EPC = Evolved packet core
E-UTRAN = Evolved UTRAN
GGSN = Gateway GPRS support node
G/MSC = Gateway MSC
GPRS = General packet radio service
GSM = Global system for mobile communications
GTP = GPRS tunnelling protocol
HLR = Home location register
HSCD = High-speed circuit-switched data
HSS = Home subscriber server
IMS = IP multimedia subsystem
IWF = Interworking function
LTE = Long-term evolution
MIMO = Multiple-input multiple-output
MSC = Mobile switching centre
MME = Mobility management entity
NFC = Near-field communications
PCU = Packet communications unit
PCRF = Policy and charging rules function
PDCP = Packet data convergence protocol
P/GW = Packet data network gateway
PLMN = Public land mobile network
RNC = Radio network controller
RLC = Radio link control
RRC = Radio resource control
SGSN = Serving GPRS support node
S/GW = Serving gateway
SMSC = Short message service centre
SCTP = Stream-control transmission protocol
TC = Trans-coder
TETRA = Terrestrial trunked radio
UTRAN = UMTS terrestrial radio-access network

UMTS = Universal mobile telecommunications system
VoLTE = Voice over LTE
VLAN = Virtual local area network
VLR = Visitor location register
WiMAX = Worldwide inter-operability for microwave access
WLAN = Wireless local area network

References

[1]   Valdar, A. and Morfett, I. 'Understanding telecommunications business'. Stevenage, UK: Institution of Electrical Engineers; *IET Telecommunications Series No. 60*, 2015, Chapter 1.
[2]   Schiller, J.H. *Mobile Communications*. Boston, MA: Addison-Wesley; 2000, Chapter 2.
[3]   Anttalainen, T. *Introduction to Telecommunications Network Engineering*. Norwood, MA: Artech House; 1999, Chapter 5.
[4]   Sutton, A. and Linge, N. 'Mobile network evolution within the UK', *Journal of the Institute of Telecommunications Professionals*, 2015;**9**(2): 10–16.
[5]   Sutton, A. and Linge, N. 'The road to 4G', *Journal of the Institute of Telecommunications Professionals*, 2014;**8**(1):10–16.
[6]   Schiller, J.H. *Mobile Communications*. Boston, MA: Addison-Wesley, 2000, Chapter 3.
[7]   Mouly, M. and Pautet, M.B. *The GSM System for Mobile Communications*. Paris: Teleco Publishing; 1992.
[8]   Schiller, J.H. *Mobile Communications*. Boston, MA: Addison-Wesley; 2000, Chapter 4.
[9]   Goleniewski, L. *Telecommunications Essentials. Boston*, MA: Addison-Wesley; 2003, Chapter 14.
[10] Walke, B., Seidenberg, P., and Althoff, M.P. *UMTS: The Fundamentals*. Chichester, UK: John Wiley & Sons; 2003, Chapter 1.
[11] Cox, C. *Essentials of UMTS*. Cambridge: Cambridge University Press; 2008, Chapter 2.
[12] Cox, C. *An Introduction to LTE*, 2nd edn. Chichester, UK: John Wiley & Sons; 2014, Chapter 4.
[13] Ofcom. (2013, February 20). Ofcom announces winners of the 4G mobile auction. [online] Retrieved from www.ofcom.org.uk. [Accessed on April, 2017].
[14] Cox, C. *An Introduction to LTE*, 2nd edn. Chichester, UK: John Wiley & Sons; 2014, Chapter 5.
[15] Dahlman, E., Parkvall, S., and Skold, J. *4G LTE/LTE Advanced for Mobile Broadband*. San Diego, CA: Elsevier Academic Press; 2011, Chapter 5.

[16] Olenewa, J.L. *Guide to Wireless Communications*, 3rd edn, International edition, Course Technology. Boston, MA: Cengage Learning, 2014, Chapter 8.

[17] Dahlman, E., Parkvall, S., and Skold, J. *4G LTE/LTE Advanced for Mobile Broadband*. San Diego, CA: Elsevier Academic Press; 2011, Chapter 9.

[18] Cox, C. *An Introduction to LTE*, 2nd edn. Chichester, UK: John Wiley & Sons; 2014, Chapter 6.

[19] Lescuyer, P. and Lucidarme, T. *Evolved Packet System (EPS): The LTE and SAE Evolution of 3G UMTS*. Chichester, UK: John Wiley & Sons; 2008, Chapter 6.

[20] Dahlman, E., Parkvall, S., and Skold, J. *4G LTE/LTE Advanced for Mobile Broadband*. San Diego, CA: Elsevier Academic Press; 2011, Chapter 17.

[21] Lescuyer, P. and Lucidarme, T. *Evolved Packet System (EPS): The LTE and SAE Evolution of 3G UMTS*. Chichester, UK: John Wiley & Sons; 2008, Chapter 3.

[22] 3GPP, www.3gpp.org.

[23] Lescuyer, P. and Lucidarme, T. *Evolved Packet System (EPS): The LTE and SAE Evolution of 3G UMTS*. Chichester, UK: John Wiley & Sons; 2008, Chapter 2.

[24] Cox, C. *An Introduction to LTE*, 2nd edn. Chichester, UK: John Wiley & Sons; 2014, Chapter 21.

[25] Dahlman, E., Parkvall, S., and Skold, J. *4G LTE/LTE Advanced for Mobile Broadband*. San Diego, CA: Elsevier Academic Press; 2011, Chapter 8.

[26] Cox, C. *An Introduction to LTE*, 2nd edn. Chichester, UK: John Wiley & Sons; 2014, Chapter 19.

[27] Dahlman, E., Parkvall, S., and Skold, J. *4G LTE/LTE Advanced for Mobile Broadband*. San Diego, CA: Elsevier Academic Press; 2011, Chapter 19.

[28] Olenewa, J.L. *Guide to Wireless Communications,* 3rd edn, International edition, Course Technology. Boston, MA: Cengage Learning, 2014, Chapter 9.

[29] Laiho, J. and Wacker, A. 'Radio network planning process and methods for W-CDMA'. In Bic, J.C. and Bonek, E. (eds.), *Advances in UMTS Technology.* London: Hermes Penton Ltd.; 2002, Chapter 6.

[30] Sutton, A. 'Small cells and heterogeneous networks', *Journal of the Institute of Telecommunications Professionals*, 2016;**10**(2):35–39.

[31] Lee, W.C.Y. *Mobile Communications Design Fundamentals*, 2nd edn. Chichester, UK: John Wiley & Sons; 1993, Chapter 2.

[32] Lee, W.C.Y. *Mobile Communications Design fundamentals*, 2nd edn. Chichester, UK: John Wiley & Sons; 1993, Chapter 6.

[33] Valdar, A. and Morfett, I. 'Understanding telecommunications business'. Stevenage, UK: Institution of Electrical Engineers; *IET Telecommunications Series No. 60*, 2015, Chapter 6.

[34] Dahlman, E., Parkvall, S., and Skold, J. *4G LTE/LTE Advanced for Mobile Broadband*. San Diego, CA: Elsevier Academic Press; 2011, Chapter 16.

[35] Clark, R. and Abernethy, T. 'Future voice services and CPE trends', *Journal of the Communications Network*, 2005;**4**(2):25–29.

[36] Cuevas, M. 'The role of standards in next generation networks', *Journal of the Communications Network*, 2005;**4**(2):16–41.

[37] Andrews, J.G., Buzzi, S., Choi, W., *et al.* 'What will 5G be?', *IEEE Journal on Selected Areas in Communications*, 2014;**32**(6):1065–1081.

[38] Dawwezeh, J., Grammenos, R.C., and Xu, T. 'Spectrally efficient frequency division multiplexing for 5G'. In Xiang, W., Zheng, K., and Shen X. (eds.), *5G Mobile Communications*. Switzerland: Springer International Publishing; 2017, Chapter 7.

[39] Darwezeh, I., Xu, T., and Grammenos, R.C. 'Bandwidth-compressed multi-carrier communications: SEFDM'. In Luo, F.-L. and Zhang, J. (eds.), *Signal Processing for 5G: Algorithms and Implementation*. Chichester, UK: John Wiley & Sons (IEEE Press); 2016, Chapter 5.

Chapter 10

Numbering and addressing

10.1 Introduction

Any form of communication between members of a population requires that each person is identified by a name or a number, and an address signifying their location. In the case of telecommunications, the set-up of telephone calls or the delivery of data packets is determined by the examination of the destination address and the application of the routeing rules for that network. Fortunately, there is universal adherence to the standards setting out the basic structures of the numbers and names for telephony and data/Internet services, respectively. Thus, every telephone line and mobile connection is identified by a unique number, potentially allowing any telephone to make unambiguous contact with any other telephone in the world; a truly remarkable achievement.

Interestingly, subscribers often become quite fond of their telephone number and there can be a pride in having certain initial digits, where these indicate a desirable area in town, perhaps. Similarly, e-mail and web page names are often considered prestigious. This means that there is value attached to certain numbers, particularly those easily remembered, giving rise to the concept of 'golden numbers'. Also, subscribers do not like having to change their telephone numbers at the whim of the national administrators of the numbering scheme. Indeed, for business customers changing their telephone numbers can be an expensive undertaking involving new signage on vehicles, reprinting advertising brochures, and so on. Consequently, the seemingly dry subject of numbering and addressing does occasionally become quite emotive.

However, this chapter confines itself to considering how numbering and addressing is defined and applied to the various types of networks (both fixed and mobile). After considering the two different approaches made in telephone and data networks, particularly IP (Internet Protocol) networks, we examine the newly introduced integrated system that allows full interworking between telephone numbers and Internet addresses. It should be appreciated that this is a surprisingly fast moving area of national regulation and there may be changes made to the establish principles of numbering and addressing, particularly as new implementation of voice services, such as voice-over IP (VoIP) (see Chapters 8 and 11) are implemented.

10.2   Numbering and addressing in telephone networks

It is important to note that, traditionally, telephone numbers are used by the exchange control system to not only route the call through the network, but also to determine the charge category for the call. Subscribers can identify from the initial digits in the number those calls which will incur long-distance, international, or special tariff rates, as described later.

Actually, in fixed telephone networks the number of each subscriber line also acts as the address of that line within the network. That is not the case, of course, for mobile networks since the mobile terminal identified by a number is not associated with any line or even a permanent location. Thus, mobile networks need to determine for every call the current address of the handset with that telephone number, as described in Chapter 9.

The internationally agreed basic format for all telephone numbers is described in recommendation E164, issued by the ITU, (see also Appendix 1) [1,2]. Each subscriber in the world has a unique international number with a maximum length of 15 digits, made up of a country code of one, two, or three digits to identify the country, followed by the national number. The length of the national number, and hence the size of the population of subscribers that can be identified, is limited by the length of the country code – thus, larger countries are allocated shorter international codes. For example, the United States, Canada and the Caribbean Islands combined have Country Code *1* (also known as the North American Numbering Plan (NANP) area); the United Kingdom has Country Code *44*; while the Irish Republic has Country Code *353*.

Figure 10.1 shows the allocation of the initial digits of all the country codes to the various regional zones of the world. It is interesting to note that Europe, with its high density of medium size countries, had the initial digits of *3* or *4* allocated to

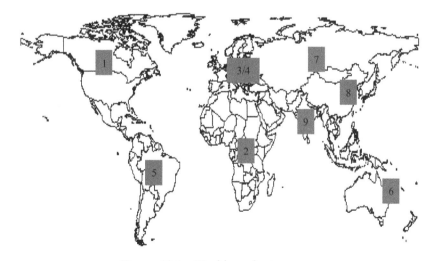

Figure 10.1   World numbering zones

its country codes. However, the allocation of country codes based on regional zones has now ceased. This is because country codes can now be assigned for global services, for example, *+800* for international freefone service, *+979* for international premium rate services and *+808* for international shared cost services. It has also been recognised that allocating the country codes strictly on a regional zone basis resulted in unbalanced utilisation. The full set of assigned country codes is listed in Appendix 2 [3].

Within a country, the national significant number can either be organised based on geographic areas, for example, for subscribers in a public switched telephone network (PSTN), or on a non-geographic basis, for example, for mobile subscribers. Figure 10.2 shows how the geographical national number for a PSTN subscriber's line is made up of two parts: the 'area code', which identifies the geographic region of the country, and the 'local number'. The latter is made up of the 'exchange code' and the 'subscriber number'. Examples of two UK local numbers are shown: London and Ipswich. The area code for London (excluding the initial *0*) is *20*, which is followed by a four-digit Exchange Code beginning with either *eight*, *seven* or *three*, and a four-digit Subscriber Number on that exchange. In the case of Ipswich, the Area Code (excluding the initial *0*) is *1473*, that is, four-digits, followed by a two-digit exchange code (capable of identifying up to 100 local exchanges units in that area) and a four-digit subscriber number. Thus, in the United Kingdom up to 10,000 lines can be identified on each exchange unit.

However, there are many variations on the way that national numbers are formatted in each country. In the United Kingdom there is a clear separation between numbers allocated to the PSTN, which have geographical significance, and mobile networks and special tariff numbers, which have no geographical significance. The first digit of the national number (excluding the leading *0*), known as

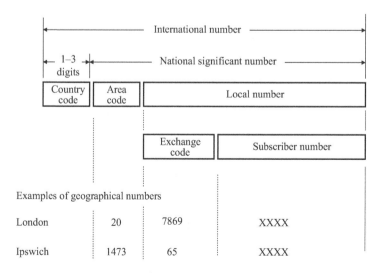

Figure 10.2   Standard format for geographical numbers (E164)

the 'Service' or 'S' digit, denotes the type of service involved. This gives users an indication of the category of the call, whether it is to a fixed-line number, a mobile number, or a special tariff number. So, all PSTN lines (i.e. geographically based) have numbers with an S-digit of *1* or *2*. (From the example of Figure 10.2, London numbers have an area code of *020*, with an S-digit of *2*; Ipswich numbers have an Area Code of *01473*, with an S-digit of *1*.) Whereas, all mobile numbers in the United Kingdom are easily distinguished by their S-digit of *7*, that is, they begin with *07*. Similarly, the specially tariffed services are in the *08*-range, and the premium rate services are clearly distinguished by beginning with *09*. Box 10.1 shows the full allocation of the S-digits in the UK National Number format. Many European countries are progressively introducing a similar numbering structure.

However, in the United States and Canada there is no use of a distinguishing S-digit, so subscribers are unable to identify whether a number is for a mobile or fixed-line phone. This mixing of geographical and non-geographical numbers in one range does, however, make maximum use of the capacity offered by the number length – unlike the UK system where the S-digit format tends to block parts

Box 10.1   The UK numbering scheme [4]

Range	Use
(0) 1	Geographic numbers
(0) 2	Geographic numbers
(0) 3	Non-geographic numbers (but charged at the geographic rate)
(0) 4	Reserved for future services
(0) 5	
55	Corporate numbers
56	Location-independent electronic communications services (e.g., voice over broadband)
(0) 6	Reserved for future services
(0) 7	'Find me anywhere' services (non-geographic)
70	Personal numbers
71–75	Mobile service
76	Radio paging service
77–79	Mobile service
(0) 8	Special Tariff services (non-geographic)
80	No charge to the caller, e.g., 'Freefone'
82	Internet for schools (no charge)
843–845	Charge, maximum rate of 5.833p pm or call
870–873	Charge, maximum rate of 10.83p pm or call
(0) 9	Premium rate services (non-geographic)
90–91	Premium rate (up to 300p pm or 500p per call)
908 and 909	Sexual entertainment services at premium rate
98	Sexual entertainment service at premium rate
pm = per minute	

of the number range from full exploitation. Nonetheless, the North American numbering scheme does contain the *800* range for special tariff (Freefone) services.

An important point to note is that the *0* usually written in front of UK national numbers is only a prefix used in national-dialling procedures, and is not part of the number. Thus, the *0* is added by a subscriber when dialling a national call, which is defined as a call between subscribers having different area codes. (In the United States and Canada, a *1* rather than *0* is used as the prefix for calls between different area codes.) The dialling procedure for a local call in the United Kingdom, that is, calls between numbers having the same area code – is to dial just the local number. International calls are made by prefixing the full international number of the foreign subscriber by the international prefix as used by the originating network – the ITU recommendation for this, which is now adopted by many countries, including the United Kingdom, being *00*. Figure 10.3 illustrates the various dialling procedures for (a) international, (b) national and (c) local calls. Correct application of the dialling procedures relies on the correct presentation of telephone numbers in correspondence, advertising, directories, etc. For example, calls between two exchanges in the London (*0*) *20* Area should use the local dialling procedure, that is, using just the Local number (*7869 XXXX*). The London number should therefore be presented as *020 7679 XXXX* (and not the often-used erroneous *0207 679 XXXX*), so that the Area Code is clearly identified.

In the case of non-geographic numbers the procedure is always for the whole number to be dialled, as illustrated in Figure 10.4 for a call to a mobile number.

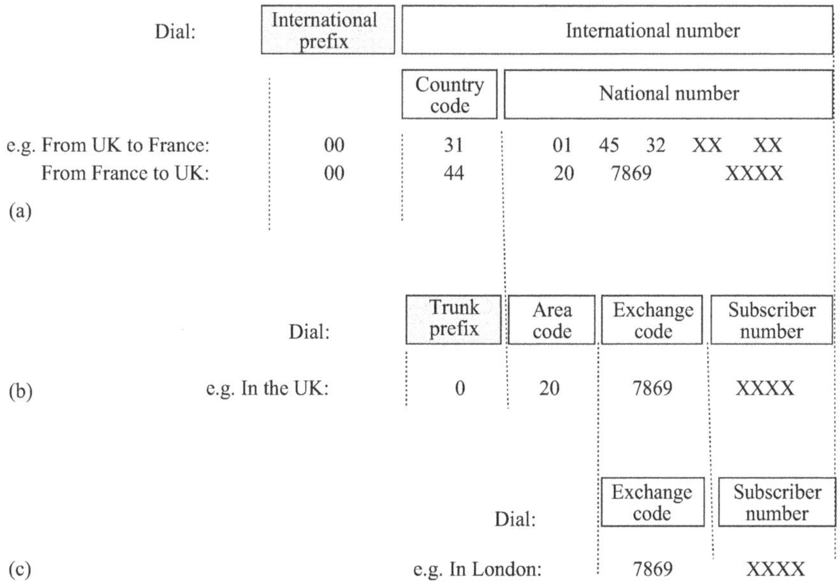

Figure 10.3   Dialling procedures for geographic numbers. (a) International call, (b) national call and (c) local call

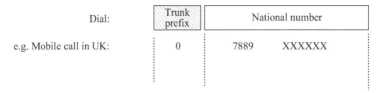

Figure 10.4   Dialling procedures for non-geographic numbers

Box 10.2   UK numbering and dialling formats

Geographic:
(01AB) CDE XXXX: 01 range with 7-digit local dialling
(01ABC) DE XXXX: 01 range with 6-digit local dialling
(01ABC) DE XXX: 01 range with 5-digit local dialling
(02A) BCDE XXXX: 02 range with 8-digit local dialling
Non-geographic:
07XXX-XXXXXX: 07 range with no local dialling
08XX-XXXXX/X: 08 range with no local dialling
09XX-XXXXX/X: 09 range with no local dialling

Box 10.2 summarises the range of telephone number formats in the UK numbering scheme and the corresponding dialling procedure. The Area codes, which can range from two to four digits, plus the trunk prefix *0*, are shown in brackets for the various lengths of geographic numbers. This mixture of six, seven and eight digit local numbers is a characteristic of the UK numbering scheme, giving as it does the ability to cope efficiently with differing densities of telephone lines within the area codes. Other countries, for example the United States, have a constant length for all numbers across the country. The three sets of non-geographic number formats are also shown; the *08* and *09* range having the option of eight or nine digits (not counting the trunk prefix) to cope with differing capacities.

We can now consider how the local numbering range can be used. As discussed above and shown in Figure 10.2, the local number is made up of the exchange code and subscriber number. Calls from other numbers in the area code are made by dialling only this local number. However, not all of the local numbering range can be used for local numbers because certain lead digits are needed for other activity. Firstly, of course, digit *0* cannot be used as the first digit of an exchange code because this could not be distinguished from the *0* or *00* used by subscribers when beginning a national or international call, respectively. Also in the United Kingdom, codes in the range *1XX* or *1XXX* are used to gain access to special features, services and indirect routeings to other operators. Thus, Exchange

Codes beginning with *1* cannot be used in the local numbering scheme. Three types of *1XX(X)* codes are used, as follows [4]:

> *Type A*: These codes are widely recognised by subscribers and used by all operators for equivalent services, for example, *100* for access to operator assistance, *123* for the speaking clock, *112* for the emergency services (the European standard code) used in addition to *999* in the United Kingdom.
>
> *Type B*: A set of codes used to enable customers to choose a service from another network operator (i.e. 124–140, 143–146, 148–149, 160–169, 181–189). Each operator is free to allocate the codes appropriately.
>
> *Type C*: This set of codes is available for independent use by network operators for their subscribers or employees (e.g. *151* for fault reporting on British Telecommunication (BT) consumer lines, *174* for BT engineers' test access).

In addition, the European standard *118XXX* range is used in the United Kingdom by the companies offering directory enquiry services. Similarly, 116XXX has been introduced as a European standard for social services help lines. Also, in the UK code *999* has for many years been used to access the emergency services, so to avoid inadvertent mimicking the network operators tend not to allocate initial digits *99* to their numbers.

Thus, only exchange codes beginning with digits *2–8* (and *90–98*) are available to a network operator – thus reducing the potential capacity to 790,000 six-digit local numbers, 7.9 M seven-digit numbers, or 79 M eight-digit numbers [5]. However, in practice, the capacity of the numbering ranges is far less than these theoretical limits due to the format of the local numbers, with the initial two, three or four-digits being used to identify the relatively small number of local exchanges within the National Code area. Thus, typically it is only the last four or five digits which can be fully used (i.e. giving a maximum capacity of 10,000 lines or 100,000 lines, respectively). The net result is that numbering ranges can in practice not support the large amount of lines that the overall number length would suggest. (As we shall see later, a similar situation has occurred with Internet or IP numbering.) Actually, occupancy of some 20%–30% of the telephone numbering ranges is generally considered good!

10.3 Administration of the telephone numbering range

Although the overall length of the significant international number and the Country Code is set by the ITU-T Recommendation E.164, as described above, the allocation of the capacity within the national number – viewed as a national resource – is governed by a national administration within each country. In the United Kingdom, it is Ofcom, the regulator of telecommunication and broadcasting, that takes this role. Similarly, the national numbering scheme for the United States is administered by the Federal Commission for Communications (FCC). The role of the administrator is to ensure that the national numbering scheme provides a framework for numbers to be allocated to all the network operators in the country, so that they can provide telephone services to their subscribers. This allocation must be in accordance with the government policy, particularly in the way that competing network operators are treated [6]. The numbering administration at Ofcom in the United Kingdom has traditionally allocated the available capacity to network operators in blocks of 10,000 numbers, that is, groups of the final four digits in the number. This relatively large minimum size means, of course, that operators requiring, say, 500–1,000 numbers use only a proportion of the allotted capacity – another reason for the utilisation factors of national numbering schemes being generally low. However, the greater decoding capabilities of modern switching systems and the increasing demands on the numbering scheme are likely to cause this procedure to change, and in the United Kingdom smaller allocations of numbers are already being assigned in some ranges. Thus, the national numbering range is progressively apportioned by the administration to all the network operators and service providers in the country; they all own their piece of the numbering range.

The national numbering scheme needs to have spare capacity to allow for the requirements of new operators in the foreseeable future, as well as providing room for expansion into new services beyond basic telephony. Ideally, to minimise the disruption to subscribers incurred through enforced telephone number changes, the national numbering schemes are designed to last for some 25 years.

However, in the United Kingdom since the 1960s there have been several changes to the numbering scheme, each requiring some level of number changes for subscribers [7]. The first, and probably the most unpopular, change occurred in 1966 with the introduction of all-figure numbers to replace the use of letters to represent the exchange names. Prior to then, the national number comprised of two letters and a number to represent the trunk exchange (e.g. *LE1* for Leicester) followed by a numerical local number. This use of letters was introduced in the late 1950s and early 1960s to enable subscribers to dial their own trunk calls (i.e. subscriber trunk dialling (STD)) using easily recognisable codes. Furthermore, in the six major metropolitan areas, that is, London, Birmingham, Manchester, Liverpool, Edinburgh, and Glasgow, it was the local exchanges that were represented by the first three letters of their name. For example, numbers within the Hampstead local exchange area were presented as *HAM* followed by the four-digit subscriber number; Swiss Cottage numbers began with *SWI*, and so on. (Similarly, to ascertain the time subscribers accessed the speaking clock by dialling *TIM*!) However, by the late 1960s this extremely convenient naming or numbering arrangement was becoming unsustainable:

- In London, the limit had been reached on the number of three-letter codes that could be used for exchange names, since groups of three or four letters correspond to the same number on the dial, and only certain combinations of letters make sense.
- With the increase in customer-dialled international calls the inconsistent allocation of letters to numbers on the dials in various countries was increasingly causing dialling errors on incoming calls to the United Kingdom.
- The London Sector plan, described in the Section 10.4, required changes to some local exchange codes which would have created unacceptable exchange-letter combinations.

It is interesting to note that the consequence of directly changing the telephone numbers from letter-number combinations to figures-only gave rise to the current allocation of trunk area codes as an apparently random spread across the country. This can be understood by considering the following examples, recognising that the leading *1* was added only recently: Area Code *1473* relates to Ipswich (in the East of England), previously coded as *IP3*; while Area Code *1472* is allocated to Grimsby (in the North of England), previously coded as *GR2*; and Area Codes *1474* is allocated to Gravesend (in the South East of England), previously coded *GR4* – whereas, area codes *1470* and *1471* cover the Isle of Skye in Scotland! In contrast, some countries, for example, Germany, have allocated the area codes on a systematic geographical regional basis.

In 1989, the trunk area code for all of London was changed from *01* to *071* for central London and *081* for the outer areas [8]. Then, beginning in 1995, there was a major restructuring of the national numbering scheme to the current structure described above and shown in Box 10.1, giving rise to the use of *0171* and *0181* for inner and outer London respectively. This was shortly followed by the re-introduction of a single code, this time *020*, for all of London [4]. These changes have been necessary to cope with the rapid increase in the number of operators (mobile and fixed), each requiring their own segregated numbering range. In addition, many new services have been introduced, most of which require non-geographic numbers and need to be differentiated from the standard geographical numbering ranges.

10.4   Routeing and charging of telephone calls

As discussed in the introduction to this chapter, numbering (and naming) of subscribers is used by the network to determine how a call should be routed and what the charge should be.

10.4.1   Numbering and telephone call routeing

In Chapter 1 we considered a simple call through the network in which the dialled number was examined by the exchange-control system at each exchange in the routeing. The structure of the geographical telephone number gives the identity of the called party in descending granularity reading from left to right, that is, country code, followed by the area code (which identifies the trunk exchange) followed by the local exchange identity, and finally the subscriber's line number on that exchange. Thus, an exchange-control system need only examine the significant dialled digits (reading from the left) depending on whether it is at the early or later stages of the call routeing, for example, only examining the area code of the dialled digits at the originating trunk exchange. Each exchange then uses the appropriate routeing table to determine how the call should be progressed.

Box 10.3 gives a brief case history of the London Sector plan to illustrate how the examination of the significant part of geographical numbers was used in the routeing of calls from the rest of the United Kingdom into London [9]. The scheme, introduced during the mid-1970s, relied on the grouping of the 350 or so local

Box 10.3   Case study of traffic routeing and numbering: The London sectorisation plan

In 1965 it was decided that a new structure was required for the handling of trunk and international calls into and out of London [9]. At that time, all such calls for the capital's 350 local exchanges were switched at trunk exchanges located in the centre. The plan provided for a new set of trunk sector switching centres (SSC) to be established in seven sectors, radiating from the 6 km radius central area. The sectors extend to the edge of the area code for London, some 20 km from the centre of London. At the start of the implementation of the project in 1970 there were about 2.2 M telephone lines within the London area. Advantage had been taken of the move to all-figure numbers (see earlier in this chapter) to ensure that the first two digits of the three-digit exchange codes for London formed into seven sectors and one central group, as shown in Figure 10.5. As mentioned earlier, this did require the introduction of some new exchange codes where simply swapping the corresponding figures for the letters was not possible.

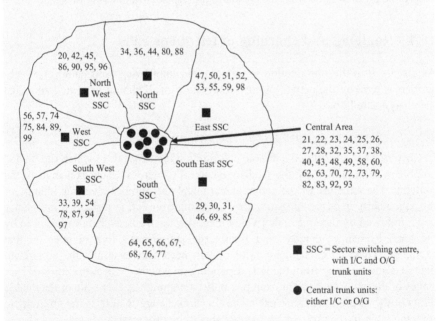

Figure 10.5   London sectorisation code allocation ('ab' digits of local exchange codes)

The national number for London subscribers was of the format: *01-abc XXXX,* where *01* was the Area Code for London and *abc* was the 3-digit local exchange code. At the distant trunk exchange, examination of just the first three digits (*1ab*) enabled the traffic to be directly routed to the serving sector switching centre or a central area trunk unit. Figure 10.6 shows an example of how two subsequent calls from a distant exchange can be sent on the appropriate route to the North-West Sector by examining *186* from dialled number *01-864 XXXX*, and to the West Sector by examining *189* from dialled number *01-894 XXXX*. This plan for London then conformed to the standard arrangement for determining the required routeing at a trunk exchange by the examination of the three digits following the 0, since at that time all trunk codes were three digits in length.

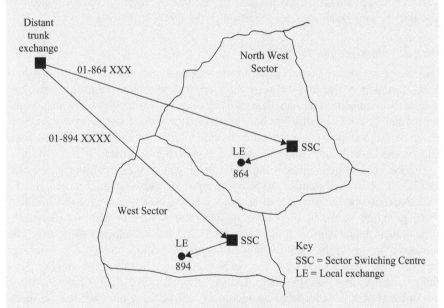

Figure 10.6   London sector routeing

In addition to limiting the growth of capacity on traffic routes into the centre of London, the Sector plan enabled other improvements to the structure of the network, and the routeing of traffic within London. The result was an improvement in transmission performance, and a build-up of the cable infrastructure in sector areas of London. The SSCs also formed the foundation for the later digitalisation of London's exchanges and transmission routes.

exchange units in London into seven sectors, each served by a trunk unit, and a central zone with several trunk units. This arrangement enabled distant trunk exchanges to determine, from the first two digits of the three-digit London local exchange code as dialled, to which of the sector or central incoming trunk units in London the call should be directly routed. A new simplified structure for the London PSTN based on digital exchanges and digital optical fibre transmission has since been introduced which reduced the number of exchanges and routes, making planning and operations easier and cheaper [10]. However, the grouping of the local exchange codes in London still conforms to the original sector plan.

In the case of calls to non-geographical numbers – for example, a mobile subscriber or a 'Freefone' call – the dialled number must first be translated into a destination network address or number, as described in Chapters 9 and 7, respectively. For the 'Freefone' call the translation of the dialled number is made by the IN (intelligent network), the control is then handed back to the trunk exchange so that the call can be routed to the new destination number through the PSTN in the normal way.

10.4.2 Number portability

The rise of the level of competition in the telephone call market created pressure from customers to be able to keep their original telephone number when they change their subscription to another network operator – a facility known as 'number portability'. This facility has been introduced into the fixed networks for both geographical and non-geographical numbering ranges, in the United Kingdom, Europe, the United States and Canada and many other countries.

In the case of geographical numbers, the extent of the portability is for subscribers to retain their number given by operator A when changing to operator B, while remaining at the same location. Thus, a new access line needs to be installed at the premises by the new operator B. (Operator A may or may not physically remove their defunct access line.) Since the subscriber's number is still part of the numbering range for the local exchange of operator A all incoming calls, wherever they originate, will be routed to that local exchange. Therefore, the number portability facility involves the detection by the so-called 'donating' exchange (on operator A's network) that an incoming call is for a number that has been ported to operator B's 'recipient' exchange. This information is held in the data store of the control system of the donating exchange, whose role is to redirect all incoming calls for the ported number to the recipient exchange. There are two main methods for routeing ported calls, as summarised below [11,12].

1. The incoming call is routed to the donating exchange in the normal way through operator A's network. On detecting that the number is ported the donating exchange sets up a ported-redirection call back to its parent trunk exchange using a prefix (*5XXXXX* in the United Kingdom) to identify the donating network, followed by the ported national number. The trunk exchange then routes the call to the point of interconnection with operator B's network. Since the call goes into and out of the donating local exchange, this method is known as 'trombone working'.

2. A more elegant and cost-effective solution now widely employed is for the call to progress only as far as the donating trunk exchange. The donating local exchange then returns a 'call drop-back' signalling message (signalling system number 7 (SSNo.7)) in response to the normal initial-address message from the trunk exchange (see Chapter 7). The call drop-back message contains the porting prefix (e.g. *5XXXXX*) plus donating number, as described above, so that the call can be sent directly from the donating trunk exchange to the point of interconnect with the recipient network. The latter then routes the call to the ported subscriber.

A further option for handling ported calls is for reference to be made to a computer data base containing the details of ported numbers, that is, the use of an IN system (see Chapter 7). This can be organised as a national data base owned and operated by an independent agency for the benefit of all network operators in the country. Individual network operators may then establish their own data bases within their own IN, accessing the national database periodically for updates of new and changed portings. However, in order to reduce the need for all calls to invoke a (relatively expensive) database enquiry, this IN method is usually limited to supporting portability of non-geographic numbers – especially those in the *08XXXXXX* Freefone, etc., range (*800* in the United States) – since such calls are easily identified by the first digit (minus the leading *0*). As the proportion of geographical numbers that are ported increases, there is an economic break-even point where the cost of having to access the data base for all geographic calls becomes equivalent to that of tromboning or call drop-back for just ported calls. At that stage the IN approach is economical for geographical as well as non-geographical fixed-network number portability.

There is also the need to provide number portability for mobile networks. The call set-up procedure for mobile networks using the home location register (HLR) and visitor location register (VLR) systems, as described in Chapter 9, lends itself to the use of an IN approach for handling number porting. As telecommunication providers look to introduce next-generation networks (NGNs) using IP-based technology, as described in Chapter 11, it is already envisaged that other ways of providing number portability, such as ENUM (described later in this chapter) come to the fore.

10.4.3 Numbering and telephone call charging

Generally, charging for calls within the PSTN is based on distance between the calling and called subscriber and the duration of the call. There may also be different tariff rates depending on the time of day. Rather than determining the distance-based charge rate for every combination of exchanges in the country – some 42 M for the 6,500 exchanges in the United Kingdom – the exchanges are instead collected into about 680 charge groups (CGs) for this purpose. By definition, calls between subscribers within a CG are at the local rate. The tariff for calls between different CGs is determined by the distance between the defined charge points in each CG. Each CG is identified by one or sometimes two area codes, there being some 840 area codes in the United Kingdom. Thus, by examining the area code in the dialled digits, the originating local or trunk exchange can determine the appropriate charge rate for the call.

These CGs vary in size, but their boundaries have been drawn recognising communities of interest and containing roughly equal numbers of subscribers within a region. Of course, due to the distribution of the population and the geography of the United Kingdom the characteristics of the CGs are varied: the largest CG being the London *(0)20* area with some 5 M subscribers, and the smallest being based on individual Scottish islands with a few hundred subscribers. However, the key requirement with telephone call charging is that it appears reasonable to customers, thus calls between subscribers in close proximity are deemed to be local and those further apart are charged at various trunk rates.

One of the major anomalies that can occur with telephone distance-charging is that calls from one end to another within a CG might be deemed local, while calls between subscribers a shorter distance apart, but on either side of the CG boundary, would be charged at the trunk rate. In the United Kingdom, this anomaly is avoided by defining the local-call fee area as being not only within a CG, but also encompassing all adjacent CGs. Only calls between non-adjacent CGs are deemed to be trunk rate. This is illustrated in Figure 10.7, where the local-fee area for a subscriber in CG-A comprises CG-A, CG-B, CG-C, and CG-D; calls to subscribers in CG-E are at the trunk or national rate.

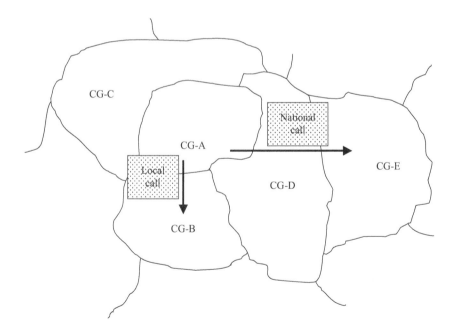

CG = Charge Group

Figure 10.7   Charging in the PSTN

Charging for international calls is determined at the originating international gateway exchange by examining the country code at the beginning of the dialled digits (after the international prefix, usually *00*). Since, as described at the beginning of this chapter, the country codes were initially grouped on a geographical-region basis the charge rates for international calls from the country can easily be grouped – thus, simplifying the billing software for the network operator and making the rates easier for customers to understand. More information on telephone call charging principles is available in Reference 11.

It is important to note that many operators depart from the principles of charging described above through the use of bundled tariffs and flat-rate charging, often without discriminating on distance or duration. Also, the introduction of VoIP over broadband, described in Chapter 8, introduces the notion of free calls, even between countries, for subscribers on the same system. This gives rise to operators and service providers developing new charging models – often based solely on subscription – to recover their network costs.

10.5 Data numbering and addressing

Data networks (see Chapter 8) also require recognised schemes for numbering/ naming and addressing. As with telephony described earlier, the numbering and addressing is used to determine the network routeing, but it is not generally used in the assessment of the charging rate, since normally no usage or distance fees are incurred – instead, the charging is subscription based.

The first number scheme standardised for public data networks is defined in the ITU-T *Recommendation X.121* [13] and was introduced in the 1970s. Figure 10.8 shows the X.121 number structure. This is based on a data network identification code

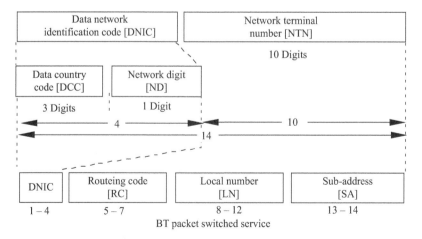

Figure 10.8 International numbering plan for public data networks (ITU recommendation X.121)

(DNIC) of four digits followed by a ten-digit network terminal number (NTN). The DNIC comprises a three-digit country code (similar in concept but different to those used for telephone numbers) and a single digit to identify up to nine networks within the country. As an example of the application of this standard, the number format of BT's packet switching service, based on the ITU-T X.25 protocol and known as packet switch stream (PSS), is also shown in Figure 10.8. For this service the NTN is split into a three-digit routeing code (RC), which identifies the parent packet switch, and a five-digit local number to identify the data line (i.e. terminating point), leaving the final two digits to act as a sub-address for use within the customer's premises [1,10]. As with telephony numbers, the country codes for X.121 addresses are defined by the ITU-T and the DNICs are allocated by Ofcom to operators in the United Kingdom, who then allocate their own-network terminating numbers.

10.6   Asynchronous transfer mode addressing

With the introduction of ATM (asynchronous transfer mode), Frame Relay and other connection-orientated data network services (see Chapter 8), new addressing schemes were specified. Public and private (i.e. corporate) ATM networks can use either the E.164 telephony numbering scheme, as described earlier, or one of the three versions of an ISO format address. The so-called ATM end-system addresses (AESA) are based on the ISO NSAP (network service access point) format, shown in Figure 10.9. The address comprises of two main parts: the initial domain part

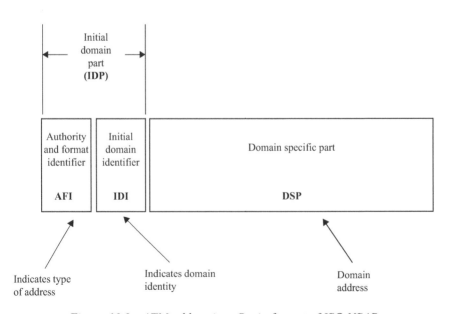

Figure 10.9   ATM addressing: Basic format of ISO NSAP

(IDP) and the domain specific part (DSP). The IDP contains a two-digit authority and format indicator (AFI), which is used to specify which type of addressing is being used, and an initial domain indicator (IDI), which gives the most significant part of the address. The remainder of the structure is known as the DSP, which is used by the network to determine the routeing to the data terminal [14]. The various formats of the ISO NSAP ATM schemes are summarised in Box 10.4 and Figures 10.10 and 10.11.

In the case where the ATM network is carrying IP packets a system is used to map the IP addresses to the ATM addresses, this is known as the ATM address resolution protocol (ARP) [14].

Box 10.4   The ISO NSAP ATM numbering scheme

(i)   *Encapsulation of E.164 numbers*: With an AFI set to *45*, the IDI is formed from an E.164 telephony-type number (although separate from the allocation of numbers for the PSTN), as shown in Figure 10.10.

Encapsulated E.164 Format

AFI = 45	E.164 number	Domain specific part

Encapsulated X.121 Format

AFI = 37	X.121 number	Domain specific part

Figure 10.10   ISO NSAP: Types of ATM addressing using encapsulation

(ii)   *Encapsulation of X.121 numbers*: Alternatively, an X.121 number can be used as the IDI; this is indicated by setting the AFI to *37*, as shown in Figure 10.10.

(iii)   *Data country code (DCC) ATM format*: With this scheme the significant part of the address is given by a DCC, allocated by ISO, which gives geographical identification of a data network – the DCC for the United Kingdom is 826. The domain specific part (DSP) is then made up of an organisation code, allocated in the United Kingdom by the Federation of Electrical Industries (FEI), which identifies the ownership of the private data network. The final part of the DSP is allocated by the organisation. Figure 10.11 shows the format of this scheme, which is indicated by an AFI of 39.

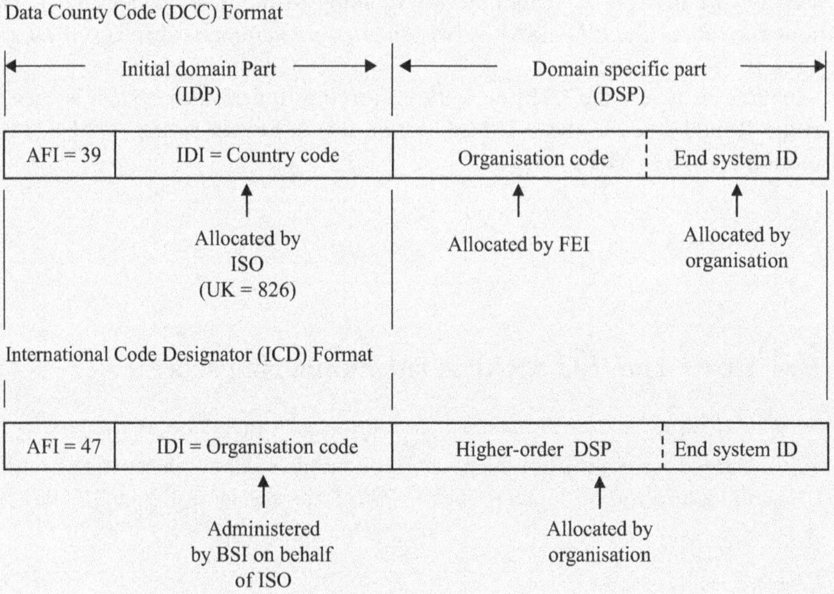

Data County Code (DCC) Format

International Code Designator (ICD) Format

Figure 10.11   ISO NSAP ATM addressing formats

(iv) *International code designator (ICD) ATM format:* The generally recommended scheme for private ATM networks is the use of the ICD format, with AFI set to *47*, in which the IDI comprises of an organisation code identifying the ownership of the private network. The British Standards Institute (BSI) administers the allocation of codes to organisations in the United Kingdom on behalf of ISO. The DSP is then allocated by the private network organisation, as shown in Figure 10.11. The alternative use of the ICD format, which is recommended for public ATM services, uses an international network designator (IND) specified by the ITU-T in place of the organisational code in the IDI.

10.7   IP numbering/naming and addressing

As with telephone numbering and addressing on the PSTN, every device (e.g. routers and data terminals or computers/hosts) on the Internet needs a unique address. Since the conventions were set by the IETF (Internet engineering task force) in 1984, 32-bit addresses adhering to the IP version 4 (IPv4) standard have been used throughout the World. A new enhanced version (IPv6) is gradually being introduced, although mass migration to this new protocol has still to occur. For convenience, we will not consider this new version until the end of this chapter. These IP 32-bit source and destination addresses specify the entry and exit points, respectively, on the Internet and are inserted in the IP packets, so that they can be routed appropriately, as described in Chapter 8. We will come back to discussing the IP addresses later in this section, after we have considered the more-familiar and user-friendly Internet names that are used in e-mails and website identification. Clearly, these names need to be translated to Internet (or intranet) addresses before the IP packets can be dispatched, which we will also consider later.

10.7.1   Internet names

The IP names and website addresses are based on the use of mnemonics that are enduring, easily identifiable and memorable. The structure of the Internet name is hierarchical, but with the granularity in the reverse order to the international telephone number, that is, moving from specific to general reading left to right. The levels of the hierarchy are in descending order as follows:

- Top-level domain (TLD)
- Domain
- Sub-domains
- Host name (The term 'host' means the device or computer attached to the Internet or private intranet – i.e. the item being addressed).

There are two categories of TLDs, namely:

1. Country Code TLD (ccTLD), of which there are some 450. For example: .ca = Canada, .es = Spain, .fr = France, .uk = the United Kingdom, .us = the United States
2. Generic TLD (gTLD): For example: .com, .edu, .gov, .int, .mil, .net, .org and .ac

Figure 10.12 shows the Internet domain name hierarchy, with an example of the analysis of the host address *ee.ucl.ac.uk* (with TLD of *uk*, domain of *ac* and subdomain of *ucl* and host machine *ee*). The base of this tree-like structure is known as the root. E-mail names are created by adding the individual name in front of this Internet name, for example, '*person's name@ee.ucl.ac.uk*'. The host in this case could be a server on which web-pages are stored, accessed by an Internet name based on the URL (universal resource locator) format: for example, *www.ee.ucl.ac. uk*/page *number*. This format can be expanded to name individual sections (i.e. pages) out of a vast collection stored on the server/host.

With the rise in popularity of the Internet and web-based services, such as online purchasing of everything from airline tickets to books and even cars, the various TLD names have become well recognised. There is much value in terms of prestige for a company in having a website name with a well-known gTLD such as . *com*. In general, short names such as *company.com* have greater marketing value

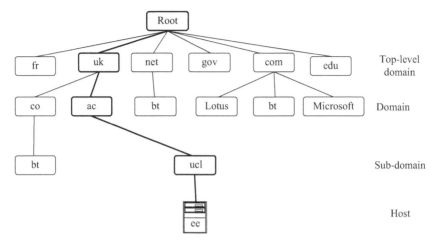

Figure 10.12   Internet domain-name system hierarchy, for example, 'ee.ucl.ac.uk'

than longer names such as *company.co.uk*. In fact, the new phenomenon of 'cybersquatting' has arisen through people registering Internet names of well-known companies in the expectation that those companies would some time later be prepared to buy the name from them at an inflated price. Finally, of course, the use of names rather than numbers for the Internet gives the opportunity for people to impersonate well-known companies by registering the same name with a different TLD or mischievously similar names, creating the phenomenon of 'cyber fraud'.

The TLD names are administered by the Internet Corporation for Assigned Names and Numbers (ICANN), which is an internationally organised, non-profit corporation that also has responsibility for IP address space allocation, and root-server system management functions. As a private-public partnership, ICANN is dedicated to preserving the operational stability of the Internet and promoting competition. In order to address some of the concerns over cybersquatting and fraudulent use of domain names it has introduced a uniform domain-name dispute resolution policy (UDRP), which is applicable across all gTLDs, and now forms the basis for resolving such issues.

10.7.2   Internet addresses

An example of an IPv4 address is shown in Figure 10.13. By convention, the 32-bit address is more conveniently shown in documentation as a set of the four decimal numbers representing each byte (8 bits). In theory, a 32-bit binary number gives:

$$2^{32} = 4,294,967,296 \text{ potential addresses.}$$

However, the original allocation of this address space in the mid-1980s has severely reduced the actual capacity available in practice. This allocation was associated with partitioning of the address space. To understand the need for

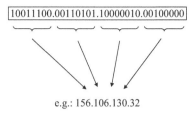

e.g.: 156.106.130.32

Figure 10.13 An IPv4 address

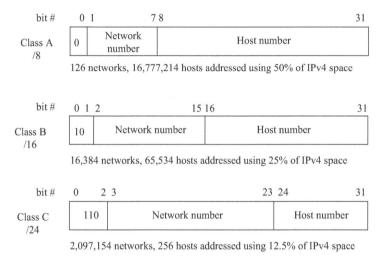

Figure 10.14 IPv4 'Classful' addressing

the partitioning we should appreciate that the (binary) IPv4 address in the destination field of an IP packet (see Figure 8.9 of Chapter 8) is used by the all the intervening routers in the Internet or intranet to determine how to onward route that packet. This is achieved by using the first part of the address to define the identity of the destination IP network, which is examined by the routers within the Internet, leaving the remainder of the address to define individual hosts on that network. The latter part is not examined until the packet reaches the ingress gateway router of the destination network, and it is then examined by any subsequent routers on that destination network.

The partitioning scheme is based on three sets of classes, each designed to accommodate different sizes of IP networks. Figure 10.14 shows the so-called 'Classful' address format [15]. Class A is designed for the very largest IP networks, having address space for some 16,777,214 hosts; Class B is designed for medium sized organisations, having address space for some 65,534 hosts; finally,

Class C, with its capacity of 256 hosts, is designed for the smaller organisations. As Figure 10.14 shows, the type of class being used is indicated by the initial binary digits of the address – *0, 10, 110*, for Classes A, B, C, respectively. For completeness, we should mention that there are also two further special classes (omitted in Figure 10.14 for clarity), namely: Class D, which is used for multi-cast addresses enabling broadcasting of an IP packet to many hosts, indicated by initial digits *1110*; and Class E, indicated by initial digits *1111*, which was reserved for future use.

One way of accommodating large organisations with several individual departmental local area networks (LANs) supporting many hosts, for example, in the case of a university campus, within the Classful structure is the use of sub-netting. This technique relies on a single network address, which is used by all routers in the Internet to reach the ingress router for the organisation, and the host address space is split into a sub-net address and its host addresses. An example of an organisation with three sub-nets associated with one network address of *156.5* is shown in Figure 10.15. Let us consider the routeing of an IP packet sent over the Internet to a host machine on sub-net A, with the destination address of *156.5.32.4*. Now, in general the control software of routers has look-up routeing tables bifurcated into addresses of all the IP networks on the Internet and the addresses of their own dependent hosts or subnetworks. For each entry, there are two fields: the IP network address, and the appropriate outgoing route, as described in Chapter 8. In our example, the network gateway router has an IP address of *156.5.0.0.* – since it is only the first two bytes that are used by the

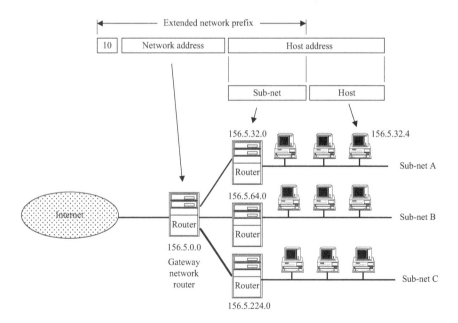

Figure 10.15   Sub-netting

Internet routers to reach that network. So, when the packet is received from the Internet with a destination address of *156.5.32.4* the network router software recognising *156.5* as its own address then examines the next bytes to determine the required sub-net – i.e. *.32*. The subnet router(s) then route to the destination host (*.4*). In general, routers within a sub-net environment use the network address together with the sub-net address when exchanging packets between hosts on the various sub-networks, this is known as 'Extended Network Prefix' working, as shown in Figure 10.15.

Unfortunately, the Classful partitioning scheme does not make good use of the 32-bit address space, since it accommodates only 126 Class A networks (which occupies 50% of the total address space), 16,384 Class B networks (25% of total address space) and 2,097,154 Class C networks (12.5% of total address space). This inherent inefficiency has been exacerbated by the allocation of addresses in the early days of the Internet without regard for the conservation of the addressing resource. Of course, the enormous rise of the popularity of the Internet, far beyond the original concept of a specialist network linking academic institutions around the world, has created unforeseeable demands for Internet addresses. It was soon realised that the Classful scheme, particularly with the Class B offering too much capacity and Class C too little capacity for hosts for most medium-sized organisations, was unnecessarily restrictive. Even as early as the beginning of the 1990s exhaustion of the IPv4 addressing space was predicted as rapidly approaching. Thus, in 1993 allocation of new Classful addresses was ceased and the alternative classless inter-domain routeing (CIDR) scheme was introduced to maximise the occupancy of the remaining available address space.

With CIDR, there is a close match between the capacity of the address space allocated and the required numbers of hosts. This is achieved through defining the position of the moveable boundary between network and host addresses. A CIDR address therefore needs two components: the full 32-bit IPv4 address, followed by an indication of the length of the network address portion. Both these pieces of information need to be held in the routeing tables of all the Internet routers. (Thus, the routeing tables need three fields of information per entry, i.e. IP network address; CIDR address length; outgoing route.) A CIDR address is written in the format: *194.34.14.0/20*, indicating that the network address occupies the first 20 bits of the full 32-bit Internet address (leaving 12 bits to identify the hosts). In practice, the routers employ a binary mask which is 'ANDed', using Boolean algebra, with the full 32-bit address to extract the portion representing the network address. Thus:

11000010 00100010 00001110 00000000 (binary of *194.34.14.0*)
ANDed with: 11111111 11111111 11110000 00000000 (20 binary 1's)
Gives
Network address: 11000010 00100010 0000

Table 10.1 gives some examples of CIDR addresses and for each length of mask the resulting number of network and hosts addresses that can be accommodated. (It should be noted that the numbers of addresses shown are derived from

Table 10.1　Examples of CIDR *Internet addressing*

CIDR prefix length	Mask	Number of network addresses	Number of hosts addresses
/13	11111111 11111000 00000000 00000000	8 k	512 k
/14	11111111 11111100 00000000 00000000	16 k	256 k
/15	11111111 11111110 00000000 00000000	32 k	128 k
/16	11111111 11111111 00000000 00000000	64 k	64 k
/20	11111111 11111111 11110000 00000000	128 k	4 k
/21	11111111 11111111 11111000 00000000	256 k	2 k

the number of binary bits in the address, and are therefore based on powers of two. Thus, 10 bits $= 1,028$ (i.e. 2^{10}), written as *1 k*; 11 bits $= 2,048$ (i.e. 2^{11}), written as *2 k*, 12 bits $= 4,096$ (i.e. 2^{12}), written as *4 k*, etc.)

10.7.3　Translating Internet names to addresses

Now we need to think about how the Internet name (e-mail or web page) is associated with an appropriate Internet address. As stated earlier, the Internet addresses and names are administered worldwide by ICANN, who in turn delegate parts of the address and name space to various regional and national bodies for allocation to Internet service providers (ISPs), companies and organisations.

In the case of e-mails, each subscriber's name, in the form of *name@ISP. com* needs to be allocated an Internet address by the ISP from their set of allocated addresses. In the case of dial-up subscribers, the ISPs usually minimise the overall need for the valuable Internet addresses by using a computer server to manage a pool of addresses, allocating them on a temporary basis to subscribers for the duration of their Internet session. However, the always-on nature of broadband subscribers means that they require a permanent allocation of Internet addresses from the ISP. Allocation of addresses to users of corporate networks (e.g. LANs) is undertaken by the company's network administrator, again using a server to allocate on a real-time temporary or permanent basis, as appropriate. The ISPs or Corporations obtain their allocation of Internet addresses from larger ISPs, who in turn obtain their allocation from one of five regional Internet registries covering Europe, North America, Asia, Latin America, and Africa, respectively.

Now to consider how the addresses for websites are determined. As mentioned above, websites are identified by the user-friendly URL mnemonic-based names. The translation between a URL and the Internet address of that site is made by a directory service known as the domain name system (DNS). This service may be provided by the subscriber's ISP or by independent DNS operators. Figure 10.16 illustrates a simple example of a user attempting to access *www.microsoft.com.*

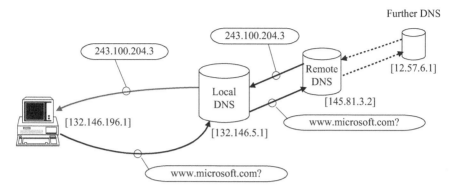

Figure 10.16   How IP terminals interrogate DNS

If an IP session already exists between the terminal and the ISP, a request for the address of Microsoft's website is sent in an IP packet with source address *132.146.196.1* (the terminal) and destination address of the ISP (*132.146.5.1*). If the DNS of the ISP does not have the required information it generates a request of the next DNS in the hierarchy (address *145.81.3.2*), as shown. In the unlikely event of this DNS not having the address, a request of a further DNS up the domain hierarchy is generated. This process continues until the definitive source of addresses, the root DNS (see also Figure 10.12), is reached. However, assuming the Remote DNS can provide a translation it sends the required address for Microsoft's website – *243.100.204.3* – in an IP packet to the ISP's Local DNS, where it is temporarily cached in case there are further requests for this address, before advising the terminal. A session can now be established with Microsoft's website from the terminal by the insertion of the Internet address in the destination field of the outgoing IP packets.

10.7.4   IPv6

Work has been undertaken by the IETF since the early 1990s to define a new version of IP, IPv6, which is designed among other things to increase vastly the capacity of Internet addresses, and improve the efficiency of network routers by better structuring of the network addresses (see also Chapter 8). A need for a significant increase in the number of Internet addresses had for some time been predicted, not only because of the current shortage in the supply of IPv4 addresses, but more importantly also because of an expected mushrooming in demand in the future. This is driven by the introduction of IP-based services, for example, 4G mobile networks and IoT (see Chapters 8 and 9, respectively).

The first thing to note about IPv6 is that it is based on 128-bit addresses, organised as eight 16-bit bytes – giving a potential capacity of 64 billion times the capacity of the IPv4 address space! Needless to say, the proposed partitioning of this 128-bit address into a rigid format significantly reduces the actual capacity available for use, but it is still enormous and it is thought unlikely to exhaust this century. The notation for IPv6 addresses is based on the representation of each 16-bit byte using hexadecimal digits (see Box 10.5). For example:

3FBE:2D00:0000:0000:0004:0377:34DE:98DC

For convenience, the leading zeros in each 16-bit byte or block can be suppressed. Also, a double colon can be used to represent consecutive zeros. So, the above example can be simplified to:

3FBE:2D00::4:377:34DE:98DC

The general format of the IPv6 address for a single interface (or host) – known as a 'global unicast address' – falls into three main parts, as shown below.

Box 10.5   Hexadecimal numbers

This table summarises the three numbering or counting systems used in telecommunication and data networking for values of 0–15. As is commonly known, the counting scheme of decimal is structured around a base of ten, that is, the use of the ten digits *0–9*, with each position to the left in the number having a weighting of increasing powers of ten. Similarly, the counting scheme of Hexadecimal is structured around a base of 16, that is, the use of 16 digits *0* to *F*, with each position to the left in the number having a weight of increasing powers of 16. While, for Binary the base is two, using digits *1* and *0*, with each position to the left in the number having a weight of increasing powers of two.

Decimal (base 10)	Hexadecimal (base 16)	Binary (base 2)
0	0	0000
1	1	0001
2	2	0010
3	3	0011
4	4	0100
5	5	0101
6	6	0110
7	7	0111
8	8	1000
9	9	1001
10	A	1010
11	B	1011
12	C	1100
13	D	1101
14	E	1110
15	F	1111

Global routeing prefix	Subnet ID	Interface ID
(*64-m*) bits	*m* bits	64 bits

The value of *m,* defining the size of the subnet identification, is commonly set to 16 bits. Thus, the global routeing prefix, which defines the network address, is commonly 48 bits.

The specification for IPv6 is reasonably immature, although some of its features are likely to be refined as experience is gained. For this reason, and the given the huge established base of IPv4 equipment across the world, the move towards IPv6 has been relatively slow so far. Also, it can be expected that the rate of adoption of IPv6 will be different within the public Internet, the corporate intranets, enterprise networks and public Wi-Fi, and within the fixed and mobile networks, depending on the commercial and service issues pertaining [16].

10.8 Interworking of Internet and telephone numbering and addresses

So far in this chapter we have considered E.164 numbering and addressing in the context of telephone and other services supported by the PSTN (fixed) and mobile networks, and Internet naming and addressing in the context of IP data services. However, the two worlds are no longer separate. We introduced the concept of providing telephony calls over an IP data network in Chapter 8, identify the move to an all-IP interface for later versions of 4G mobile networks in Chapter 9, and we investigate the eventual large-scale replacement of the PSTN by VOIP in so-called NGNs in Chapter 11. These are all examples of the much-heralded convergence of voice and data services. Clearly, an important requirement of this convergence is the smooth translation between telephone numbers and Internet addresses.

In Chapter 8 the sequence of a voice call between two telephones on different PSTNs carried over an intermediate IP network (the Internet or an intranet) was described (see Figure 8.17). The interworking of the PSTN and IP network require the originating PIG (PSTN-to-IP Gateway) to seek the IP destination address of the distant PIG so that the voice IP packets can flow between gateways to create a call connection. This involves the originating PIG obtaining from a domain name system (DNS) the IP address of the PIG serving that numbering range of the PSTN. Given that the DNS also holds name-to-address information for the Internet, as described above, the addition of translation details for PSTN numbers creates a significant extra storage load. This need for extra capacity not only increases the cost of the DNS, but it also affects performance due to the greater interrogation times involved. Consequently, as the quantity of VoIP calls increases, a more elegant solution for the translations of PSTN numbers to Internet addresses is needed. Such a solution is offered by the ENUM system specified by the IETF in collaboration with the ITU-T [17].

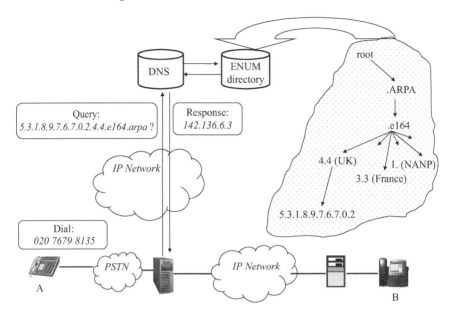

Figure 10.17   Use of ENUM system

The ENUM is a DNS-based system that converts a E.164 telephone number into an Internet name or address for routeing through an IP network. The design of the system reflects the fact that the use of telephone numbers is so universally accepted that even when a high proportion of phones are IP-based, and circuit-switched PSTNs are replaced by VOIP equivalents, people will be still be using such numbers.

Figure 10.17 shows an example of the use of ENUM, in which PSTN subscriber A wishes to call a LAN-based subscriber B who has an IP phone. However, subscriber B is still recognised as a telephony subscriber with an E.164 number *020 7679 8135*. On dialling this number the call from subscriber A is routed through the PSTN to the appropriate IP gateway (i.e. PIG). An enquiry for an IP address for this number is then sent from the PIG to its DNS. ENUM has a dedicated TLD – expected to be *.arpa* – encompassing all the translations for the world's E.164 numbers. The enquiry message to the ENUM system is created as follows:

- Dialled number: *020 7679 8135*
- Convert to standard international number: *+44 20 7679 8135* (i.e. a UK number)
- Eliminate spaces and non-digits: *442076798135*
- Reverse and insert dots: *5.3.1.8.9.7.6.7.0.2.4.4*
- Add TLD and domain name: *5.3.1.8.9.7.6.7.0.2.4.4.e164.arpa*

This standard DNS enquiry format is then passed to the ENUM directory for translation into an Internet address using a decoding tree, as shown in Figure 10.17.

The call is then routed through the IP network to the destination subscriber phone B.

A variant on this example is for the case of a SIP-controlled call (see Chapter 7), where the ENUM directory translates the enquiry into an IP name rather than an address, for example, *name@ISP.com*. This is then subsequently converted at a DNS to an Internet address for routeing through the IP network in the normal way. Actually, the ENUM system is more complex than described above because it also copes with the many forms of services that use E.164 numbers other than telephony, for example, text and e-mail. Thus, a single E.164 number can be translated within the ENUM system into several outputs: Internet numbers or names, depending on how the node in the Internet dealing with that service is to be accessed.

10.9 Summary

In this chapter, we have examined the structure of the internationally recognised telephone number scheme, and noted how the dialled number is used by the control systems in the exchanges to determine the appropriate call charges and how it should be routed through the network. The application of similar numbering schemes was applied to Connection-orientated data services such as X.25, frame Relay and ATM. The chapter then looked at the Internet, noting that the names are used in place of numbers and that there is a clear distinction between numbers/names and addresses – unlike the arrangement for the PSTN, where the subscribers' numbers are also used as the network addresses. Finally, we looked at how ENUM helps manage the interworking between networks using either form of addressing, thus allowing increasing convergence of services in the future.

References

[1]   McLeod, N.A.C. 'Numbering in telecommunications', *British Telecommunications Engineering*, 1990;**8**(4):225–231.

[2]   ITU-T Recommendation E.164. 'Numbering for the ISDN Era' (ITU, Geneva).

[3]   Complement to ITU-T Recommendation E.164: 'List of ITU-T Recommendation E.164 Assigned Country Codes' (Annex to ITU Operational Bulletin No. 805, Geneva).

[4]   OFCOM. 'The National Telephone Numbering Plan', 2015 (http://www.ofcom.gov.uk).

[5]   OFCOM. 'A user's guide to telephone numbering', originally published by Oftel and subsequently adopted by Ofcom (www.ofcom.gov.uk).

[6]   Buckley, J. 'Telecommunications regulation'. Stevenage: The Institution of Electrical Engineers; *IEE Telecommunications Series No. 50*, 2003, Chapter 6.

[7]   Osborn, B., Aucock J., Banks, M., Dunning, L., Hall, B., and Underwood, H. 'National code and number change: Technical solutions for BT's network', *Journal of the Communications Network*, 2002;**1**(1):107–113.

[8]   Banerjee, U., Rabindrakumar K., and Szczech, B.J. 'London code change', *British Telecommunications Engineering*, 1989;**8**(3):134–143.

[9]   Wherry, A.B. and Birt, J.F. 'The London sector plan: Background and general principles', *Post Office Electrical Engineers Journal*, 1974;**67**(1):3–9.

[10]   Wood, S. and Winterton, R. 'A new structure for London's public switched telephone network', *British Telecommunications Engineering*, 1994;**13**(3):192–200.

[11]   Flood, J.E. 'Numbering, routing and charging'. In Flood J.E. (ed.), *Telecommunications Networks*, 2nd edn, *IEE Telecommunications Series No. 36*, 1997, Chapter 9.

[12]   Buckley, J. 'Telecommunications regulation'. Stevenage: The Institution of Electrical Engineers; *IEE Telecommunications Series No. 50*, 2003, Chapter 7.

[13]   ITU-T Recommendation X. 121. 'International numbering plan for public data networks' (ITU, Geneva).

[14]   Ross, *J. Telecommunication Technologies*. Indianapolis, IN: Prompt Publications; 2001, Chapter 8.

[15]   Tanenbaum, A.S. *Computer Networks*, 4th edn. Amsterdam, The Netherlands: Prentice Hall PTR; 2003, Chapter 5.

[16]   Davidson, R. 'IPv6: An opinion on status, impact and strategic response', *Journal of the Communications Network*, 2002;**1**(1):64–68.

[17]   Neustar. 'ENUM: Driving convergence in the Internet age', www.enum.org.

Chapter 11

Putting it all together

11.1　Introduction

In the first ten chapters of this book we considered various aspects of tele-communication networks: their components, how they are constructed, and the way that voice and data is carried. Now, in this final chapter we take a holistic view of how all the pieces are put together, how the end-to-end service provided to the users is managed, and how the networks are progressively enhanced and developed to take advantage of new technology. Specifically, the so-called next-generation network (NGN) concept is examined. It is hoped that by the end of this chapter readers will have gained an insight into the intriguing question raised in the Fore-word to the book, that is, whether it is right to assume that all communications will eventually move onto the Internet.

11.2　Architecture

As we have seen in earlier chapters there are many components used in tele-communication networks, which require appropriate linking in order to provide services for customers. This situation is shown conceptually in Figure 11.1, with the network components of access, circuit-switching, core transmission, intelli-gence and signalling shown as pieces in a jigsaw puzzle, residing within the domain of the applicable commercial and regulatory environment, which together provides services. The classical method of understanding in a systematic way how all the pieces fit together is through the use of architectures. In everyday life we under-stand how the use of architecture helps to describe the structure of a house, giving the relationships between the walls, windows, floors, roof, etc. It also shows the deployment of the various services, such as water, drainage, heating, air-conditioning, electricity and gas. Architecture is applied in much the same way in telecommunication networks.

However, architecture in telecommunication networks is more than just a diagram showing how all the 'boxes' are joined together, important though this is! An architecture also needs to encompass the commercial and regulatory conditions applying to the network, and the services it receives and supplies to other networks/ service suppliers, as well as the services provided to its users (customers). The architecture of an operator's network, for example, must comply with the strategic

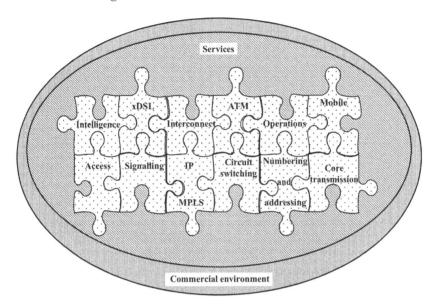

Figure 11.1   Putting it altogether

policy of the company in terms of the business objectives and services offered [1,2]. Therefore, typically four architectural views are required to adequately define a network operating in today's environment, as shown in Figure 11.2 which is described below.

11.2.1   Commercial or service model view

This view shows the various roles involved in getting a service to a customer, and the players that undertake those roles. In addition, most importantly this view needs to identify the flow of money between the roles. One might characterise this view as showing who does what to whom and who is paying for it! A simple example is that of a subscriber to BT's network accessing '*Tyres you need*' – a (fictional) nation-wide tyre-fitting company's call centre using a 'Freefone' (0800) number (see Chapter 7). The subscriber pays a line rental to BT for the local line, which is also used for normal telephony calls (with appropriate call charges to BT). Since *Tyres you need* has provided the subscriber with free-call access to the company's call centre by virtue of the Freefone number, BT does not receive any money from the subscriber for such calls; instead, *Tyres you need* pays BT a rental for the Freefone number. Thus, there is a set of commercial relationships between the subscriber, *Tyres you need*, and BT covering services and payment.

We can elaborate the simple example of the dial-up call to *Tyres you need* by considering what happens if the subscriber later makes a purchase over the Internet via the web page of the trader *Tyres you need*. There is then a further commercial

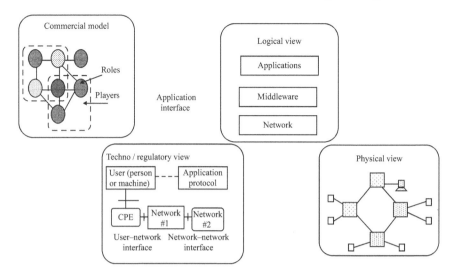

Figure 11.2   The many architectural views

relationship between the trader and the subscriber. However, the situation could become even more complex if a search engine or web portal is used by the subscriber when seeking the trader on the Internet, because the trader may then need to pay the portal service provider for the customer referral. Such use of the Internet (World Wide Web) for purchasing, often called 'e-purchase', is part of a general phenomenon known as 'e-commerce' or 'e-business'. This itself is part of the so-called bigger 'information industry', which embraces the provision of information (e.g. share prices and local weather forecasts) over the Internet or corporate networks, broadcast television via satellites, etc., as well as e-business applications. During the mid-1990s the terms 'Global Information Highway' and 'Information Superhighway' were coined to describe the route to this world of electronically-accessible information.

In 1995, the European Telecommunication Standard Institute (ETSI) undertook a strategic review of the implications of the emerging infrastructure for the Information industry, and any consequent need for new standards. They produced a generic reference diagram for the roles in the information industry, separated into 'structural roles' providing the information routeing, supported by 'Infrastructural roles' providing the supporting infrastructure of terminal and network equipment. Network operators provide the infrastructural role of 'Telecommunication Service provision'. This ETSI Strategic Review reference diagram is shown in Figure 11.3; it represents a good example of a commercial or service architectural view.

The subject of commercial models in telecommunications is examined in more detail in Reference 27.

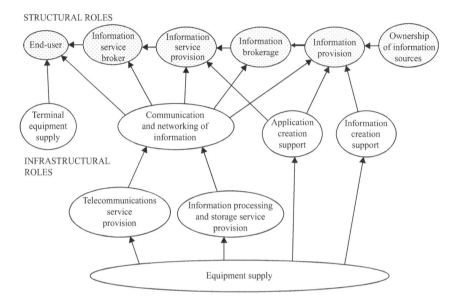

Figure 11.3   ETSI SRC6 roles in information industry

11.2.2   Techno-regulatory view

The second architectural perspective shown in Figure 11.2 is the Techno-regulatory view. This indicates the boundary conditions of the network: the user–network interfaces on one side and the network–network interfaces, where interconnections to other networks are made, on the other side. Generally, these two sets of inter-faces are defined under the control of the national or state regulator to ensure competitive openness. Thus, users can employ any make of terminal devices pro-vided they comply with the user–network interface specification. Similarly, other network operators and independent service providers will be assured of technical interoperability when connecting to the network.

In addition to the physical and electrical interface characteristics, there may be application – or software – levels of interfacing that need to be specified. Examples of these for network–network interfaces are the message sets for signalling between the networks, for example, signalling system Number 7 (SSNo.7) or SIP and API (application programme interfaces), as described in Chapter 7. These APIs may also apply as software/interfaces at the user–network boundary.

11.2.3   Functional or logical view

The functional or logical view (see Figure 11.2) describes contributions made by the various elements of a network. In this perspective, the elements are deemed to provide a ''service' to other elements of the network. Generally, each element is supported by the service of a lower-level element and, in turn, it provides a service to a higher-level element. Therefore, these elements tend to be considered as residing in layers. (Engineers love to analyse complex situations into multilayered functional architectural views!) Examples of this form of architectural view that are described in this book are the OSI 7-layer model (Chapter 8 and Figure 8.4) and the 4-layer model of SSNo.7 (Chapter 7 and Figure 7.8).

11.2.4   Physical view

The fourth view is that of the physical perspective, as shown in Figure 11.2, which is probably the most commonly understood manifestation of architecture. This view shows the physical joining of the various boxes of equipment. Most of the block-schematic diagrams used throughout this book are presenting physical views.

These views are just different ways of looking at the same thing. They may be likened to looking at a mountain from four directions, say, North, East, South and West, each view being a different way of seeing the same mountain. We return to the use of architectural views several times in this chapter.

11.3   A holistic view of a telecommunication network

We will now use two architectural perspectives, logical and physical, to understand how all the component networks considered in this book fit together to form a public switched telephone network (PSTN). Such holistic pictures are important for the understanding of how the overall network performs, how it provides services, and how it can be developed in the future.

11.3.1   Logical multi-layered view of a Telco's networks

Figure 11.4 presents a multi-layered logical architectural view of the components of a typical set of an operator's networks [3]. The lowest Layer, the Transmission-Bearer network, provides the transmission capability used by all higher layers – thus, acting as a transmission utility. This layer contains the copper and optical fibre cables, as well as any radio links forming the fixed Access Network as well as the cell sites and backhaul links of the mobile networks. It also contains the junction and trunk circuits between the fixed, mobile and data network nodes of the Core Transmission Network. As described in Chapter 5, the nodes of the Core Transmission Network (i.e. the Core Transmission Stations) provide points for interconnection of circuits or groups of multiplexed circuits (this is known as 'transmission flexibility'). About half of the Core Transmission Stations are co-sited with the switching equipment, that is, both are in the exchange building; the remainder are transmission-only nodes, housed in transmission equipment buildings. This is reflected in Figure 11.4; when tracking the nodes in the bottom two layers it can be seen that there are more nodes in the Transmission Bearer Network than in the switching layers above.

The circuit-switching layer contains all the switching nodes: remote subscriber concentrators, local exchanges (LEs) with co-sited subscribed concentrators,

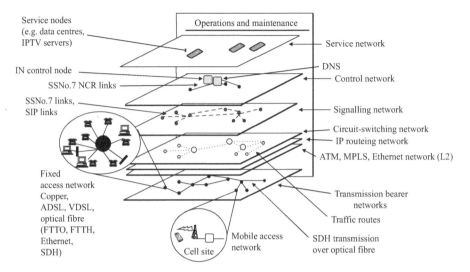

Figure 11.4   The multilayer view of a set of telecommunication networks

junction tandem exchanges, trunk exchanges, international exchanges and the mobile switching centres. Between the nodes, shown as dotted lines in Figure 11.4, are the traffic routes. Of course, the transmission paths for these traffic routes are physically carried in the bottom layer; although logically the traffic routes are in the Switching Layer. This separation of physical and logical functionality is an important architectural concept. Actually, the switching layer is closely associated with two other layers: the IP routers serving the data networks and the Layer 2 networks providing ATM, MPLS and Ethernet, as shown in Figure 11.4.

The next layer up is that of the Signalling Network. Between the subscribers and the serving concentrator (whether co-sited or remote) are the various subscriber-signalling systems (see Chapter 7). The SSNo. 7 links run between the fixed and mobile exchange nodes. Again, logically the SSNo.7 links are in this third layer, even though physically their transmission is within the time slot 16 s of 2 Mbit/s digital systems carried in the Transmission-Bearer Network Layer. Similarly, this layer includes the SIP signalling for the IP network.

At the fourth layer of Figure 11.4, we have the Synchronisation Network, comprising of synchronisation nodes located at all the digital exchanges and data nodes, linked by a set of synchronisation links. The role of this network layer is the control of the speed of the digital clocks at each of the nodes. As described in Chapter 4, digital time-division multiplexing (TDM) networks require all participating switching and multiplexing nodes (voice or data) to be synchronised. The actual arrangements for synchronising may differ between national networks, ranging from disseminating the timing from a central atomic clock through a mesh of synchronisation links and control nodes [4,5] to the reception at each network node of the GPS (Global Positioning Satellite) system timing pulses. Again, the transmission of the synchronisation links or satellite links is physically carried in the Transmission utility of the bottom layer.

Intelligence for advanced call control, in the form of IN SCPs (intelligent-network service control points), and the control links to the fixed and mobile switching nodes is provided by the fifth layer of Figure 11.4 (see Chapter 7). This layer also includes the home location register (HLR) and IMS (IP multimedia subsystem), etc., of the mobile networks. The control links are provided by SSNo.7 non-circuit-related signalling, carried physically in time slot 16 s of 2 Mbit/s systems in the Transmission Bearer Network of Layer One.

Finally, overseeing all these component networks shown in Figure 11.4 is the Administration Network at Layer Six. This network comprises of the set of network-management centres and their communication and control links, which perform the function of managing the operation of all the lower network layers. We will consider the network management function in a little more detail later in this chapter.

The importance of this logical model of a typical set of networks run by an operator is that it clearly demonstrates the role of the component networks and their functional inter-relationships. In practice, operators tend to manage the component networks separately, with different planning and operating organisations and cultures dealing with each level. However, of course, it is the combined operation of all the component networks – i.e., across all six layers – that produce the services provided to the customers.

11.3.2 *Physical view of the set of a Telco's networks*

Figure 11.5 presents a physical view of all the networks associated with a typical telecommunication network operator, that is, a so-called 'Telco'. The view shows on the left-hand side the various forms of subscribers: mobile, fixed residential and small business, and medium and large business fixed customers – with the first vertical line indicating the user–network boundary for each of the categories. The second and third vertical lines indicate the boundary between the Access Network and the Core networks, i.e., the location of the aggregator nodes. In the case of the PSTN, the subscriber concentrator unit takes this role, being either remotely located or co-sited with the parent LE. Within the Core Network there are two sets of nodes located at the third and fourth vertical line. For the PSTN, the first group is the junction tandems (JT) or trunk exchanges (TE); in the second group are just TEs. Finally, the TEs are connected to international exchanges. At the extreme right-hand side is the final vertical line indicating the important network–network boundary, the interface between the Telco's network and the other public network operators (known as communication providers (CPs)) and service providers (e.g. ISPs). The PSTN comprises the network terminating point (at the user–network boundary) the Access Network, LEs, JT, TE and international exchanges.

As Figure 11.5 shows, the parent trunk exchange acts as a gateway to many other networks and non-PSTN exchanges. These include the CPs, the mobile

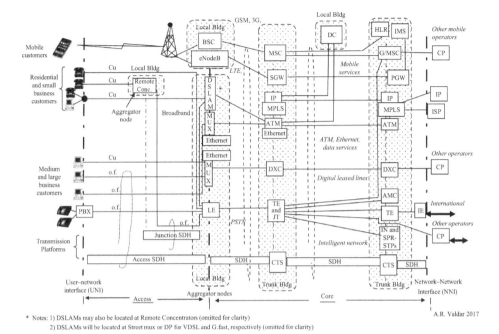

Figure 11.5   Typical set of Telco networks

switching centres (MSC), auto-manual centres (which provide operator-assistance, including emergency [*999* or *112*] and directory enquiry services), as well as the Intelligent Network (IN) service control points (SCPs). Within the Core Network all links between nodes are carried over digital synchronous digital hierarchy (SDH) transmission on optical fibre cables or microwave point-to-point radio. These forms of transmission are also used where high-capacity links are required in the Access Network, for example, for high-speed leased lines, integrated services digital network (ISDN) links to digital PBX (private branch exchanges) or high-speed data services – all provided to business customers' offices.

Also shown are the ATM and Ethernet switches and IP routers providing the core data network capacity, used by a variety of access methods: 'broadband' over xDSL (digital subscriber loop) (with DSLAM at the broadband aggregator node), optical fibre and copper cables. High-speed digital leased lines are directly carried over SDH transmission links on the Access and Core Networks with manual patching or automatic cross-connecting at the Core Transmission Stations. Lower-speed digital leased lines are automatically patched via digital cross-connect (DXC) units. Access to the Telco's data centres is via the Data Core network which is usually housed in separate buildings.

Figure 11.5 shows the typical locations of the various network nodes. It is important to note that the local and trunk buildings accommodate many different network nodes, with the Core Transmission network links between buildings across the country.

For clarity, some of the other nodes that might be in a Telco's existing set of networks have been omitted from Figure 11.5, for example, special business networks for Centrex and virtual private network (VPN) services (see Chapter 2). However, despite the inevitable simplifications, this physical view does convey the complexity and variety of the set of Telco's networks serving a range of customer types, with a wide portfolio of services. We will consider this complexity and variety when we look at the issues around enhancing and modernising the networks, later in this chapter.

11.4 Quality of service and network performance

11.4.1 Introduction

A primary objective for a network operator or service provider is to provide its customers or users with the required quality of service (QoS). Of course, users want as high a QoS as possible, but the operators need to balance the manpower and equipment costs of meeting certain standards against the price that customers are prepared to pay. Usually, there is a clear statement by the operators in the contracts for service describing the quality that customers can expect – there may even be penalty clauses or promises of compensation when these levels are not met. Broadly, QoS covers both network-related as well as non-network-related aspects. Examples of the latter include: the time between receiving an order and the provision of a service to a new customer, time to repair faults, and accuracy and convenience of billing. Network-related aspects (e.g. congestion, clarity of a call connection) are dependent on the way that the network performs. As indicated earlier in this chapter, the performance of the network perceived by the user results from the contribution of all the component networks involved in providing the service end-to-end.

Figure 11.6 illustrates the interrelationships of the factors affecting QoS, showing the non-network and network-related factors [6]. The network performance is dependent firstly on the extent that the network is available for service, known as the 'availability', measured as a proportion of a year (e.g. 99.99%). This is influenced by the inherent reliability of the equipment used in the network and the extent of redundancy in the network structure and the use of protection switching (see Chapter 5). The availability is also dependent on the speed with which service can be restored following equipment failure, that is, the average time to repair faults or provide an alternative way of carrying the traffic.

The other major factor influencing network performance is the so-called 'trafficability', which is a measure of the ease with which telephone calls, data packets or even permanent transmission paths for leased lines, can be routed through the network. The level of trafficability for telephone service is set by measures such as grade of service (GoS), giving the probability of calls failing because of network

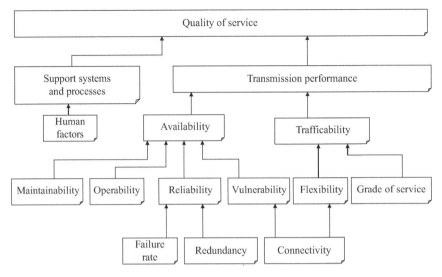

Figure 11.6   Factors affecting quality of service

congestion (see Chapter 6). Trafficability depends on the extent of interconnectivity of the network – i.e., the degree of meshing between network nodes and the opportunities for patching or switching-in alternative paths for leased lines or transmission routes between nodes.

We now briefly consider the main parameters used in measuring network performance [7].

11.4.2   Transmission loss and loudness in a PSTN

Probably the most obvious performance parameter for a PSTN is the loudness of the telephone call connection, as perceived by the users. The loudness of a telephone call is set by the characteristics of the telephone instruments at either end or the transmission loss between them through the network connection. Figure 11.7 shows the arrangement, with the two directions of transmission separated. Each type of telephone terminal has a characteristic inherent loudness as generated by its microphone – the sending level, and as generated by its earpiece – the receiving level. Since loudness is a subjective measure for humans, panels of observers are used in a testing environment to evaluate the sending and receiving loudness ratings for each type of telephone terminal – known as Send loudness rating (SLR) and Receiver loudness rating (RLR), respectively.

Referring to Figure 11.7, it can be seen that the overall loudness rating as heard by subscriber B is made up of the sending rating of telephone A, *SLR(A)*, the network loss across the network, *x* dB (see Box 4.1 in Chapter 4 for an explanation of dB units), and the receiving rating of telephone B, *RLR(B)*. In the reverse

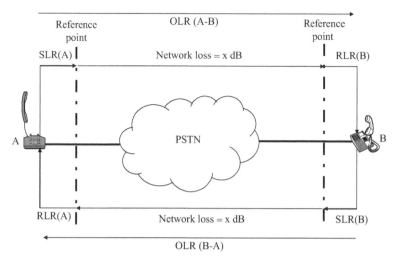

Figure 11.7   Transmission loss in the PSTN

direction, the loudness rating as heard by subscriber A is made up of: $SLR(B) + x + RLR(A)$ dB.

Operators design their networks (whether fixed or mobile) to work within an acceptable range of overall loudness, recognising the range of loud and soft speakers in the populations of users, as well as the range of telephone terminals available. Again, based on the views of panels of speakers and listeners the acceptable range of overall loudness is determined to be between the upper levels, where 50% of customers have difficulty because of it being too loud, and the lowest level, where 50% of customers would have difficulty because of it being too quiet.

11.4.3   Transmission stability

Figure 11.8 shows simplified view of a PSTN, with two-wire copper local loops at each end of a four-wire (i.e. separate Go and Return paths) switch and Core Transmission Network. At the point of the two-to-four-wire conversion (provided by the 'hybrid transformers', as described in Chapter 1) there is, in practice, due to mismatches in balancing the line impedance, the possibility of some of the send power breaking into the receive direction, and vice versa. If the four-wire network is operated at too low a loss overall, the breakthrough electrical signals would circulate in the loop created between the two two-to-four-wire conversion points. Subscribers would hear an annoying screech, confusingly known technically in electrical engineering parlance as 'ringing' and the circuit would be deemed unstable.

Stability in a PSTN – or any network using a mixture of four- and two-wire circuits – is ensured by operating the four-wire portion of the network at a 6 dB loss (i.e. −6 dB). This ensures that any circulating breakthrough signals are reduced,

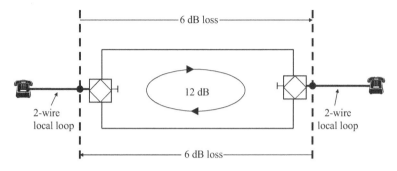

Figure 11.8   Stability

(attenuate) by 12 dB per revolution, rapidly eliminating any ringing. This 6 dB loss is introduced at the subscriber line card in the serving digital LE (see Chapter 6). Similarly, if two-wire circuits are used with voice-over-IP (VoIP) equipment 6 dB of loss needs to be inserted by the codec in the media gateway device (see Chapter 8).

11.4.4   Echo and delay

Apart from the screeching effect (i.e. ringing) caused by unattenuated circulating signals, there is the problem of echo paths also being set up. Echo becomes annoying for the telephone users when the propagation delay between the directly transmitted and breakthrough circulating voice signals becomes appreciable. Two sets of paths are established: the talker echo-path and the listener echo-path, as shown in Figure 11.9. Normally, the loss inserted at the subscriber line card for stability purposes, described above, also reduces the level of both types of echo signals sufficiently to give satisfactory quality for telephony service.

All transmission of an electrical current incurs a propagation delay, the extent depending on the type of medium and the distance involved. Table 11.1 shows details of typical transmission system propagation delays. In addition, nodal equipment such as multiplexors, digital circuit switches, packet routers, etc., introduces their own delays. However, if the level of propagation delay through the network is above that of a standard digital PSTN, say due to the use of ATM packet switching for voice (see Chapter 8) or satellite transmission, then special echo-control equipment is normally required. This is usually located with the offending systems in the network.

It should be noted that codecs used in mobile handsets and many of the VoIP terminals use special computer processing systems to reduce the required bit rates below the normal standard of 64 kbit/s. Such processing introduces extra delay which is perceptible to telephony users. With certain combinations of mobile handsets and VoIP networks the resulting total delay can cause annoyance to users – although, it will not introduce echo if no two-wire copper is used in the connection since two-to-four-wire conversion is avoided.

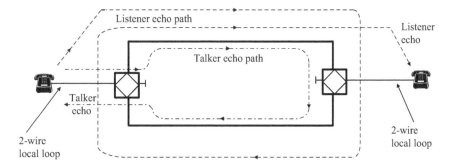

Figure 11.9 Echo paths

Table 11.1 Contributions to network delay [8]

Equipment	Delay
2 Mbit/s pair cable	4.3 μs/km typical
140 Mbit/s optical fibre	4.9 μs/km typical
140 Mbit/s microwave radio link	3.3 μs/km typical
Digital multiplexor–demultiplexor	1–2 μs typical
Digital trunk exchange	450 μs (International recommendation)
Digital local exchange	1.5 ms (International recommendation)
Geostationary satellite	260 ms

11.4.5 Digital errors

An important performance characteristic of any digital system used in tele-communication networks, whether it deals with data or encoded voice, is that of digital errors. These can result from a variety of causes, including minor timing and sampling glitches, impulsive noise causing distortion to the signal, and equipment malfunction. Digital errors may occur in just one bit, that is, a binary *0* being corrupted to a binary *1*, or vice versa – or a cluster of bits being in error. The latter is often referred to as an 'error burst'. There are techniques using extra information inserted at the sending end to the data packets of data systems or the digital stream of transmission systems, which enable induced errors to be detected, and even corrected, at the receiving end, as described in Chapters 4, 7 and 8. However, these error-detecting and correcting systems can cope with only a limited range of errors. When the error rate exceeds these limits, whole packets need to be resent in the case of packet switched networks, or the whole system must be shut down in the case of TDM digital switching and transmission systems. For example, the latter would occur when the error rate on a digital transmission system over an optical fibre degraded to one error in every 100,000 bits (i.e. 1 in 10^5) or worse.

Table 11.2   Digital network error standards

Performance parameter	Definition	G821 recommendation
Errored seconds	A time period of 1 s with 1 or more errors	<1.2% of 1 s intervals to have any errors
Severely errored seconds	A time period of 1 s with 65 or more errors	99.935% of 1 s periods with error ratio better than 10^{-3}
Degraded minutes	A time period of 1 min with 5 or more errors	98.5% of 1 min periods with error ratio better than 10^{-6}

Table 11.2 presents the ITU international recommendation (or standard) G821 on the values of various digital error performance parameters at which the networks should operate. Specific causes of error resulting from timing problems between the sender and receivers on a digital transmission system include:

• *Jitter*. The very short-term variation in timing (equivalent to 'flutter' in an audio HiFi system).
• *Wander*. The medium-term variation in timing (equivalent to 'wow' in an audio HiFi system).
• *Slip*. Where complete TDM frames (see Chapter 4) are dumped or repeated (i.e. slipped) to compensate for the accumulated long-term differences in timing between the node and the rest of the network [8].

There are many other parameters of network performance used by operators to help manage the network at an acceptable quality level. For example, the performance of call processing by an exchange control system is measured by the delay to dial tone and connection-establishment delay. The performance measurement of ATM packet switched service uses several parameters, for example, cell transfer delay (CTD) and cell-error ratio (CER).

Finally, we come to the, perhaps strange, concept of apportioning during the design stage the performance impairments that each of the various network elements may incur. The network is then operated and maintained at this performance level. This apportionment must be based on the eternal balance between acceptable cost and required quality. The overall network performance offered by a network operator (which may or may not match customer's expectations!) is usually based on international standards and historical levels of achievements. Each overall performance parameter is then allocated to the contributing network elements in a way that minimises the total capital cost of equipment and the operational cost of the technician manpower to maintain the network to that level of performance. This usually means that the Core Network, with its higher-loaded circuits and more centralised operations, is required to perform to higher levels than the disparate more voluminous Access Network (e.g. 20–30 million lines).

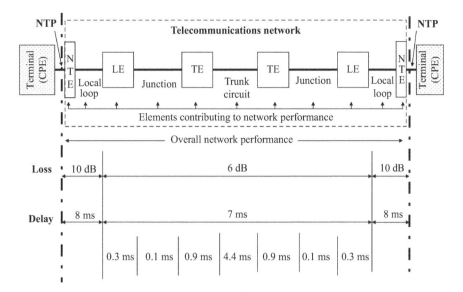

Figure 11.10   Apportionment of performance parameters

Figure 11.10 illustrates the apportionment of two important parameters, loss and delay, to the PSTN, as discussed below.

Loss. The overall reference equivalent (ORE) for acceptable telephone calls is around 37 dB, of which the two telephone terminals contribute a total of 11 dB. As explained earlier, the overall loss of a four-wire digital PSTN is set at 6 dB to ensure stability and to minimise echo problems. This leaves some 20 dB maximum loss for the sum of the copper local loops at each end. The local loop contains a wide range of cable lengths and gauges, with subscribers at differing distances from the exchange (or remote concentrator). Typically, the maximum loss of a local loop is set at 10 dB – although, many subscribers lines incur lower losses. With telephones terminals typically having a sending reference equivalent of 10 dB SRE and a receiving reference equivalent of 1 dB RRE, this causes the worst case loss end-to-end to be $20 + 6 + 1 + 10 = 37$ dB RE. However, most telephone calls would incur some 10 dB less overall loss than this worst case [8].

Delay. The overall limit of acceptable end-to-end delay for a telephone call through a PSTN is 23 ms (i.e. 23 thousandths of a second). The allowed contribution of all the network elements for the most complex call (i.e. highest number of components incurred in the call) is shown in Figure 11.10 [8].

It must be emphasised that the situation shown in Figure 11.10 assumes just one network involved in the call connection. However, many calls are carried over several different networks, for example, from a mobile (cellphone) via an alternative network carrier and terminated on the incumbent operator's PSTN and local loop. The apportionment of performance to the various participating networks, of course, should total to an overall level which is acceptable to users. In practice, given the wide variety of mobile and fixed telephone terminals and network operators, there may occasionally be combinations that cause difficulty for certain users (e.g. too much delay or not loud enough for listening clarity).

11.5 Operations

Undoubtedly, the main attribute that differentiates one network operator from another is the quality of their management of the operations of the networks and support systems, and the day-to-day activity of dealing with customer-service interactions. This is because all network operators can use the same range of technology in their equipment, so it is the way that the networks and services are operated that makes the difference. Curiously, this seemingly obvious factor is missed by successive generations of managers working for network operators who tend to concentrate on the choice of network technology and neglect the consideration of the less glamorous aspects of how a new service or network is to be managed. This has resulted many times in cases where new services or network enhancements have been launched with much publicity, only to be temporarily withdrawn shortly afterwards for some months whilst the operational aspects are sorted out. By the same token, the lead time (see Chapter 6) to introduce a new service, network, or new technology in an existing network, can be dominated by the time taken to develop the necessary new operations-support systems.

Clearly, we need to include a view on operations in this chapter's consideration of putting all the pieces together. Not surprisingly, there are many books and technical journals covering this vast subject [8–13] and we can provide only a brief overview here.

Operations covers all the day-to-day activities involved in running a public or private telecommunication network. This includes the planning, design and installation of equipment into the network and its upgrades and expansion – generally referred to as 'design and build' functions. It also includes the providing of service to new customers (known as 'provisioning') and the management of the performance of the network – known as 'provide and maintain' functions. These two sets of functions, which are focused on the network, are normally considered to provide the so-called role of 'network management'.

For any national operator, network management is a vast undertaking. It involves the control of many thousands or millions of lines, multiplexors, subscriber concentrator switches, digital transmission line systems, LEs, junction tandems, trunk exchanges, IP routers, Ethernet switches, etc. Each piece of equipment must be configured, assigned to a customer or for common use in the network, its performance monitored, faults must be repaired and the equipment brought back into service. Just keeping an up-to-date inventory of all the equipment identities, their location, their status, etc., is a huge task. Although historically the records used were paper-based with manual tracking, now operators use computer-based network management systems. Thus, the network management function is delivered through a range of large computer systems running programs that enable technicians at several dispersed centres within the country to control remotely whole regions of the network. Such systems require large databases to hold the inventory and status information of the equipment in the catchment area. Real-time monitoring and remote control is provided by the extensive deployment of control links from the various network elements (i.e. equipment) to network-management centres. Typically, the latter are referred to as operations and maintenance centres' (OMCs). The control-links from the equipment and the OMCs are deemed to reside in the top layer (Administrative Layer) of the multilayered model of Figure 11.4, described earlier in this chapter.

The OMCs provide a range of functions, typically including:

(i)   Remote monitoring of alarms from exchanges and Core transmission systems.
(ii) Remote access to the exchange-control systems to:
   •   Change the status or features of a subscriber's line;
   •   Initiate a new subscriber's line;
   •   Monitor a subscriber's line;
   •   Change the telephone number of a subscriber's line;
   •   Change the contents of the exchange routeing codes and tables;
   •   Set software changes and upgrades;
   •   Manage software restoration actions;
   •   Install software builds.
(iii) Monitoring of unmanned exchange and Core-transmission buildings for intruder and fire alarms, etc.
(iv) Remote collection of traffic usage information (to be used for dimensioning and forecasting of growth in demand, traffic dispersion, etc.).
(v)   Remote collection of call-record data from the exchanges for forwarding to separate billing centres.

In addition, there are separate network management centres which monitor the whole national network, the links to other operators and the international links to other countries. These national or regional control centres have the responsibility of overall control of the network performance. Importantly, it is the technicians at these centres that can initiate remedial action across the range of transmission systems and exchanges to cope with major breakdowns or traffic overloads in the network. Where there is advanced warning of likely telephone traffic surges, for example, as a results of televised telephone voting, where a massive number of calls can be expected to a single number during a short time of day – technicians at the network control centres can initiate re-routeing of calls and other measures, such as call gapping, to limit the extent of overload and ameliorate the effects on the QoS for the customers. In the case of call gapping, the control systems of the offending LEs are set to switch only a limited proportion of calls to the overloaded destination, for example, one call every 5 s [14].

In practice, the management of the network is undertaken at two levels. The first level is that of the management of individual pieces of network equipment or 'elements', for example, cables, multiplexors, line systems, cross-connects, exchanges, signalling systems, and IN databases. So-called 'element managers' are control systems, usually computer based, that are specific to the elements' technology. For example, an SDH add-drop multiplexor (ADM) controller is used to configure the ports on all the ADMs supplied by a particular manufacturer. Normally, element controllers can extend remote control to all the many elements within an area – typically a region within a country, as set by practical constraints or organisational boundaries of the network operator. In addition to managing the configuration of the equipment, element managers usually also monitor one or more performance parameters (e.g. digital error rate) and any fault alarms or system error messages. Element managers are usually located in operational buildings, such as exchanges or Core Transmission stations.

The second level of network management is at the overall network level, having end-to-end control for that particular network. Examples include: the full network view of leased lines, telephone calls and ATM cell routeing. Generally, network managers, which are also computer-based systems, coordinate the outputs from all the element mangers involved in the network so that a total overview is obtained. It is these network management systems that are in the Network Management Centres described above.

So far, we have considered only the management of the network itself, but there is a further range of operations associated with managing the interactions with customers – usually referred to as 'customer service' or 'service management'. There are four main areas of service management, namely: order taking, fault management, provisioning and billing. The key aspect of service management is that it involves providing an interaction with customers. This is provided by service centres which are contacted by customers through telephone calls, e-mails, websites, or even in person. The support systems for service management employ large-scale computing with massive databases.

Figure 11.11 presents a summary top-level view of the widely accepted logical architecture for operations management, which is structured as a five-layer hierarchy. At each layer a distinct set of operational activities are undertaken by groups of people, using dedicated computer support systems, associated with the relevant databases, following prescribed processes, and providing outputs for different recipients. At the base of the hierarchy are all the network elements, that is, the network itself. As described above, these are managed by technicians at network buildings using element controllers, shown as the second layer in the architecture. The element controllers are, in turn, managed on an end-to-end basis by the network-management centres, which are deemed to sit in the third or network-control layer. Above this is the service-management layer providing the interface to the customers of the network service. Finally, there is a top layer which comprises of all the activities associated with managing the operator's business. This includes functions such as budget build, financial tracking of expenditures within the organisation, particularly expenditure on network equipment!, human-resource management, payment of salaries, invoicing and treasury functions, etc. Generally,

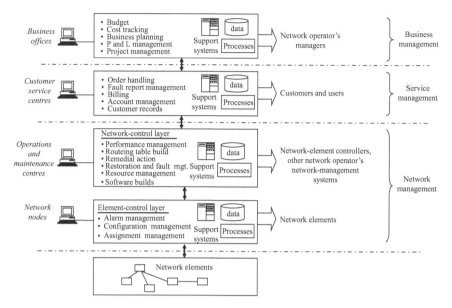

Figure 11.11   Operations management hierarchy

the Element-Control layer and the Network-Control layer are assumed to act as a combined Network-Management function.

Whilst the five-layer architectural view of Figure 11.11 helps define the various categories of activities involved in managing a telecommunications-network-operator business, it does not provide a structure for the design of the vast range of support systems and their data bases and the process associated with execution. However, the Telecommunication Management Forum (TMF), which includes representation from network operators and equipment manufacturers worldwide, has addressed this problem. The TMF have developed the so-called FAB model to help the industry agree on how the set of activities or processes involved in providing network services should be structured. Figure 11.12 presents the model, which identifies three sets of processes: those associated with customer care, service development and operations, and network and systems management. Its name is derived from the three fundamental categories of action that the processes support:

- Fulfilment, covering all that is involved in provisioning service to a customer
- Assurance, covering the management of the QoS received
- Billing, including all that must be done in the network and the support systems.

The TMF have incorporated the FAB model into a comprehensive functional view covering all the processes of a network operator, known as the Extended Telecommunications Operations Map (eTOM). This gives an all-embracing view of the processes associated with element management, network management,

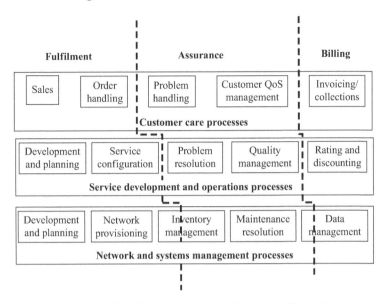

Figure 11.12   Tele management forum's FAB model

service management and business or enterprise management – covering not only the operational (day-to-day) aspects, but also the long-term strategy, planning and development aspects. Figure 11.13 presents the generic picture of eTOM, with three distinct domains identified: the first is Strategy, Infrastructure and product, the second is Operations, and the third is Enterprise (or business) management. For the first two domains, the processes are grouped into those that specify what is required in interfacing to customers (shown as the top horizontal in Figure 11.13); those processes associated with service management, that is, day-to-day interactions with customers – as the second horizontal; those processes associated with managing the computer support equipment and the network equipment (third horizontal); and finally the bottom horizontal contains the processes associated with managing the relationship with equipment suppliers and operational partners and internal management of the operator's business.

All the relevant processes in the horizontals are fully described in the eTOM (although for clarity not shown in Figure 11.13). For example, service management and operations processes for billing are in the intersection of the second horizontal and final column of eTOM. It is through commonly agreed and understood models like eTOM that the network operators can buy from a number of suppliers standardised off-the shelf software support systems for their business, service, and network-management processes, rather than incur the cost and complexity of developing their own bespoke systems. As a result, most network operators have undertaken the huge task of rationalising their operations processes and support-system architectures so as to conform to eTOM and there is a general preference for the use of off-the-shelf operational support systems.

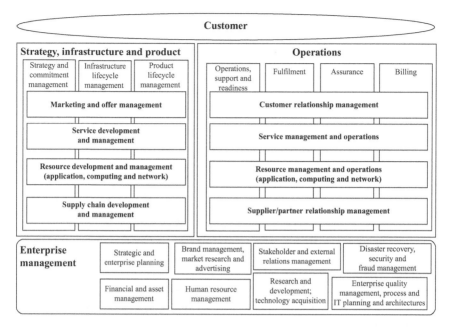

Figure 11.13   Extended telecommunications operations map (eTOM)

11.6   Network evolution

In looking at how all the pieces fit together we need to consider the way that all the technologies used for the network elements go through a lifecycle from inception and small-scale deployment, through progressive growth to the mass deployment of a mature technology, ending with obsolescence and withdrawal from the network. The shape of these life cycles is similar to that of the well-known products life cycle [15]. As with products, the useful life of a technology as an element within the network is usually extended by progressive enhancements to keep it competitive compared to the newer technology alternatives constantly becoming available to the industry. Eventually, of course, one of the emerging new technologies becomes so attractive in terms of what it can do (i.e. its 'functionality') or its costs, that it starts to be adopted by network operators in preference to continuing to install established technology for growth in capacity. Once the new technology becomes mature it is also used to replace the old (i.e. earlier generations of technology) equipment in the network.

Figure 11.14 shows a succession of lifecycles for different generations of telephone switching systems, showing how each is eventually superseded by a new technology, and the overlap period when the established technology is replaced by the new technology. For illustration purposes the relevant years are shown in the Figure 11.14, but these are indicative only. Each of the technologies – manual switch-boards, two-motion selector automatic analogue switching (using electromechanical technology), semi-electronic analogue (using reed-relay switches with

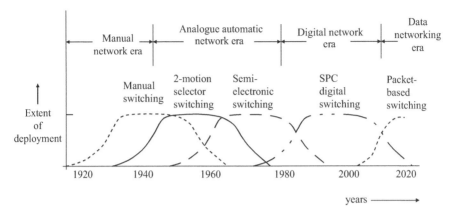

Figure 11.14 Technology product lifecycles

electronic control), and digital SPC switching (using semiconductor electronic switching and control) – has undergone significant enhancements to perpetuate their useful lives [16]. A corresponding set of technology lifecycles exists for all the different network elements: Access Network and Core Transmission systems, signalling systems, IN systems, etc. The set of lifecycles for all the network elements applying simultaneously for a period leads to the concept of 'network eras'. For example, during the periods of analogue two-motion selector and semielectronic switching the transmission systems were predominantly also analogue (e.g. FDM) with multi-frequency tone signalling between exchanges – hence this is referred to as the 'Analogue Network Era'. The range of technologies described in this book fall into either the established 'Digital Network Era' or the rapidly emerging 'Data Network Era', as shown in Figure 11.14. Thus, the technology used in the World's telecommunication networks is currently being changed from that of one Era to the next, as examined in the section below.

Telecommunication networks are constantly in a state of evolution, as new technology is progressively used to improve the cost, quality and functionality of the services offered to the operator's customers. Of course, wholesale changes of the network equipment are highly disruptive to the day-to-day network operations and the conversion costs can outweigh the possible advantages in using the new technology. Thus, there is normally only gradual introduction of new technology, creating a slowly evolving network, which is interrupted periodically by major shifts in technology, when there are sufficient reasons for a radical change as a new era is started [17].

11.7   Next-generation networks

11.7.1   The NGN concept

The term commonly used to describe the new Data-Network Era is next-generation network (NGN). Most network operators are in the process of changing or planning to change their networks progressively towards an NGN. The generally accepted reasons for making this transition are as follows.

- The adoption of the Internet by people at work and at home has radically changed the communications needs and expectations of the network operators' customers. Most day-to-day business transactions are now conducted via e-mail, and the World Wide Web, with the readily available information found by search engines, provides research facilities, online purchasing and transactions, as well as sources of recreation. This combination of computing – or information technology (IT) – and communications functions are often referred to as information and communications technology (ICT). Descriptions of the way that the industry and business has changed as a result of ICT include such expressions as 'The Internet Economy' and the 'Digital Networked Economy'. An NGN is expected to be better suited to supporting ICT services for the predominantly data-networking needs of customers in this new economy.
- Credible IP-technology-based alternatives to a digital telephone switching systems, as used in the PSTN, have become available. (These are described in Chapter 8.) An NGN based on this technology therefore offers the advantages of conveying voice and data directly over a common platform, giving convergence (i.e. ICT) as well as cost-saving opportunities for network operators.
- The digital circuit-switched telephone exchanges of the PSTN, described in Chapters 1 and 6, are reaching the end of their economic lives. Network operators therefore need to replace this obsolescent equipment. Thus, an NGN is viewed also as a way of modernising the PSTN.

The consensus within the industry on the necessary attributes of an NGN may be summarised as follows:

(i) Voice and data traffic are supported on the same packet-based network.
(ii) Voice and data traffic are integrated at the user's premises, ideally using multimedia terminals.
(iii) Broadband access
(iv) Customer services are accessible everywhere, that is, 'any time, any place'.
(v) Capable of supporting several service providers on a single NGN, in a multioperator environment.
(vi) Overall capital and operational cost savings compared to existing generation networks.

Although the details of the technical design of the various versions of NGN differ between network operators and the equipment manufacturers, there is one overarching technology that dominates – namely, the use of VoIP as the switching technology for telephony [18] (see Chapter 8). This use of the major alternative to the existing PSTN technology of circuit switching enables the NGN to be structured around a network of IP routers which carry not only voice but also the wide range of data services on the one 'platform'. (The latter term is frequently used to describe a deployment of equipment and its network management systems that act as a self-contained and complete piece of network infrastructure.) Network operators therefore have the opportunity to replace the PSTN exchanges as well as some of the specialised data network nodes by deploying an NGN. The resulting reduction in the number of network platforms, with the consequent savings in operating expenses, meets necessary attribute (i) listed above.

It should be emphasised that the IP platform used in an NGN to support voice and data services must provide an adequate quality of conveyance for all the services carried. This means that the basic IP network must have additional equipment that controls the flow of packets so that those services that do not tolerate delay or loss are given priority over those that do. In practice, this flow and quality management is provided using virtual paths at Layer 2 of the OSI model underpinning the IP platform – examples being MPLS, ATM and Ethernet (see Chapter 8). A further important point is that the IP/MPLS, ATM or Ethernet network is dedicated to the operator's NGN and does not form part of the uncontrolled set of IP routers on the (public) Internet.

Figure 11.15 shows a generic architectural view for the way that voice is carried over a fixed NGN. The top of the figure shows how the existing PSTN carries telephony calls (see Chapter 1). It also shows how xDSL provides a separate data path bypassing the PSTN, which can be used for Internet access via an ISP or, perhaps more profoundly, it provides an alternative to the PSTN by allowing voice in the form of IP packets to be carried over an ATM or Ethernet data network. However, the NGN itself comprises of the access media gateways, typically located at the LE buildings (physically next to the PSTN subscriber concentrator switchblocks that they will eventually replace), and the enhanced edge and core IP

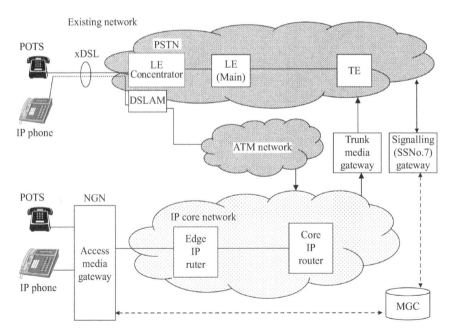

Figure 11.15   Generic fixed NGN architecture for voice calls

routers, as described above. The subscribers are permanently connected to the media gateways. In the case of subscribers using existing conventional telephone terminals, the media gateway needs to provide the analogue-to-digital conversion (using a codec) as well as the packetisation of the voice. Of course, neither of these functions is required for subscriber lines from so-called 'IP phones', that is, terminals that include codec and packetisation [19]. But for the many years during the transition to the new era, an NGN needs to cope with customers using existing phones as well as the new IP phones.

Also shown in Figure 11.15 are the media gateway controllers (MGCs), which provide call control on the NGN. Any calls to or from existing subscribers on the PSTN are passed via a trunk media gateway which converts the media flow between IP and TDM. The Trunk SSNo.7 signalling gateway provides the signalling interconnection between the MGC of the NGN and the PSTN (A fuller description of all the above is provided in Chapter 8.).

Finally, the alternative data path for voice calls routed over the broadband (i.e. xDSL) link of the subscriber on the existing network can extend through the ATM/Ethernet network to the enhanced IP network of the NGN, thus providing a further way of delivering IP voice packets from the existing network to the NGN. Care is required by the network operator to ensure that such hybrid routeings of voice calls over the new and old networks do not introduce QoS problems, in the form of delay and distortion, for the users.

The required broadband access for a fixed NGN's customers (i.e. attribute (iii) above) is provided by a variety of technologies, depending on the geographical terrain, the local availability of installed optical fibre, and the customer base being served (see Chapters 4 and 5). In the case of serving the residential and small business customers, the use of high-speed electronics over already-installed copper cable between exchange and subscriber, that is, ADSL – using all copper, or in combination with optical fibre and street electronics – VDSL or G.fast is favoured by most incumbent network operators. Alternatively, PON and some fibre to the premises systems are used to provide even higher speeds for all business customers and an increasing range of residential subscribers. In other cases, fixed microwave radio access systems give an alternative means of broadband delivery to the NGN.

It is expected that the ready availability of multimedia and ICT services at high speeds over the NGN will encourage users to adopt ever-more 'bandwidth-hungry' communications services from the network operators. This means that the capacity of transmission between network nodes will progressively need to increase, making it worthwhile to deploy new very high capacity optical fibre systems, using dense wave-division multiplexing (DWDM) – that is, carrying many different colours of light. Furthermore, the Core Transmission networks could be converted to optical-only cross connects and ADMs, totally eliminating the use of electronics. This will give rise to the establishment of wavelength or λ (lambda)-routed Core Transmission networks providing the required huge digital capacities between major towns for the NGN (see Chapter 5).

Although not shown in the simplified Figure 11.15, NGNs include call and service control systems that supply the range of ICT services, including customer authentication. Typically, this also involves the use of APIs for third-party service providers (see Chapter 7). There may also be control links between the NGN control systems and the HLR of 2G, 2.5G or 3G and IMS of 4G mobile networks (see Chapter 9), enabling some new location-based services. The latter are examples of converged services derived from fixed and mobile networks working closely together, that is, the so-called 'fixed-mobile convergence (FMC)'.

We can now pull together the various technology options described so far for an NGN and create a generalised logical architecture, as shown in Figure 11.16. This presents an architectural view of the relationships between the different functional layers in an NGN and the possible choices of technology for the components. The architecture shows the different choices for the Access and Core parts of the NGN, to reflect the differences in traffic densities and network structure, and hence economics in the two areas. As Figure 11.16 shows, the platform layers correspond loosely to the OSI 7-Layer model (see Chapter 8), with the cables, optical fibres and wireless links forming the basic Level 0, on which all the communications are carried. At the top of the architectural view the range of services are characterised as voice, real-time data and private circuits – all of which are delay or latency intolerant, and the latency-tolerant data services. All services are carried either over IP packets (at OSI Layer 3) or over cells or frames at OSI layer 2. In the case of leased lines, which in a conventional network are provided over dedicated transmission paths, a system is required which will manage the spasmodic

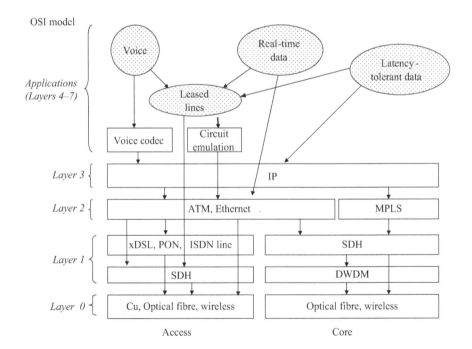

Figure 11.16   Generalised logical architecture for NGN

nature of the conveyance of the packet or cell layers of an NGN in such a way that the customer receives a close approximation to the required continuous transmission – a technique known as 'circuit emulation'. The leased lines will be used by the customers to link the nodes of the corporate private networks, carrying voice and data, as appropriate.

The relationships between the services and platforms or between one platform and another are indicated by the arrowed lines in the architectural view of Figure 11.16. It should be noted that, where appropriate, upper layers may be supported directly by layers two or three levels below. The way that the contents of one platform are translated into a lower platform is often referred to as 'mapping' or 'adaptation'. Thus, for example, bits forming a large IP packet are 'adapted' into the many smaller cells/frames of an ATM or Ethernet stream.

11.7.2   Implementation of an NGN

It should be emphasised that there is no inherent cost difference between the IP packet switching and the circuit-switching technology for voice – both are digital and use very large-scale semiconductor integrated (VLSI) circuits. Rather, the cost advantage of an NGN is due to two architectural advantages [20,21], namely:

- *First advantage*: The use of a single multi-service platform for all services (see Chapter 8).
- *Second advantage*: The shift of subscriber interface costs (BORSCHT, see Chapter 6) from the network to the subscriber's terminal.

Chapter 9 describes how the architecture of the 4G, long-term evolution (LTE) and LTE Advanced mobile system creates an all-IP platform for all the services carried, including telephony – i.e., it meets the first advantage. Furthermore, mobile networks in general have the economic advantage of the BORSCHT functions being provided in the handset, which is purchased by the subscriber, so meeting the second advantage. These two advantages make the operator's business case for moving to an NGN architecture much more compelling than for a fixed network still supporting subscriber's copper local lines [20,21]. Therefore, LTE Advanced can be viewed as a mobile version of an NGN.

It is likely that further developments of mobile systems (4G and 5G) will replace much of the earlier generations of equipment, including many of the PSTN lines, so furthering the adoption of an NGN approach. However, as mentioned earlier in Section 11.7.1, operators will eventually need to replace the ageing circuit-switched PSTN exchanges and this is likely to be a version of NGN. In order to achieve the second architectural advantage (to reduce user-interfacing costs) most incumbent operators are likely to encourage the subscribers' adoption of IP phones as part of their NGN implementation strategy.

Box 11.1 gives a description of the early version of an NGN originally designed by BT (named 21st CN), which was intended to replace the PSTN and several other network platforms. Even though implementation of the original 21st CN programme was not pursued, mainly due to the commercial and technical

Box 11.1   An example of a fixed network NGN physical architecture

Figure 11.17 presents a physical architectural view of BT's original design for their fixed network NGN, known as 21st CN. (This view is derived from publicly available initial design information and may differ in some details from the final design.) Its logical architecture corresponds to the generalised view given in Figure 11.16. The main component is the aggregator node, located at some 5,500 local or concentrator exchange buildings, that is, at the focal points of the copper Access Network. The aggregation is provided by multiservice access nodes (MSAN), on which all subscriber lines are terminated. The MSAN contains several types of line card:

(i)   A combination POTS and xDSL (digital subscriber loop) card for (a) converting telephone voice on copper lines to digital (PCM) samples, which are converted to IP packets by a media gateway (MG); and (b) for terminating the broadband ADSL signals on the copper loop, serving residential and small business subscribers.

(ii)   A reduced functionality line card for interfacing to ISDN lines (copper or optical fibre based).

(iii)   Ethernet cards for business subscribers that have their services terminated on lines using the LAN protocol.

(iv)   Circuit-emulation cards for medium speed leased lines. The higher-speed private circuits are directly carried over SDH.

An important factor in the MSAN design is its ability to terminate lines from subscribers using conventional telephones as well as IP phones and LAN equipment.

The various services are carried over a set of Ethernet virtual LANS on SDH or Gigabit Ethernet optical fibre transmission systems from the MSANs to parent Metro Nodes, of which there will be some hundred around the country. It is these Metro Nodes that contain the IP Edge routers, which provide the gateway to the MPLS-enhanced IP Core Network at the heart of 21st CN. The Metro Nodes also contain the Trunk Media Gateway and Signalling Gateway, as described in Chapter 8, providing the interconnection back to the existing PSTN and other operators' networks. Call control and service intelligence is provided at one of several Data Centres by IN servers – SCPs – and media gateway controllers (MGCs), as described in Chapters 7 and 8, respectively. Call sessions are set up across the network between the appropriate MSAN line cards using SIP signalling [22–32].

The BT's 21st CN was intended to replace several of the existing network platforms, including: the PSTN, digital private circuit network, frame-relay network, X25 packet switched network and the PDH transmission network.

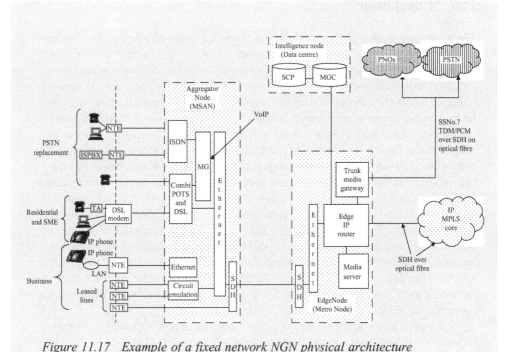

Figure 11.17   Example of a fixed network NGN physical architecture

complications of not having the second architectural advantage, it does nicely give
an example of the characteristics of a fixed network implementation of an NGN.

11.8   Summary

This chapter attempted to piece together all the various elements of a telecommunication network. To this end, the concept of the four architectural views –
commercial model, techno-regulatory, functional (or logical) and physical – was
introduced. The first application of an architectural perspective, viewing the various
component networks of a PSTN as a set of layers, was then used to give a holistic
description. Users-perceived QoS and the measures of a network's performance were
then introduced. These measures cover telephony services as well as digital transmission and data packet services. This led to the concept of network operations, and
how an operator needs to use support systems to manage the network equipment on a
day-to-day basis, generally encompassed by the term network management. We also
looked briefly at service management, that is, customer interactions, such as order
taking, billing, provisioning and fault reporting. The chapter finished by presenting a
view of the way that telecommunication networks evolve and the concept of an NGN.

11.9   Conclusion

The Foreword to this book raises the intriguing question of whether the existing well-established networks providing switched telephone service are to be replaced by the Internet. We have learnt how the existing network operates, using digital circuit-switched technology and digital transmission over copper, optical fibre or radio, all under the control of exchange-based processors. However, we have also learnt how the newer packet-based data platforms can provide the telephone switched service. One such platform is the public Internet, which is really a constellation of many thousands of loosely linked IP networks made available through a widely adopted naming and addressing system. Whilst VoIP can be successfully carried over the Internet, the quality is inevitably dependent on the instantaneous loading by other users and so large-scale reliable use of VOIP requires the provision of a dedicated IP network. Such networks, with suitable quality enhancements, will undoubtedly form the basis of the next generation of networks. However, as this book has demonstrated, NGNs will still use much of the existing infrastructure, such as the copper Access Network and Core Transmission systems. Thus, whilst IP, the technology of the Internet, will form part of the solution for voice calls in the future there is a continuing need for dedicated managed network platforms to form the public telecommunications infrastructure.

Perhaps, a bigger question for the future, however, is how will mobile networks, wireless Wi-Fi and fixed networks interrelate? Early work on 5G Mobile Technology suggests that they might all be replaced by *ad hoc* groupings of wireless-based networks owned and operated by the customers themselves. So, perhaps the more relevant question is: Whither the network operator?

References

[1]   Stewart, H. 'Building an architectural framework', *British Telecommunications Engineering*, 1994;**13**:186–191.

[2]   Wright, T. 'An architectural framework for networks', *British Telecommunications Engineering*, 1995;**14**:141–148.

[3]   Valdar, A. and Morfett, I. 'Understanding telecommunications business'. Stevenage: Institution of Electrical Engineers; *IET Telecommunications Series No. 60*, 2015, Chapter 6.

[4]   Redmill, F.J. and Valdar, A.R. 'SPC digital telephone exchanges'. Stevenage: Institution of Electrical Engineers; *IEE Telecommunications Series No. 21*, 1995, Chapter 10.

[5]   Bregni, S. *Synchronization of Digital Telecommunications Networks*. Chichester, UK: John Wiley & Sons; 2002, Chapter 4.

[6]   Oodan, A., Ward, K., Savolaine, C., Daneshmand, M., and Hoath, P. 'Telecommunications quality of service management: From legacy to emerging services', *IEE Telecommunications Series No. 48*, 2003, Chapter 2.

[7]   Cook, G.J. 'Network performance'. In Flood, J.E. (ed.), *Telecommunications Networks*, 2nd edn. *IEE Telecommunications Series No. 36*, 1997, Chapter 17.

[8]   Oodan, A.P., Ward, K.E., and Mullee, A.W. 'Quality of service in telecommunications', *IEE Telecommunications Series No. 39*, 1997, Chapter 10.

[9]   Furley, N. 'The BT operational support systems architecture framework', *British Telecommunications Engineering*, 1996;**15**:114–121.

[10]   Garrison, R., Spector, A., and de Groot, P.C. 'The BT network traffic management system: A window on the network', *British Telecommunications Engineering*, Special Issue on Network Management Systems, 1991;**10**(3): 222–229.

[11]   Fairley, J.M. 'Network management'. In Flood, J.E. (ed.), *Telecommunications Networks*, 2nd edn, *IEE Telecommunications Series No. 36*, 1997, Chapter 18.

[12]   Aidarous, S. and Plevyak, T. (eds.). *Telecommunications Network Management into the 21st Century, IEE Telecommunications Series No. 30*, 1996.

[13]   Valdar, A. and Morfett, I. 'Understanding telecommunications business'. Stevenage: Institution of Electrical Engineers; *IET Telecommunications Series No. 60*, 2015, Chapter 8.

[14]   Flood, J.E. 'Numbering, routing and charging', *IEE Telecommunications Series No. 30*, 1996, Chapter 9.

[15]   Valdar, A. and Morfett, I. 'Understanding telecommunications business'. Stevenage: Institution of Electrical Engineers; *IET Telecommunications Series No. 60*, 2015, Chapter 9.

[16]   Valdar, A.R. 'Circuit switching evolution to 2012', *Journal of the Institute of Telecommunications Professions*, 2012;**6**(4):27–31.

[17]   Valdar, A.R., Newman, D., Wood, R., and Greenop, D. 'A vision of the future network', *British Telecommunication Engineering*, 1992;**11**(3): 142–152.

[18]   Philpott, M. 'Migration strategies to an all-IP network', *Journal of the Telecommunications Network*, 2002;**1**(3):47–51.

[19]   Crane, P. 'A new service infrastructure architecture', *BT Technology Journal*, 2005;**23**(1):15–27.

[20]   Valdar, A. and Morfett, I. 'Understanding telecommunications business'. Stevenage: Institution of Electrical Engineers; *IET Telecommunications Series No. 60*, 2015, Chapter 5.

[21]   Valdar, A.R. 'The economics of packet versus circuit voice switching', *Journal of the Institute of Telecommunications Professions*, 2016;**10**(1): 37–43.

[22]   Wardlaw, M. 'Intelligence and mobility for BT's next generation networks', *Journal of the Telecommunications Network*, 2002;**1**(3):28–47.

[23]   Beal, M. 'Evolving networks for the future: Delivering BT's 21st century network', *Journal of the Communications Network*, 2004;**3**(4):4–10.

[24]   Reeve, M.H., Bilton, C., Holmes, P.E., and Bross, M. 'Networks and systems for BT in the 21st century', *BT Technology Journal*, 2005; **23**(1):11–14.

[25] Levy, B. 'The common capability approach to new service development', *BT Technology Journal*, 2005;**23**(1):48–54.

[26] Strang, C.J. 'Next generation systems architecture: The matrix', *BT Technology Journal*, 2005;**23**(1):55–68.

[27] Brazier, R.W. and Cookson, M.D. 'Intelligence design patterns', *BT Technology Journal*, 2005;**23**(1):69–81.

[28] Catchpole, A.B. 'Corporate multimedia communications', *BT Technology Journal*, 2005;**23**(1):98–107.

[29] Nunn, A. 'Voice evolution', *BT Technology Journal*, 2005;**23**(1):120–133.

[30] Ingham, A., Graham J., and Hendy P. 'Engineering performance for the future of intelligence', *BT Technology Journal*, 2005;**23**(1):134–146.

[31] Stretch, R.M. and Adams, P.M. 'Standards for intelligent networks', *BT Technology Journal*, 2005;**23**(1):154–159.

[32] Valdar, A. and Morfett, I. 'Understanding telecommunications business'. Stevenage: Institution of Electrical Engineers; *IET Telecommunications Series No. 60*, 2015, Chapter 1.

Appendix 1

Standards organisations

A1.1 Introduction

By its very nature, telecommunications depend on interaction between distant systems, and for this to be successful, some degree of commonality is required in specification of the equipment interfaces and protocols used. There are several organisations addressing the areas of standards for telecommunication networks. These are organised on a global, regional or national basis. In addition, there are a range of other organisations, often consortium or forums of interested parties in the industry that set *ad hoc* standards, many of which become adopted later by the more formal regional and global organisations. The key organisations referred to throughout this book are summarised below.

A1.2 Global organisations

A1.2.1 International Telecommunication Union

International Telecommunication Union (ITU) is a specialist agency of the United Nations, comprising of some 200 member countries, responsible for telecommunication standardisation. The organisation is split into two main areas of interest:

(i) ITU-T covering telecommunications (ITU-T was previously known as CCITT (*Comité Consultatif International de Télégraphique et Téléphonique*))

(ii) ITU-R covering radio (ITU-R was previously known as CCIR (*Comité Consultative International des Radio communications*)).

A1.2.2 International Organisation for Standardisation

International Organisation for Standardisation (ISO) is a joint organisation with the International Electrical Commission (IEC) responsible for standardisation of information technology (IT). The ISO 7-layer model is probably the most famous output from this organisation.

A1.3   Regional organisations

A1.3.1   *European Telecommunications Standards Institute*

European Telecommunications Standards Institute (ETSI) is the main standard organisation covering the European Union countries. It follows on from the work of the predecessor organisation CEPT (*Conférence Européenne des Administrations des Postes et des Telecommunications*). A notable output of CEPT and ETSI is the GSM standard for second-generation mobile systems, now adopted worldwide.

A1.3.2   *American National Standards Institute*

American National Standards Institute (ANSI) is the overarching standard organisation for the United States covering many areas in addition to telecommunications.

A1.3.3   *Institute of Electrical and Electronics Engineers*

The Institute of Electrical and Electronics Engineers (IEEE) is one of the largest professional societies in the world, which among its many activities is standardisation for the electrical and electronic engineering field, including telecommunication. One of the most notable standards is the IEEE 802.11 series covering local area networks (LANs).

A1.4   Other organisations

A1.4.1   *Internet Engineering Task Force*

The Internet Engineering Task Force (IETF), which is composed of manufacturers, operators, service providers and academics, is the main body setting the technical standards for the Internet. Its most famous output is the TCP/IP protocol.

A1.4.2   *Internet Corporation for Assigned Names and Numbers*

The Internet Corporation for Assigned Names and Numbers (ICANN) controls the assignment of top-level domain names and other parts of the naming and numbering scheme for the Internet worldwide.

A1.4.3   *Network Management Forum*

Network Management Forum (NMF) industry forum sets the standards for service and network management of telecommunication networks.

Appendix 2

List of ITU-T recommendation E.164 assigned country codes

Country code	Country or geographical area
1	United States, Canada and the Caribbean Islands
20	Egypt
212	Morocco
213	Algeria
216	Tunisia
218	Libya
220	Gambia
221	Senegal
222	Mauritania
223	Mali
224	Guinea
225	Cote d'Ivoire
226	Burkina Faso
227	Niger
228	Togolese
229	Benin
230	Mauritius
231	Liberia
232	Sierra Leone
233	Ghana
234	Nigeria
235	Chad
236	Central African Republic
237	Cameroon
238	Cape Verde
239	Sao Tome and Principe
240	Equatorial Guinea
241	Gabonese Republic
242	Congo
243	Democratic Republic of the Congo
244	Angola
245	Guinea-Bissau
246	Diego Garcia
247	Ascension

(Continues)

(*Continued*)

Country code	Country or geographical area
248	Seychelles
249	Sudan
250	Rwandese Republic
251	Ethiopia
252	Somali Democratic Republic
253	Djibouti
254	Kenya
255	Tanzania
256	Uganda
257	Burundi
258	Mozambique
260	Zambia
261	Madagascar
262	Reunion
263	Zimbabwe
264	Namibia
265	Malawi
266	Lesotho
267	Botswana
268	Swaziland
269	Comoros
269	Mayotte
27	South Africa
291	Eritrea
297	Aruba
298	Faroe Islands
299	Greenland
30	Greece
31	Netherlands
32	Belgium
33	France
34	Spain
350	Gibraltar
351	Portugal
352	Luxemburg
353	Ireland
354	Iceland
355	Albania
356	Malta
357	Cyprus
358	Finland
359	Bulgaria
36	Hungary
370	Lithuania
371	Latvia
372	Estonia
373	Moldova

(Continued)

Country code	Country or geographical area
374	Armenia
375	Belarus
376	Andorra
377	Monaco
378	San Marino
379	Vatican City State
380	Ukraine
381	Serbia and Montenegro
385	Croatia
386	Slovenia
387	Bosnia and Herzegovina
388	Group of countries, share code
389	The Former Yugoslav Republic of Macedonia
39	Italy
40	Romania
41	Switzerland
420	Czech Republic
421	Slovak Republic
423	Lichtenstein
43	Austria
44	United Kingdom
45	Denmark
46	Sweden
47	Norway
48	Poland
49	Germany
500	Falkland Island
501	Belize
502	Guatemala
503	El Salvador
504	Honduras
505	Nicaragua
506	Costa Rica
507	Panama
508	Saint Pierre and Miquelon
509	Haiti
51	Peru
52	Mexico
53	Cuba
54	Argentine Republic
55	Brazil
56	Chile
57	Colombia
58	Venezuela
590	Guadeloupe
591	Bolivia
592	Guyana

(Continues)

(Continued)

Country code	Country or geographical area
593	Ecuador
594	French Guiana
595	Paraguay
596	Martinique
597	Suriname
598	Uruguay
599	Netherlands Antilles
60	Malaysia
61	Australia
62	Indonesia
63	Philippines
64	New Zealand
65	Singapore
66	Thailand
670	Democratic Republic of Timor-Leste
672	Australian External Territories
673	Brunei Darussalam
674	Nauru
675	Papua New Guinea
676	Tonga
677	Solomon Islands
678	Vanuatu
679	Fiji
680	Palau
681	Wallis and Futuna
682	Cook Islands
683	Niue
684	American Samoa
685	Samoa
686	Kiribati
687	New Caledonia
688	Tuvalu
689	French Polynesia
690	Tokelau
691	Micronesia
692	Marshal Islands
7	Russian Federation and Kazakhstan
800	International Freephone Service
808	International Shared Cost Service (ISCS)
81	Japan
82	Korean (Republic of)
84	Viet Nam
850	Democratic People's Republic of Korea
852	Hong Kong, China
853	Macao, China
855	Cambodia
856	Lao People's Democratic Republic
86	People's Republic of China

(*Continued*)

Country code	Country or geographical area
870	Inmarsat SNAC
871	Inmarsat
872	Inmarsat
873	Inmarsat
874	Inmarsat
878	Universal Personal Telecommunication Service (UPT)
879	Reserved for national non-commercial purposes
880	Bangladesh
881	International Mobile, shared code
882	International Networks, shared code
886	Taiwan
90	Turkey
91	India
92	Pakistan
93	Afghanistan
94	Sri Lanka
95	Myanmar
960	Maldives
961	Lebanon
962	Jordan
963	Syrian Arab Republic
964	Iraq
965	Kuwait
966	Saudi Arabia
967	Yemen
968	Oman
971	United Arab Emirates
972	Israel
973	Bahrain
974	Qatar
975	Bhutan
976	Mongolia
977	Nepal
979	International Premium Rate Service (PRS)
98	Iran
991	Trial of a proposed new international telecommunication public correspondence service, share code
992	Tajikistan
993	Turkmenistan
994	Azerbaijani Republic
995	Georgia
996	Kyrgyz Republic
998	Uzbekistan
999	Reserved for possible future use within the Telecommunications for Disaster Relief (TDR) concept

Index